VOLUME FIFTY THREE

ADVANCES IN INSECT PHYSIOLOGY

Insect Epigenetics

VOLUME FIFTY THREE

ADVANCES IN INSECT PHYSIOLOGY

Insect Epigenetics

Edited by

HELEEN VERLINDEN
Department of Biology,
University of Leuven, Belgium

Series Editor

RUSSELL JURENKA
Department of Entomology,
Iowa State University, Iowa, United States

ELSEVIER

ACADEMIC PRESS
An imprint of Elsevier

Academic Press is an imprint of Elsevier
125 London Wall, London, EC2Y 5AS, United Kingdom
The Boulevard, Langford Lane, Kidlington, Oxford OX5 1GB, United Kingdom
525 B Street, Suite 1800, San Diego, CA 92101-4495, United States
50 Hampshire Street, 5th Floor, Cambridge, MA 02139, United States

First edition 2017

ISBN: 978-0-12-811833-7
ISSN: 0065-2806

Publisher: Zoe Kruze
Acquisition Editor: Kirsten Shankland
Editorial Project Manager: Alina Cleju
Production Project Manager: Vignesh Tamilselvvan
Cover Designer: Matthew Limbert

Typeset by SPi Global, India

CONTENTS

CONTRIBUTORS

Amir Ayali
Tel-Aviv University, Department of Zoology, Tel-Aviv, Israel

Margaret J. Beaton
Mount Allison University, Sackville, NB, Canada

William G. Bendena
Queen's University, Kingston, ON, Canada

Kevin Berry
University of Alaska Anchorage, Institute of Social and Economic Research, Anchorage, Alaska, United States

Jerome Buhl
University of Adelaide, School of Agriculture, Food and Wine, Adelaide, Australia

Warren W. Burggren
University of North Texas, Denton, TX, United States

Jian Q. Cao
School of Life Sciences, Simon F.S. Li Marine Science Laboratory and Centre of Soybean Research, Partner State Key Laboratory of Agrobiotechnology, The Chinese University of Hong Kong, Shatin, Hong Kong

Arianne J. Cease
School of Life Sciences; School of Sustainability, Arizona State University, Tempe, Arizona, United States

Andrew Cridge
Laboratory for Evolution and Development, Genetics Otago, University of Otago, Dunedin, New Zealand

Darron A. Cullen
KU Leuven (University of Leuven), Zoological Institute, Leuven, Belgium

Rien De Keyser
University of Leicester, Department of Neuroscience, Psychology and Behaviour, Leicester, United Kingdom

Peter Dearden
Laboratory for Evolution and Development, Genetics Otago, University of Otago, Dunedin, New Zealand

Bert Foquet
Texas A&M University, Department of Entomology, College Station, Texas, United States

Joleen C. Hadrich
Colorado State University, Department of Agricultural & Resource Economics, Fort Collins, Colorado, United States

Tom Harrop
Laboratory for Evolution and Development, Genetics Otago, University of Otago, Dunedin, New Zealand

Jerome H.L. Hui
School of Life Sciences, Simon F.S. Li Marine Science Laboratory and Centre of Soybean Research, Partner State Key Laboratory of Agrobiotechnology, The Chinese University of Hong Kong, Shatin, Hong Kong

Alexander K. Koliada
D.F. Chebotarev Institute of Gerontology, NAMS, Kyiv, Ukraine

Iryna A. Kozeretska
Taras Shevchenko National University of Kyiv, Kyiv, Ukraine

Alexandre V. Latchininsky
University of Wyoming, Ecosystem Science and Management, Laramie, Wyoming, United States

Mackenzie Lovegrove
Laboratory for Evolution and Development, Genetics Otago, University of Otago, Dunedin, New Zealand

Tom Matheson
University of Leicester, Department of Neuroscience, Psychology and Behaviour, Leicester, United Kingdom

Ruth Müller
GM Mosquitoes Confined Release Facilities, PoloGGB, Terni, Italy; Environmental Toxicology and Medical Entomology, Institute of Occupational, Social and Environmental Medicine, Goethe-University, Frankfurt am Main, Germany

Ann-Marie Oppold
Molecular Ecology Group, Senckenberg Biodiversity and Climate Research Centre, Frankfurt am Main, Hessen; Institute for Ecology, Evolution and Diversity, Goethe-University, Frankfurt am Main, Germany

Swidbert R. Ott
University of Leicester, Department of Neuroscience, Psychology and Behaviour, Leicester, United Kingdom

Mario A. Poot-Pech
Yucatán State Plant Protection Committee (Locust Program), Mérida, Yucatán, México

Emily Remnant
Behaviour and Genetics of Social Insects Laboratory, School of Life and Environmental Sciences, University of Sydney, Sydney, NSW, Australia

Julie A. Reynolds
The Ohio State University, Columbus, OH, United States

Brian E. Robinson
McGill University, Department of Geography, Montreal, Quebec, Canada

Stephen M. Rogers
Arizona State University, School of Life Sciences, Tempe, Arizona, United States; University of Cambridge, Department of Zoology, Cambridge, United Kingdom

Svitlana V. Serga
Taras Shevchenko National University of Kyiv, Kyiv, Ukraine

Jonathan M. Smith
University of Leicester, Department of Neuroscience, Psychology and Behaviour, Leicester, United Kingdom

Hojun Song
Texas A&M University, Department of Entomology, College Station, Texas, United States

Gregory A. Sword
Texas A&M University, Department of Entomology, College Station, Texas, United States

Stephen S. Tobe
University of Toronto, Toronto, ON, Canada

Wai S. Tong
School of Life Sciences, Simon F.S. Li Marine Science Laboratory and Centre of Soybean Research, Partner State Key Laboratory of Agrobiotechnology, The Chinese University of Hong Kong, Shatin, Hong Kong

Alexander M. Vaiserman
D.F. Chebotarev Institute of Gerontology, NAMS, Kyiv, Ukraine

Jozef Vanden Broeck
KU Leuven (University of Leuven), Zoological Institute, Leuven, Belgium

Rik Verdonck
KU Leuven (University of Leuven), Zoological Institute, Leuven, Belgium

Heleen Verlinden
KU Leuven (University of Leuven), Zoological Institute, Leuven, Belgium

Andreas Vilcinskas
Institute for Insect Biotechnology, Justus Liebig University of Giessen; Fraunhofer Institute for Molecular Biology and Applied Ecology, Giessen, Germany

Marcin W. Wojewodzic
University of Birmingham, School of Biosciences, Birmingham, United Kingdom; Centre for Epigenetics, Development and Evolution, University of Oslo, Oslo, Norway

Hiu Y. Yu
School of Life Sciences, Simon F.S. Li Marine Science Laboratory and Centre of Soybean Research, Partner State Key Laboratory of Agrobiotechnology, The Chinese University of Hong Kong, Shatin, Hong Kong

INTRODUCTION TO ADVANCES IN INSECT PHYSIOLOGY: INSECT EPIGENETICS

Heleen Verlinden
Zoological Institute, KU Leuven (University of Leuven), Leuven, Belgium

It is a great honour to introduce Volume 53 of *Advances in Insect Physiology*. This volume focuses on the impact of environmental factors on individuals and populations. Biological systems have the crucial capacity to adapt to their environment. Natural selection acts upon heritable genetic diversity (for instance caused by mutation, recombination, transposition, duplication) in a population. Eventually this leads to well-adapted organisms. The genes of an individual with higher fitness in the current environment are more likely to be passed on to the next generation. However, if the environmental conditions change faster than the genetic evolutionary adaptations take place, the demand for developmental adjustments without genetic mutations arises. Under these circumstances, individuals with flexible phenotypes, where one genotype can give rise to different phenotypes according to the environmental cues, are favoured (Pigliucci, 2001). Since the mechanisms that create these rapid modifications are critical for survival, they also contribute to evolution (Cardoso et al., 2015; Frank, 2011).

Nongenomic modifications to chromatin structure (alterations that do not affect the nucleotide sequence of a gene) are encompassed by the term "epigenetics", which was initially used to describe how cells differentiate from their original totipotent state during embryonic development (Waddington, 1942). The epigenome determines the way individuals use their static genome to live in a dynamic environment. Depending on the time lag and strength of response to dynamic environmental cues, some traits are more plastic than others. Variable selective pressure leads to genetic systems underlying hyper-adjustable traits. In this context, behaviour represents the primary means of response to environmental challenges and thus exhibits a faster and stronger plasticity than most other, physiological or morphological characteristics (Jensen, 2015; Pigliucci, 2001).

Modifications can be intragenerational and/or even transgenerational in case the epigenetic changes have occurred in the germ line (Burggren, 2014).

Important is that these traits also disappear when no longer preferable. The most prominent epigenetic mechanisms are methylation of cytosine in DNA, posttranslational histone modifications (histone PTMs) and regulation by noncoding RNA (ncRNA). In addition, these different mechanisms can also influence, or be influenced by, each other directly or indirectly (Ho and Burggren, 2012; Hunt et al., 2013; Sato et al., 2011; Zhou et al., 2014). For example, polycomb group and trithorax group proteins can interact with histone modifications and with long ncRNAs to alter the accessibility of chromatin, therefore representing an intriguing intersection between different epigenetic mechanisms (Geisler and Paro, 2015).

In this volume, different environmental factors are discussed that influence the phenotypical outcome of individuals, entire populations and future generations. In Chapter 1, Dr Warren Burggren introduces the different epigenetic mechanisms and their effects on phenotypes. Changes in epigenetic marks can be triggered by environmental stress or are part of a natural process during development, metamorphosis and ageing. Inevitably, they will have important ecological and evolutionary implications.

An important factor that influences epigenetic marks is nutrition. In Chapter 2, Dr Peter Dearden and coworkers discuss the influence of nutrition and nutritional stress on epigenetic marks. They highlight how the diet during early life stages influences the phenotype throughout the rest of life and even in future generations. Nutrition affects gene expression in an individual, its hormone levels and its phenotype. Many studies indicate epigenetic changes, but the authors righteously emphasise that hardcore evidence of direct interactions and functional relevance remains scarce in insects.

In Chapter 3, Dr Jerome Hui, Dr William Bendena and coworkers dig deeper into the effects of some of the most important insect hormones, insulin and ecdysteroids and their functional interactions with microRNAs. Intriguing relations between nutritional state, epigenetic mechanisms and insect hormones are described. These hormones are crucial for the regulation of important physiological processes, such as ecdysis and metamorphosis. Therefore, these interactions provide pathways by which the development of insects can be adjusted to the available resources.

In Chapter 4, Dr Irina Kozeretska, Dr Alexander Vaiserman and coworkers give an overview of epigenetic changes affecting ageing. Ageing is a complex process influenced by a large number of environmental and genetic factors. Differences in longevity of social insects are an excellent example of how different nutritional factors influence epigenetic mechanisms and thus also gene expression. Also, studies of epigenetic marks in other insects provide crucial knowledge concerning healthy ageing.

When conditions are suboptimal for survival, many insects undergo some form of diapause. In Chapter 5, Dr Julie Reynolds describes the altered gene expression profiles mediating developmental arrest, metabolic depression, increased stress resistance and other physiological and biochemical changes associated with this dormant state. The influence of epigenetic processes on gene expression before, during and after diapause is explored.

Another environmental stressor influencing gene expression in insects is the presence of parasites. The presence of parasites is correlated with specific epigenetic marks, which may alter expression of genes involved in pathogen recognition and immunity-related signalling. However, parasites have also evolved strategies to directly affect the gene expression in their host to their own benefit. In Chapter 6, Dr Andreas Vilcinskas digs deeper into the impact of parasites on the epigenetic regulation in their hosts.

Chapter 7 is a special one. In this chapter, members of the Global Locust Initiative (founded at the 25th International Congress of Entomology 2016 in Orlando, Florida, USA) cover the literature about locusts, since the seminal review by Pener and Simpson (2009). When considering phenotypic plasticity, locusts are of special interest. They display one of the most extreme forms of phenotypic plasticity observed in the animal kingdom. Their phenotype is highly dependent on the population density. Under low population densities, locusts are usually cryptic in both behaviour and appearance and will actively avoid conspecifics (solitary phase). Environmental factors have a huge impact not only on the overall population size but also on the distribution of individual locusts. High population densities cause locusts to undergo a switch to a more active, non-cryptic phenotype, the gregarious phase. Locusts in the gregarious phase are a major threat to agriculture, and the environment in general, in large parts of the world. A better understanding of all factors influencing the phenotype of locusts, as well as the switch between phases, is crucial to develop efficient ways of combating swarms. In addition, locusts provide an excellent model to study the reversibility of epigenetic marks. In recent years, several studies point at a role of epigenetics in the phenotypic plasticity of locusts, as was thoroughly reviewed by Ernst et al. (2015). This chapter gives a very broad overview on different environmental and molecular aspects affecting phenotype in locusts.

In Chapter 8, written by Dr Marcin Wojewodzic and Dr Margaret Beaton, the presence of predators, prey and toxins is considered as influencers of phenotype. For this chapter, we move to a non-insect arthropod genus that serves as a model for environmental epigenetics, *Daphnia*. This chapter also provides a nice overview of the currently available techniques to study methylation patterns, together with their advantages and disadvantages.

In the final chapter, Ann-Marie Oppold and Dr Ruth Müller dig deeper into the effects of toxins, more specifically insecticides, on epigenetic marks. Insects and humans interact in many ways. Some insects are beneficial, but many are not. Humans try to fight detrimental insect pests in many different ways, ideally without harming the beneficial ones. In the last couple of decades, different types of insecticides have been developed. However, insects seem very capable in gaining resistance over a few generations, which makes it a constant battle to find new, effective, but also selective products. It should not be surprising that insecticides induce stress in insects and it should also not be surprising that epigenetic mechanisms may be at play to generate changes in gene expression and in phenotype that can cause this resistance. This chapter provides an overview of which insecticides have been used in the past and how different strategies of resistance to these compounds are acting in insects. Where possible, involvement of epigenetic marks is described. Interestingly, also the possibility of manipulating epigenetic marks in the control of insect pests is brought forward.

In conclusion, the ability to cope with changing environmental conditions is crucial for survival of all organisms on Earth. In the last decades, advancements have been made in understanding how adaptive phenotypic changes can take place through natural selection and/or epigenetic regulation. Although several key mechanisms underlying phenotypic plasticity seem to show conservation in the animal kingdom, these do not always seem to function in the same way in different (groups of) organisms. Environmental influences and life conditions play a major role in the development of diseases as well. Hence, a better understanding of how epigenetic mechanisms perpetuate or can be reversed will not only provide important targets for selective pest management, but it will also be important for avoiding or changing epigenetic marks that have detrimental consequences for human health.

REFERENCES

Burggren, W.W., 2014. Epigenetics as a source of variation in comparative animal physiology—or—Lamarck is lookin' pretty good these days. J. Exp. Biol. 217, 682–689.

Cardoso, S.D., Teles, M.C., Oliviera, R.F., 2015. Neurogenomic mechanisms of social plasticity. J. Exp. Biol. 218, 140–149.

Ernst, U.R., Van Hiel, M.B., Depuydt, G., Boerjan, B., De Loof, A., Schoofs, L., 2015. Epigenetics and locust life phase transitions. J. Exp. Biol. 218, 88–99.
Frank, S.A., 2011. Natural selection. II. Developmental variability and evolutionary rate. J. Evol. Biol. **24**, 2310–2320.
Geisler, S.J., Paro, R., 2015. Trithorax and Polycomb group-dependent regulation: a tale of opposing activities. Development 142, 2876–2887.
Ho, D.H., Burggren, W.W., 2012. Parental hypoxic exposure confers offspring hypoxia resistance in zebrafish (*Danio rerio*). J. Exp. Biol. 215, 4208–4216.
Hunt, B.G., Glastad, K.M., Yi, S.V., Goodisman, M.A., 2013. Patterning and regulatory associations of DNA methylation are mirrored by histone modifications in insects. Genome Biol. Evol. 5, 591–598.
Jensen, P., 2015. Adding 'epi-' to behavior genetics: implications for animal domestication. J. Exp. Biol. 218, 32–40.
Pener, M.P., Simpson, S.J., 2009. Locust phase polyphenism: an update. Adv. Insect Physiol. 36, 1–272.
Pigliucci, M., 2001. Phenotypic Plasticity: Beyond Nature and Nurture. Johns Hopkins University Press, Baltimore, MD, pp. 1–344.
Sato, F., Tsuchiya, S., Meltzer, S.J., Shimizu, K., 2011. MicroRNAs and epigenetics. FEBS J. 278, 1598–1609.
Waddington, C.H., 1942. The epigenotype. Endeavour 1, 18–20.
Zhou, H.L., Luo, G., Wise, J.A., Lou, H., 2014. Regulation of alternative splicing by local histone modifications: potential roles for RNA-guided mechanisms. Nucleic Acids Res. 42 (2), 701–713.

CHAPTER ONE

Epigenetics in Insects: Mechanisms, Phenotypes and Ecological and Evolutionary Implications

Warren W. Burggren
University of North Texas, Denton, TX, United States

Contents

Abstract

Insects are being extensively investigated in the context of both intra- and transgenerational epigenetic phenomena and mechanisms. DNA methylation, histone modification and noncoding RNAs are all involved in phenotypic modification in insects. Changes in patterns of these epigenetic markers may arise from stressors, including physicochemical factors, nutrition, predation and crowding. Special cases of metamorphosis, social societies, solitarious to gregarious phase changes, diapause and

Advances in Insect Physiology, Volume 53
ISSN 0065-2806
http://dx.doi.org/10.1016/bs.aiip.2017.04.001

insect–microbe interactions are all being actively explored in the context of the insect epigenome. Epigenetic inheritance emerges as a rapid way of modifying phenotype in large proportions of a population to bridge across an environmental stressor for a few generations, and can be both widespread in a population and capable of being rapidly "sunsetted", both advantages over genetic inheritance. Finally, much emphasis has been placed on epigenetic mechanisms, perhaps at the expensive of fully understanding the role of epigenetics in altering phenotype. Correlations are abundant, while demonstration of cause and effect relationships between epigenetic markers and insect phenotype remains elusive.

1. INTRODUCTION

1.1 Scope of This Chapter

By design, this chapter is broad in scope, as it aims to introduce the topic of insect epigenetics to the reader. In some instances, the lack of treatment of a particular topic in the material below is because others have already written extensively on the topic, and the reader is directed to some of these references for an entry into the literature. In other instances, it is simply because we know relatively little about the specific topic—either we do not know or understand an aspect of the epigenetic phenomena or mechanisms in *any* organism, or just as likely, we do not understand how they operate in insects, specifically.

Before moving directly into insect epigenetics, a short diversion to discuss semantics will help avoid subsequent confusion.

1.2 Semantics in Epigenetics

Epigenetic phenomena in insects have long been a source of interest to biologists. In fact, a search of the PubMed database reveals that the oldest of several hundred papers on some form of insect epigenetics was published in 1964 (Gill, 1964), literally decades before most biologists had ever even heard of "epigenetics".[a] That particular study on dietary effects on adult phenotype is illustrative because it uses the term "epigenetics" in what has become, statistically, by far the most common use of that word—namely, changes in gene expression (but not gene sequence) ultimately leading to modified phenotype *during the course of an organism's life span*. Such modified phenotypes emerging during an individual's development, which reflect *intragenerational epigenetics*, essentially is one form of developmental

[a] Remarkably, Gill (1964) is the fourth oldest paper containing the word "epigenetics" in the PubMed database. Even so, Waddington explained the "epigenotype" using *Drosophila* as an example in 1942.

phenotypic plasticity. This use of epigenetics is most often in the context of gene expression gone wrong, as in the "epigenetics of cancer", with much less emphasis on ecological or evolutionary perspectives (Burggren, 2016). Certainly, insects and their epigenetics have figured prominently in our understanding in the medical literature of disease states in humans (Dietz et al., 2015; Mukherjee et al., 2015) (Fig. 1).

A far smaller subset of researchers—in fact, <5% according to a meta-analysis based of PubMed publications (Burggren, 2016)—use the word "epigenetics" to refer to *transgenerational epigenetic inheritance*, essentially the transfer of a phenotype from the P_0 to F_1 generation (or beyond) without changes in gene sequence. Essentially, intra- and transgenerational epigenetics are related but distinct phenomena, but the descriptions, definitions and uses of epigenetics often confuse, conflate or ignore the two phenomena

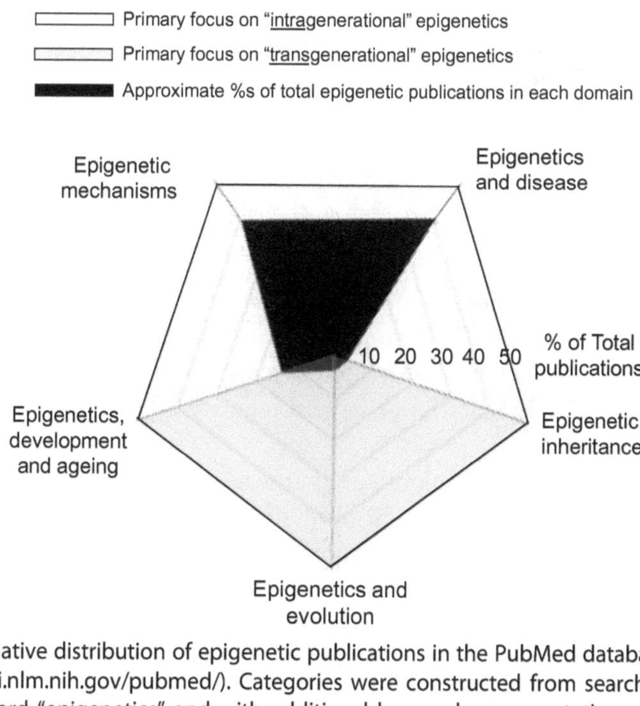

Fig. 1 Relative distribution of epigenetic publications in the PubMed database (http://www.ncbi.nlm.nih.gov/pubmed/). Categories were constructed from searches pairing the keyword "epigenetics" and with additional keywords representative of each five general categories shown in this radar diagram. There is a heavy preponderance of epigenetic papers dealing primarily with "intragenerational" aspects of epigenetics (e.g., disease, mechanism, development), with a very small percentage dealing with "transgenerational" aspects (e.g., inheritance, evolution). After *Burggren, W.W., 2016. Epigenetic inheritance and its role in evolutionary biology: re-evaluation and new perspectives. Biology 5, e24,* which describes methodology for this meta-analysis.

and their relationships—see Burggren (2016) and Burggren and Crews (2014) for an introduction to the oft-confusing semantics of epigenetics. Noteworthy is that the insect literature is not devoid of these semantic liberties, and this chapter will attempt to employ careful, clear terminology.

1.3 Beyond Model Systems

Insects (most notably the fruit fly *Drosophila melanogaster*) have served as highly tractable animal models for studying a wide range of biological processes and systems, and so it is no surprise that insects are also being employed as models for studying epigenetics. Such studies often explore basic mechanisms that can be revealed through experimental manipulation—an example is epigenetic modulation of the relationship between dietary restriction and ageing, where *D. melanogaster,* the honey bee *Apis mellifera* and other insects have been used to probe these interactions (Ford, 2013; Lian et al., 2015; Peleg et al., 2016) (see also chapter "Epigenetic regulation of longevity in insects" by Kozeretska et al.). In toxicology research, the red flour beetle *Tribolium castaneum* and other insects are also being studied in the context of transgenerational epigenetic inheritance of maladaptive phenotypes resulting from pharmaceutical exposure (Bingsohn et al., 2016; Vandegehuchte and Janssen, 2014). Development of pesticide resistance may in some instances involve transgenerational epigenetic inheritance of altered methylation patterns, and the tiger mosquito *Aedes albopictus* is being studied in this regard (Oppold et al., 2015) (see also chapter "Epigenetics: How the environment can have impact on genes and regulate phenotypes" by Oppold and Müller).

Yet, as tractable and important as insects can be as models for other organisms (ultimately human, in most studies), the tremendous diversity of insect biology, ranging from metamorphosis to complex societies, makes an understanding of insect epigenetics important in its own right, as amply evident in this volume! We now consider molecular epigenetic mechanisms in insects, and how they resemble and differ from other organisms.

2. MOLECULAR EPIGENETIC MECHANISMS IN INSECTS

A great deal of effort has focused on the molecular mechanisms underlying both intra- and transgenerational epigenetics in both plants and animals. The literature is too vast to review here (PubMed contains more than 15,000 papers on the subject), and the topic has been extensively treated elsewhere (and will be discussed in other chapters in this volume). Briefly, several different, but not mutually exclusive, molecular mechanisms for epigenetic modification of gene expression have been identified, including most commonly DNA methylation, histone modification and microRNAs. These epigenetic markers are highly conserved across taxa and are well represented in insects, as we will now explore.

2.1 DNA Methylation in Insects

Compared to other organisms, insects exhibit relatively low levels of DNA methylation, typically in the range of 0%–3% of cytosines showing methylation (Ernst et al., 2015; Field et al., 2004; Lyko et al., 2010; Mukherjee et al., 2015). By comparison, mammals and birds show ~5% DNA methylation, rising to ~10% in fish and >30% in some amphibians and many plants (Globisch et al., 2010; Head, 2014; Kamstra et al., 2015; Rai et al., 2008; Vidalis et al., 2016). Some insects have been reported to have no detectable levels of DNA methylation—e.g., *T. castaneum* (Zemach et al., 2010). Others, such as the ants *Camponotus floridanus* and *Harpegnathos saltator*, have DNA methylation rates of 0.2%–0.3% (Bonasio et al., 2012). In embryos of the honey bee, an emerging model within insect studies of DNA methylation and other epigenetic markers, less than 1% of the 5′-C-phosphate-G-3′ (CpG) islands (typically at the transcription initiation site for genes), is methylated (Remnant et al., 2016). The methylation levels are consistent with other estimates of low DNA methylation in adult honey bees (Drewell et al., 2014; Lyko et al., 2010). Classically (if anything is "classic" in the field of insect epigenetics), the Diptera as a whole have been viewed as devoid of DNA methylation (Bewick et al., 2016). *D. melanogaster*, as a specific example, was long viewed as devoid of methylation (Tweedie et al., 1997). However, more sensitive detection techniques for methylation are revealing both detectable global levels and altered levels associated with phenotype modification (Ashapkin et al., 2016; Boffelli et al., 2014; Glastad et al., 2015a; Guo et al., 2011; Redchuk et al., 2012).

Against the backdrop of vertebrate levels of DNA methylation, these typically low values in insects have led some to question the function, if not presence, of low DNA methylation levels in insects—see Boffelli et al. (2014) and Kim et al. (2016) for review. Noteworthy, however, is that vertebrates tend to have more globally methylated DNA, whereas in insects DNA methylation is primarily confined to the transcription initiation site for genes, where nearly one half of DNA may be methylated (Bewick et al., 2016; Glastad et al., 2011). For example, single base-level resolution analysis of DNA methylation in the genome of the silk moth *Bombyx mori* has indicated very low methylation rates of approximately 0.11%, but all of this methylation occurs in CpG islands (Xiang et al., 2010). Loci related to small RNAs (sRNAs) within gene bodies, especially of moderately expressed genes, were more extensively methylated in *Bombyx* and other species (Xiang et al., 2010; Zemach et al., 2010).

While global or even gene-specific DNA methylation data can be instructive, of course, the absolute levels are less important than the *change* in level under specific circumstances. Of key importance is whether an essential gene or gene group is expressed, repressed or silenced by changes in methylation. Thus, changes in overall methylation rate of the genome may be very small, but extensive for certain regions of the genome. Put differently, the *overall* levels of DNA methylation in insects may suggest a very low signal-to-noise ratio, but key genes appear to both be methylated and affect phenotype in or across generations. As Field et al. (2004) wisely cautioned more than a decade ago at the infancy of DNA methylation studies in insects, declarations of "no methylation" must be tempered by the fact that there are significant developmental changes in patterns and extent of epigenetic markers (see Section 4.2.1), and also that DNA methylation in particular may be limited to specific sequences/genes. Thus, while global DNA methylation levels may be low to the point of being almost undetectable, this does not preclude highly targeted methylation and subsequent phenotypic modification.

Beyond measurements of patterns or extent of DNA methylation is this epigenetic marker associated with known environmental conditions and the insects ability to adapt to them, for example? The simple answer is "Yes". Reviewing the literature is beyond the scope of this chapter (and is discussed in other chapters in this volume). However, by way of a few examples, changes in DNA methylation extent and location have been associated with viral infection (Galbraith et al., 2015) or nutrition (Strachecka et al., 2015) in the bee *A. mellifera*. Unusually high levels of DNA methylation occur in the desert locust *Schistocerca gregaria*, which also has an unusually high level of phenotypic plasticity, suggesting DNA methylation involvement in insect polyphenism (Falckenhayn et al., 2013). As a final example, in females of the wasp *Nasonia vitripennis*, DNA methylation patterns are affected by photoperiod, and either knocking down DNA methyltransferases or pharmacologically blocking DNA methylation disrupts photoperiodic diapause (Pegoraro et al., 2016). Notably, a recent review article has compiled a comprehensive list of phenotypic traits—developmental, physiological, morphological, behavioural, etc.—that are associated with degree of DNA methylation in insects (Kim et al., 2016).

A final note on DNA methylation is that few studies investigating DNA methylation have focused on individual variation in methylation levels as opposed to population or species level patterns. There is robust DNA methylation in the clonal raider ant *Cerapachys biroi* but, interestingly, there is great variation between ants. Contrary to expectations, the variation does not coincide with queen- and worker-specific DNA methylation (Libbrecht et al., 2016). This led these authors to make the provocative claim that previous studies of DNA methylation patterns specifically associated with queen and worker castes in social insects may be statistical artefacts. We return to epigenetics and castes in Section 4.2.2.

2.2 Histone Modification in Insects

Chromatin configuration (and thus the level of gene transcription) is relaxed or condensed according to the degree of acetylation, methylation, phosphorylation and/or ubiquitylation of associated histones. An entry into this literature is available through recent reviews (Izzo and Schneider, 2010; Janssen et al., 2017; Onder et al., 2015; Rothbart and Strahl, 2014; Yan et al., 2016; Zhang et al., 2015). Well understood in many organisms, histone modification is also being explored in insects and associated with altered gene expression and subsequent altered phenotypes. In fact, more than 160 histone modifications have been documented in insects (see Ernst et al., 2015). As examples, histone acetylation/deacetylation specifically has been implicated in regulation of pupal diapause in the flesh fly *Sarcophaga bullata* (Reynolds et al., 2016). Histone acetylation also regulates caste-specific behaviours in the carpenter ant *Camponotus florida*, where foraging and scouting behaviours specific to minor and major worker castes were disrupted when the ants were fed of inhibitors of the acetyltransferase CBP (Simola et al., 2016). Patterns of histone modifications (both methylation and acetylation of lysine) differ between queens and 96-h-old larvae of *A. mellifera*, though a function has not yet been ascribed to these histone modifications (Dickman et al., 2013). Although a field in its infancy, neural development in insects is also being examined in the context of histone modifications, where they are associated with (not necessarily the cause of) neural differentiation in *D. melanogaster* and other insects (Ye et al., 2016, 2017). As a final example, histone-modifying enzyme systems are differentially expressed in brain tissues of solitarious and gregarious migratory locusts *Locusta migratoria*, although these changes are at this point only correlated with phase changes in phenotype (Guo et al., 2016).

While the study of the patterns and role of histone modifications in insects is not as developed as for DNA methylation, there is clearly a solid base emerging for future studies of this epigenetic marker.

2.3 Noncoding RNAs as Epigenetic Markers in Insects

Noncoding RNAs are transcribed from DNA but, rather than being translated into protein, instead regulate gene expression at both transcriptional and posttranslational levels. Numerous categories of noncoding RNA exist, including microRNAs (miRNA), small interfering RNAs (siRNA), PiWi-interacting RNAs (piRNA), long noncoding RNAs (lncRNA), architectural RNAs (arcRNA) and circular RNAS (circRNA) (Abu and Jamal, 2016; Beermann et al., 2016; Chujo et al., 2016; Hanan et al., 2016; Ng et al., 2016; Sas-Chen et al., 2017). More forms are being discovered yearly, if not monthly.

Noncoding RNAs are proving to be important to insect epigenetics. Paramutation, where one allele affects another allele at the same locus and thereby induces a heritable change in the second allele, can result from the actions of both noncoding RNAs and DNA methylation. Paramutation resulting from both PiwiRNA and siRNA repression of mobile DNA in the germline occurs in *D. melanogaster* (Ronsseray, 2015). Noncoding RNAs (siRNA, miRNA and PiRNA) are involved in antiviral defence in insects (Brutscher and Flenniken, 2015; Fablet, 2014; Nandety et al., 2015), as well as other eukaryotes. Silk yield in the domestic silkworm (*Bombyx mori*) also appears to be modulated by differentially expressed lncRNAs, given that three different silk gland lnRNA genes are differentially expressed in domestic and wild silkworms (Zhou et al., 2017). MicroRNAs are also involved in insulin regulation in *D. melanogaster* (see chapter "The role of microRNAs in *Drosophila* regulation of insulin-like peptides and ecdysteroid signaling: Where are we now?" by Cao et al. of the current volume). As a final example, it was mentioned earlier that histone modification is involved in the developmental trajectory of the flesh fly, *S. bullata*. So, too, are siRNAs, including piRNAs (Reynolds et al., 2013).

While most work on noncoding RNAs in insects has focused on intragenerational epigenetic effects, transgenerational epigenetic inheritance is implicating noncoding RNAs, as well. For example, maternal piRNAs in *D. melanogaster* can be transmitted in germ cell cytoplasm from P_0 to F_1, where they subsequently induce gene silencing (Brennecke et al., 2008; Grentzinger et al., 2012) that can continue for 50 or more generations (de Vanssay et al., 2013). In *S. bullata,* siRNA-related genes show reduced transcript abundance in the first instar larvae when their mothers have undergone diapause compared to those larvae whose mothers have no diapause history, suggesting a role for siRNAs in not only development but in epigenetic inheritance (Reynolds et al., 2013).

New methodologies for detecting and assessing noncoding RNAs continue to emerge (Hu et al., 2015), and insects will surely be at the forefront of the emerging story of gene expression modification by noncoding RNAs.

2.4 The Epigenetic Marker Options: Must It Be Either/Or?

Important to note is that different mechanisms for epigenetic modification of gene expression are not mutually exclusive. Thus, the genome of organism, including insects, can be *concurrently* modulated by DNA methylation, histone modification and the action of noncoding RNAs. Interestingly, in the ant *C. floridanus*, DNA methylation is strongly predicted by histone modifications in genes (Glastad et al., 2015b). Indeed, these authors suggest that histone modification and DNA methylation "... are more predictive of gene expression when considered together than when considered independently". As will be discussed below (Section 4.2.2), caste determination in social insects appears to potentially involve *all* of the major epigenetic markers. These observations should direct us to undertake more concurrent analysis of multiple epigenetic markers, their interactions and their predictive power in insects.

Finally, it is important to emphasize that there can be important interactions between the various epigenetic mechanisms. For example, RNAs can actually direct the extent of DNA methylation in both plants and animals (Cuerda-Gil and Slotkin, 2016; Matzke et al., 2015; Wendte and Pikaard, 2017). Little wonder that some workers in the area refer somewhat cynically refer to epigenetic regulatory mechanisms as "dark matter".

3. EPIGENETIC DYNAMICS

Though often not emphasized in epigenetic studies, epigenetic markers and the phenotypic effects they create are rarely in a steady state. This is hardly surprising, given that molecular processes like methylation and acetylation of the genome are opposed by demethylation and deacetylation processes, as an example. So-called epigenetic dynamics have been revealed at many organizational levels in plants, animals and microbes, and have been implicated in both intra- and transgenerational epigenetic effects—for an introduction, see Bou Kheir and Lund (2010), Burggren (2015), Dhar et al. (2014), Sasai et al. (2013) and Zhao et al. (2015).

Epigenetic phenomena, even when viewed as in nonsteady-state conditions, are often viewed as digital—simply put, modified phenotypes and the epigenetic markers causing them are present (1) or they are absent (0). Increasingly, phenotypes associated with epigenetic markers are being discovered to be graded (i.e. analogue), occurring to varying degrees under different conditions and situations and even under different doses of the stressor that is triggering changes in degree of DNA methylation, for example. A "wash in" (essentially a delayed or slowed appearance) or "washout" (a slow disappearance) of an epigenetically modulated trait can occur during the life span of a single individual (Burggren, 2015). In the context of transgenerational epigenetic inheritance, "wash in" and "washout" can also occur across a few generations or even across successive broods within the F_1 generation. A clear demonstration of this comes from the toxicological literature, where exposure of the P_0 generation of zebrafish *Danio rerio* to the polycyclic aromatic hydrocarbon benzo[*a*]pyrene ultimately results in larval pericardial oedema that, in a dose-dependent fashion, may take two to three generations to first appear (Corrales et al., 2014). Then, the effects may take multiple generations to disappear. Washout of nongenetic inheritance also occurs in the water flea *Daphnia magna*. In this crustacean, a reduced body mass in the F_1 generation as a result of P_0 hypoxic exposure washes out as F_1 individuals develop, but also washes out across reproductive events (subsequent broods) in the same mother (Andrewartha and Burggren, 2012). Water fleas from Broods 1 and 2 from the same female show reduced body mass early in their development, but this phenotypic modification had disappeared in Brood 3 of this female (Fig. 2). Additional information on the uses of the water flea for epigenetic studies is provided in chapter "The future of Environmental Epigenetics: insights using the clonal water-flea model" by Wojewodzic and Beaton in this volume.

While the generation- and brood-related epigenetic dynamics described above are likely to be broadly applicable to most, if not all, organisms, scant evidence exists for insects. One experimental complication in following phenotype across multiple broods is that many insects show sperm storage (which may or may not be from a single mating), clumping, dumping and other related reproductive phenomena (Avila et al., 2011; Orr and Zuk, 2012). Against this backdrop, the use of clonal, asexually reproducing insects (Fournier et al., 2016; Matsuura, 2017; Simon et al., 2010; Wenseleers and Van Oystaeyen, 2011) may be the most efficient manner of demonstrating epigenetic dynamics in insects.

4. EPIGENETIC PHENOTYPIC MODIFICATION

Understanding the molecular mechanisms underlying epigenetic phenomena, as described above in Section 3, is vital for understanding epigenetics in insects (or any other organism). However, it could be argued

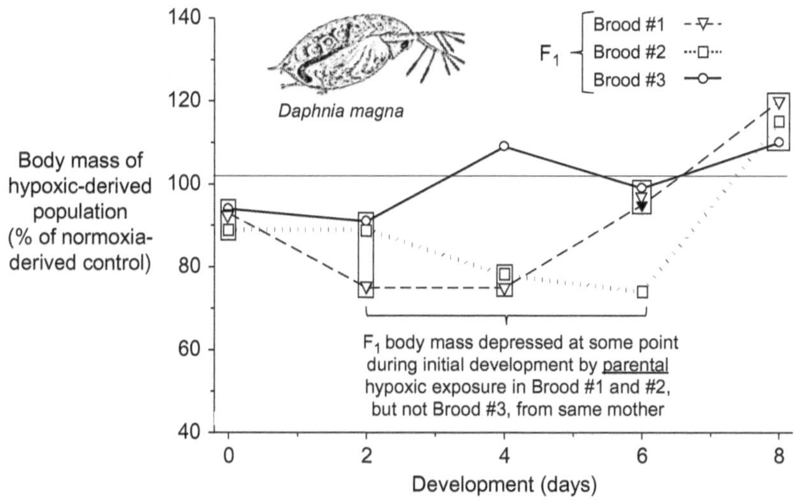

Fig. 2 Transgenerational influence of hypoxic exposure in the water flea *Daphnia magna*. The parental generation (F_0) was exposed to chronic hypoxia, and the effects on phenotype (body mass) assessed in three successive F_1 generation broods. The epigenetic inheritance of a depressed body mass in the first 8 days of development induced by parental hypoxic exposure is evident in the first two broods, but is "washed out" in the third and final brood. *Rectangles* enclose mean values not significantly different from each other. *After Andrewartha, S.J., Burggren, W.W., 2012. Transgenerational variation in metabolism and life-history traits induced by maternal hypoxia in* Daphnia magna. *Physiol. Biochem. Zool. 85, 625–634.*

that the intense focus on molecular mechanisms has occurred at the partial expense of determining the actual epigenetic effects on phenotype and inheritance. Thus, we often know how epigenetic markers have changed in an organism without fully understanding the implications of such changes at organismal, population and species levels (Burggren, 2016).

Consequently, we now turn to the phenotypic implications of changes in molecular epigenetic markers in insects.

4.1 Environmental Factors Triggering Epigenetic Responses

4.1.1 Physicochemical Factors

Surprisingly (and disappointingly), there is little information about how changes in natural physicochemical environmental factors, such as temperature, oxygen level, pH and light, evoke phenotypic changes in insects through epigenetic regulation of gene expression—as opposed to, for example, direct effects on enzymes and metabolic pathways. Yet, a few striking examples exist. For example, temperature affects chromatin structure in

D. melanogaster, resulting in changes in abdominal colouring (Gibert et al., 2007). Phenotypic modifications such as these may derive from the effects of temperature on heat shock protein expression (Chen et al., 2015; Feliciello et al., 2013; Pezer and Ugarkovic, 2012), though, of course, heat shock protein expression is directly stimulated by many different environmental stressors. Another example of a physicochemical factor evoking changes in epigenetic markers comes from *S. bullata* where short-day conditions induce entry in diapause (Reynolds et al., 2016). Light levels also affect methylation levels of genes, and subsequent neuromorphological phenotype, in the honey bee *A. mellifera ligustica* (Becker et al., 2016).

Generally speaking, the effects of basic environmental factors on epigenetic markers and their effect on gene regulation and phenotype—within or across generations—are very poorly understood and warrant further investigation.

4.1.2 Nutrition

Nutritional status, reflected in the quality and/or quantity of the food, has profound effects on insects, and there are many examples of epigenetic markers underlying phenotypic changes that derive from nutritionally induced changes in DNA methylation, histone modification and/or noncoding RNAs (Cridge et al., 2015; Kucharski et al., 2008; Maleszka, 2008; Snell-Rood et al., 2013). Among the most interesting of these are the phenotypic differences that emerge in social insects, where changes in morphology, behaviour and physiology rely upon nutritional characteristics. We resist here the temptation to further explore this fascinating topic of nutrition and epigenetic changes in insects, as it is treated in depth in chapter "Nutrition and epigenetic change in insects" by Daerden in this volume.

4.1.3 Predation

Phenotypic responses (especially of morphology and behaviour) of insects in response to predation pressures are well documented (Kishida and Nishimura, 2006; Kunert and Weisser, 2003; Kunert et al., 2008; Mikolajewski et al., 2015). In the aphids *Aphis gossypii, A. fabae*, and *Megoura viciae*, exposure of the parents to predators results in a greater proportion of winged offspring, in an apparent classic example of epigenetic inheritance (Kunert et al., 2008; Mondor et al., 2005).

Interestingly, the effects of radiation released from the Fukushima Daiichi Nuclear Power Plant on the pale grass blue butterfly *Zizeeria maha* have been investigated in the context of transgenerational transfer of phenotype (Taira et al., 2015). Ingestion of radioactive food materials by *Z. maha* as larvae results in high mortalities in the subsequent F_1 generation when they are fed radioactively contaminated foods, which could suggest epigenetic transgenerational transfer of decreased radioactivity resistance. However, this does not necessarily evoke epigenetic markers so much as maternal effects of direct transfer to the offspring. Indeed, in many species of insects the males produce a large "nuptial gift" of sperm that can range from 5%–20% of the male's body mass (Alexandre et al., 2011; Fedorka and Mousseau, 2002). Thus, "male provisioning", akin to female provisioning through yolk and its constituents, may be involved, rather than (or in addition to) any radiation-induced change in epigenetic marker configuration.

4.1.4 Crowding

Population density appears to be another environmental factor that has implications for epigenetic markers and their regulation of gene expression in insects. In locusts, crowding (along with other factors) triggers stark changes in morphological, physiological and behavioural phenotype associated with transition from solitarious to gregarious forms (Ernst et al., 2015), as well as increasing egg weight in gregarious offspring. Crowding also alters hatchling body size, coloration and other phenotypic traits in both swarming and nonswarming grasshoppers (Chen et al., 2015; Gotham and Song, 2013; Maeno and Tanaka, 2011; Tanaka and Maeno, 2006). Heat shock protein expression has been implicated in some of these changes (Badisco et al., 2011; Chen et al., 2015; Ernst et al., 2015; Maeno and Tanaka, 2011), but whether heat shock protein expression itself is regulated by a classic epigenetic mechanism (DNA methylation, histone modification, noncoding RNAs) has not been determined. Another target of epigenetic regulation is the genes involved in expression of a variety of hormones, which underpin the dramatic phase changes in locusts. A comprehensive schema for epigenetic mechanisms involved in the regulation of transition between solitarious and gregarious insects has been given by (Ernst et al., 2015)—see also chapter "The molecular physiology of locust swarming behaviour" by Cullen et al. in this volume.

4.2 Special Cases

4.2.1 Development, Metamorphosis and Ageing

Insects, like other arthropods, represent an interesting case study in alternative forms of development, with holometaboly (complete metamorphosis), hemimetaboly (partial metamorphosis) and ametaboly (absence of metamorphosis) all abundantly represented (Minelli et al., 2006). Insect development involves not only metamorphosis but also hyperplastic and hypertrophic growth within discrete developmental stages, with some insects growing several log cycles of body mass from egg to adult. Yet, as Nijhout et al. (2014) comment generally, but also specifically for insects, "The mechanisms that control the sizes of a body and its many parts remain among the great puzzles in developmental biology". The genetic and molecular basis of metamorphic growth and control in insects is still emerging (see Borras-Castells et al., 2017; Li et al., 2016; Ma et al., 2016; Morandin et al., 2015; Nijhout and Callier, 2015; Urena et al., 2016). Reviewing this broad topic is beyond the scope of this review. Briefly, however, microRNAs (see Section 2.3) have been intimately associated with the metamorphic process in insects (Asgari, 2013; Belles, 2017; Cristino et al., 2011; Lucas et al., 2015). While these and other authors discuss the regulation of gene expression by a large array of microRNAs, rarely is the term "epigenetics" specifically evoked. Yet, microRNAs as epigenetic markers are closely associated with gene expression regulation in the context of both intra- and transgenerational epigenetics in developing insects. The role of microRNAs in epigenetic regulation of development and ageing in insects is additionally explored in chapters "The role of microRNAs in *Drosophila* regulation of insulin-like peptides and ecdysteroid signaling: Where are we now?" by Cao et al. and "Epigenetic regulation of longevity in insects" by Kozeretska et al. Additional introduction to the development and ageing literature on epigenetic markers (including DNA methylation, histone modification and microRNAs), their clearance and reprogramming can be gained from Boija and Mannervik (2015) and Ernst et al. (2015).

4.2.2 Social Insects

One of the richest and most active areas of epigenetic research in insects involves epigenetic programming of caste phenotype in social insects. Nutritional cues contained with the "royal jelly" direct development of larvae into queens. While the full profile of regulatory substances in royal jelly has not been made, up to 15% of royal jelly consists of so-called major royal jelly proteins (Albert et al., 1999; Buttstedt et al., 2014). Hydroxy fatty acids in royal jelly also apparently have regulatory functions beyond mere nutritional value (Kodai et al., 2011; Noda et al., 2005), as do glycosides and sterols (Kodai et al., 2007).

Considerable evidence suggests that epigenetic mechanisms are involved in caste determination in social insects—for an entry into the literature, see Ford (2013), Foret et al. (2012), He et al. (2017), Holman et al. (2016), Kim et al. (2016), Spannhoff et al. (2011) and Vaiserman (2014). For example, in honey bees a silencing of the expression of the DNA methyltransferase DNMT3 redirects the larval developmental trajectory towards queens, mimicking the effect of royal jelly (Kucharski et al., 2008). These authors suggest that the gene *dynactin p62* plays a key role in regulating alternate developmental trajectories, based on different methylation patterns in queens compared to workers. A CpG island in the gene protein kinase C-binding protein 1 also shows differential methylation as a function of social interaction in similarly aged worker and nurse bees (Lockett et al., 2012). DNA methylation is similarly involved in caste development in the carpenter ant *C. floridanus*, through actions on ant body size and other phenotypic traits and, consequently, the division of labour within the colony (Alvarado et al., 2015; Bonasio et al., 2012). DNA methyltransferase expression (and presumable the ultimate pattern of DNA methylation) is also involved in differential development in the fire ant *Solenopsis invicta* (Kay et al., 2016).

While DNA methylation certainly appears to be involved in caste creation and its maintenance in social insects, histone deacetylation has also been implicated in caste switching in the honey bee *A. mellifera* (Huang et al., 2012; Spannhoff et al., 2011) and the ant *C. floridanus* (Simola et al., 2016). To further complicate our understanding of phenotypic regulation and caste determination during development in social insects, noncoding RNAs also appear to be implicated (Ashby et al., 2016; Guo et al., 2013; Li et al., 2014; Macedo et al., 2016). Overall, then, it would seem that *all* of the major epigenetic mechanisms for modifying gene regulation during development have been implicated in some way in caste determination in social insects. This underscores the point made earlier (Section 2.4) that complex phenotypic changes may possibly—even likely—result from complex combination of concurrent changes in epigenetic markers.

As a final note on the topic of the role of caste determination in social insects, a recent study investigated whole genome DNA methylation neural tissue of the clonal raider ant *Ooceraea biroi* (Libbrecht et al., 2016). These authors did *not* find any difference in DNA methylation patterns between queens and workers. They call for a more cautious analysis of DNA methylation data, pointing out misinterpretations that could spring from possible sampling artefacts and from statistical analysis. Surely provoking more discussion and experimentation, Libbrecht et al. (2016) consequently opine that "there is currently no evidence that genome-wide variation in DNA methylation is associated with the queen and worker castes in social insects".

4.2.3 Gregarious Insects

Insects such as migratory locusts that as individuals experience morphological, physiological and behavioural changes as they move between solitarious and gregarious phases are a particularly interesting case study in insect epigenetics. In locust species such as *Schistocerca gregaria*, the behavioural transition from solitarious to gregarious behaviour can occur within hours, while morphological and physiological traits changes can take several moults or even generations. The result is a difference in phenotype so striking that for years naturalists thought that the gregarious and solitarious forms were different species. This phenotypic switch has been broadly attributed to neuroendocrine mechanisms, especially involving serotonin in *S. gregaria* (Anstey et al., 2009) and catecholamine pathways in *L. migratoria* (Ma et al., 2011; Shi et al., 2014). However, not all aspect of this transition can be explained "simply" through hormones and neurotransmitters, leading to an increasingly favoured view that epigenetic mechanisms involving DNA methylation, histone modification and/or microRNAs are involved (Ernst et al., 2015). Intriguing are two observations that, when combined, suggest epigenetic mechanisms are involved. First, morphological changes once triggered can span several generations (Burrows et al., 2011; Simpson and Miller, 2007; Tanaka and Maeno, 2006). Second, these effects are reversible, suggesting a "washout" of an epigenetically inherited phenotype (Burggren, 2015).[b] Rapid reversibility is a strong indicator of a phenotype that is epigenetically inherited (Burggren, 2014, 2016).

The reader is directed to chapter "The molecular physiology of locust swarming behaviour" by Cullen et al. in this volume that further explores the environmental influence on the phenotype of locusts.

4.2.4 Diapause

Many insect species exhibit various forms of diapause, characterized by highly reduced feeding and movement and, ultimately, reduced metabolic rate and energy expenditure. Diapause is typically induced by environmental factors such as photoperiod, thermal changes or chemical signals derived from plants upon which the insects feed (Hand et al., 2016; Schiesari and O'Connor, 2013). Yet, as for so many aspects of insect biology, the underlying molecular mechanisms of insect diapause remain enigmatic.

Certainly, microRNAs and their role on histone positioning have been implicated in diapause regulation broadly speaking (Hand et al., 2016; Lu

[b] Maternal (parental) effects are sometimes considered to be a subset of modes of epigenetic inheritance (Burggren and Crews, 2014).

et al., 2013; Reynolds et al., 2013). What remains unclear is whether epigenetic markers directly regulate entry into and exit from diapause, or whether they are "merely" associated with phenotypic modifications necessary for successful diapause. This topic is explored more fully in chapter "Epigenetic influences on diapause" by Reynolds in the current volume.

4.2.5 Insect–Microbe Interactions and the Epigenome

One of the most exciting and, at times, mystifying areas emerging in animal biological research involves the symbiotic role of microbes with their hosts organisms, and especially captivating is the role of microbes in the gut–brain axis (Bauer et al., 2016; Dinan and Cryan, 2017). Few journal article titles capture the status quo, and the imagination, as "The Brain's Geppetto—Microbes as Puppeteers of Neural Function and Behaviour?" (Stilling et al., 2016). Understandably, most focus in the burgeoning field has been in mammals generally and humans specifically. Yet, within the insects are some of the most extensive and intimate metabolic and ecological relationships with gut flora (Brune and Dietrich, 2015; Douglas, 2015; Kim et al., 2016).

Symbiotic relationships mediated by DNA methylation in insects appear to be lacking, with only a few examples in those insects showing significant methylation "signals". One such relationship involves the symbiotic relationship between *Wolbachia*, an intracellular bacterium, and the mosquito, *Aedes aegypti*. Infection with *Wolbachia* of the mosquito results in changes in host methylation due to effects on tRNA methyltransferase DNMT2 and subsequent immune system changes affecting antiviral defences (Ye et al., 2013; Zhang et al., 2013).

Likely to be of greater significance in insect–microbe interactions are the large number of noncoding regulatory RNAs—especially sRNAs in archaea and bacteria (Babski et al., 2014; Kim et al., 2016). sRNAs produced by bacteria can target eukaryotic genes, and Kim et al. (2016) hypothesize that these sRNAs can either be coopted into the RNAi pathway or actually mimic eukaryotic sRNAs. While evidence for direct epigenetic influences by microbes on organismal level phenotype in insects is scant, the potential for such regulation is enormous. While the gut diversity of microbes residing symbiotically with insects overall is thought to be lower than in mammals, for example, some insects (especially social insects) harbour enormous numbers of microbial species (Engel and Moran, 2013), each of which can potentially have some influence on complex regulatory functions via the gut–brain axis (or perhaps more correctly, gut–ganglion axis).

5. ECOLOGICAL AND EVOLUTIONARY IMPLICATIONS

5.1 Epigenetics and Population-Level Survival

Much has been written about mechanisms of epigenetic inheritance, and how epigenetically inherited traits may become fixed in the genome (Lowdon et al., 2016; Morris, 2015; Prendergast et al., 2014; Sharma, 2015; Skinner, 2015). Less, however, has been expressed about the role that epigenetic inheritance may play at the level of the individual or population, especially in the context of dynamic environmental stressors. The benefits to population-level survival are best demonstrated by comparing and contrasting genetic with epigenetic inheritance (Table 1), which reveals the "epigenetic advantage" of employing epigenetic markers to effect short-term change (in an evolutionary context).

Inheritance by mutation is a somewhat random and, even if the resulting trait is advantageous, any given mutation only affects a very small proportion (perhaps even just a single individual) of any given population of a species. On the other hand, epigenetically inherited traits can (and often do) concurrently affect large proportions of the population (Burggren, 2016). Typically there is significant epigenetic diversity/variability in animal and plant populations (Burggren, 2016; Giuliani et al., 2016; He et al., 2011; Quintana-Murci, 2016; Seymour and Becker, 2017; Stuart et al., 2016), including insects (Chen et al., 2015; Vieira et al., 2012; Wang et al., 2015). Yet, many individuals in that diverse population will nonetheless simultaneously show a similar if not identical epigenetically mediated phenotypic response to a common environmental stressor (crowding, hypoxia, hypothermia). Consider the migratory locust as a stark example. Even though there is diversity in the epigenome of migratory locusts (Chen et al., 2015), environmental stressors induce a simultaneous transition from the solitarious to gregarious morph in tens of millions of locusts (Fig. 1), with this phenotype then potentially epigenetically inherited for several generations. If a gregarious form of locust for survival of environmental change required *genetic mutation* to occur, and then natural selection for it to become prominent in the population...well, it would not happen. Thus, epigenetic inheritance of an advantageous phenotype can be quickly evoked, allowing a large proportion of a population to survive a rapidly appearing environmental stressor that might otherwise threaten that population (Burggren, 2014, 2016; Kim et al., 2016).

Table 1 A Comparison of the Difference Implications of Genetic and Epigenetic Inheritance

	Form of Inheritance		
Parameter	**Genetic**	**Epigenetic**	**Implications**
Mechanism	Mutation through insertion, deletion, duplication, frameshift, repeat expansion, etc.	Modification of existing gene expression through action of epigenetic markers	Gene mutation is random and infrequent; gene expression typically easy to modify via epigenetic markers
Value of inherited trait	Positive or negative	Positive or negative	Both genetic and nongenetic inheritance may be random, resulting in a mixture of maladaptive and adaptive traits
Breadth of initial impact in population	Very small percentage of population initially affected by any given mutation	Potentially most, if not all, of population can be immediately affected in response to environmental stressor	When environmental stressor is widespread, large proportions of population with epigenetically inherited trait may be better positioned for survival
Time span and reversibility of impact	Indefinite; not reversible	Short term (within a few generation); reversible upon removal of epigenetic repression of gene expression	Epigenetically inherited trait can be rapidly "sunsetted" when environmental stressor disappears or washes out

While a rapid "on" response to a threatening environmental situation may be important to population survival, so too might be the relatively rapid "off" response, either in parallel with the diminution of the environmental stressor, or through washout of the epigenetic effect. Again, returning to the example of migratory locusts, the gregarious phenotype serves the locust well as crowding begins to occur, but is not an ideal phenotype under other conditions—if the gregarious phenotype were always optimal, it would have been selected for as the permanent, singled phenotype of evolutionary time. Thus, the ability of an epigenetic phenotype to be "sunsetted" allows an organism (and populations) to quickly shed costly or otherwise disadvantageous phenotypes as environmental conditions change (Burggren, 2016).

From an ecological perspective, then, epigenetic phenotype modification and its transgenerational inheritance provide a rapid response to environmental change. This may prove to be especially important in insects, most of which are small and unable to move large distances, and thus may be especially vulnerable to short-term environmental change. This will doubtlessly be an interesting area for future investigation in insect ecology.

5.2 Epigenetics in Evolution

Epigenetically derived modifications of phenotype directly affect an organism's fitness. As a result, individuals with advantageous phenotypes resulting from epigenetic phenomena will be selected for, and those with disadvantageous phenotypes will be selected against (Burggren, 2016; Jablonka et al., 1998; Laland et al., 2015; Pigliucci, 2009; Skinner, 2015). Given that diversity coexists in the genome and epigenome, epigenetically derived phenotypes with more or less advantageous phenotypes will doubtlessly drive a population's genome (Klironomos et al., 2013). The largely unanswered question, however, is..."Does the diversity in the epigenome correlate with diversity in the genome in such a pattern that drives phenotypic change over evolutionary time?" The answer to this question is only slowly emerging for any organism, and in insects is especially hampered by the lack of information on *both* the genome and epigenome of a single species. However, studies are beginning to characterize both "-omes" in single species—e.g., the small carpenter bee (*Ceratina calcarata*) (Rehan et al., 2016).

A key issue regarding epigenetics and evolution, and one of the most active and potentially far-reaching aspects of epigenetic research, is understanding whether and to what extent epigenetically derived phenotypes can become permanently fixed in the genome. The extent to which "Lamarckian evolution" influences "Darwinian" and "new-Darwinian" evolution is being vigorously debated, as are the mechanisms by which epigenetics influences evolution through fixed (permanent) genome modifications. Regretfully, a deep dive into this topic is beyond the scope of this chapter, and the reader is directed to the following evolving literature for additional information: Asano et al. (2013), Burggren (2016), Dall et al. (2015), Damiani (2007), Kim et al. (2016), Klironomos et al. (2013),

Noble (2015), Pilpel and Rechavi (2015), Sharma (2015), Skinner (2015) and Springer and McGinnis (2015).

6. CONCLUSIONS AND FUTURE DIRECTIONS

Epigenetic mechanisms have been "implicated", "suspected" and "associated with" many phenotypic changes in insect morphology, biochemistry, physiology and behaviour. However, the words "demonstrated", "shown" or even "proven" do not emerge in summary statements as often as one would like. Typically, changes in the state of DNA methylation, histone modification or noncoding RNA profile—sometimes to quite striking degrees—are "associated with" sometimes equally striking changes in phenotype not just in insects, but in many other organisms. Unfortunately, the definitive experiments "proving" an epigenetic basis for a phenotype within a generation, or epigenetic inheritance across generations, are often difficult, expensive and complex. But, as experimentalists, we have to bite that bullet. As Maleszka (2016) so aptly notes, "To achieve rapid progress, insect epigenetics needs to focus on mechanistic explanations of epigenomic dynamics and move beyond low-depth genome-wide analyses to cell-type specific epigenomics". More experiments designed to determine cause-and-effect relationships are also required. To this end, the emergence of pharmacological blockers of DNA and histone methylation, particularly useful in so-called epigenetic therapy (Afshinnekoo and Mason, 2016; Momparler and Cote, 2015), as just one example, may add to our currently somewhat limited toolkit for demonstrating direct linkage of epigenetic markers to specific phenotypes.

Ultimately, insects can be highly tractable models for exploring the epigenetic basis of social behaviours, developmental regulation, toxicological effects, etc., for animals generally (Bingsohn et al., 2016; Bonasio, 2015; Mukherjee et al., 2015; Yan et al., 2014). This does not mean that insects should be relegated to being a model for other animals, but rather that they may in fact be representative of relatively conserved epigenetic mechanisms. Doubtlessly, given the burgeoning intensity of research into epigenetic mechanisms and outcomes (Burggren, 2014, 2016), the vitally important role of epigenetics in insect biology will gain ever-increasing recognition.

ACKNOWLEDGEMENT

The author was supported in part by NSF grant IOS-1543301.

REFERENCES

Abu, N., Jamal, R., 2016. Circular RNAs as promising biomarkers: a mini-review. Front. Physiol. 7, 355.

Afshinnekoo, E., Mason, C.E., 2016. Epigenetic therapy in a new era of medicine: creating and integrating molecular profiles of patients. Ann. Transl. Med. 4, 436.

Albert, S., Bhattacharya, D., Klaudiny, J., Schmitzova, J., Simuth, J., 1999. The family of major royal jelly proteins and its evolution. J. Mol. Evol. 49, 290–297.

Alexandre, D., Linhares, R.T., Queiroz, B., Fontoura, L., Uchôa, A.F., Samuels, R.I., Macedo, M.L., Bezerra, C.S., Oliveira, E.M., Demartini, D.R., Carlini, C.R., Silva, C.P., 2011. Vicilin-derived peptides are transferred from males to females as seminal nuptial gift in the seed-feeding beetle *Callosobruchus maculatus*. J. Insect Physiol. 57, 801–808.

Alvarado, S., Rajakumar, R., Abouheif, E., Szyf, M., 2015. Epigenetic variation in the egfr gene generates quantitative variation in a complex trait in ants. Nat. Commun. 6, 6513.

Andrewartha, S.J., Burggren, W.W., 2012. Transgenerational variation in metabolism and life-history traits induced by maternal hypoxia in *Daphnia magna*. Physiol. Biochem. Zool. 85, 625–634.

Anstey, M.L., Rogers, S.M., Ott, S.R., Burrows, M., Simpson, S.J., 2009. Serotonin mediates behavioral gregarization underlying swarm formation in desert locusts. Science 323, 627–630.

Asano, M., Basieva, I., Khrennikov, A., Ohya, M., Tanaka, Y., Yamato, I., 2013. A model of epigenetic evolution based on theory of open quantum systems. Syst. Synth. Biol. 7, 161–173.

Asgari, S., 2013. MicroRNA functions in insects. Insect Biochem. Mol. Biol. 43, 388–397.

Ashapkin, V.V., Kutueva, L.I., Vanyushin, B.F., 2016. Dnmt2 is the most evolutionary conserved and enigmatic cytosine DNA methyltransferase in eukaryotes. Genetika 52, 269–282.

Ashby, R., Foret, S., Searle, I., Maleszka, R., 2016. MicroRNAs in honey bee caste determination. Sci. Rep. 6, 18794.

Avila, F.W., Sirot, L.K., LaFlamme, B.A., Rubinstein, C.D., Wolfner, M.F., 2011. Insect seminal fluid proteins: identification and function. Annu. Rev. Entomol. 56, 21–40.

Babski, J., Maier, L.K., Heyer, R., Jaschinski, K., Prasse, D., Jager, D., Randau, L., Schmitz, R.A., Marchfelder, A., Soppa, J., 2014. Small regulatory RNAs in Archaea. RNA Biol. 11, 484–493.

Badisco, L., Ott, S.R., Rogers, S.M., Matheson, T., Knapen, D., Vergauwen, L., Verlinden, H., Marchal, E., Sheehy, M.R., Burrows, M., Vanden Broeck, J., 2011. Microarray-based transcriptomic analysis of differences between long-term gregarious and solitarious desert locusts. PLoS One 6, e28110.

Bauer, K.C., Huus, K.E., Finlay, B.B., 2016. Microbes and the mind: emerging hallmarks of the gut microbiota-brain axis. Cell. Microbiol. 18, 632–644.

Becker, N., Kucharski, R., Rossler, W., Maleszka, R., 2016. Age-dependent transcriptional and epigenomic responses to light exposure in the honey bee brain. FEBS Open Bio 6, 622–639.

Beermann, J., Piccoli, M.T., Viereck, J., Thum, T., 2016. Non-coding RNAs in development and disease: background, mechanisms, and therapeutic approaches. Physiol. Rev. 96, 1297–1325.

Belles, X., 2017. MicroRNAs and the evolution of insect metamorphosis. Annu. Rev. Entomol. 62, 111–125.

Bewick, A.J., Vogel, K.J., Moore, A.J., Schmitz, R.J., 2016. Evolution of DNA methylation across insects. Mol. Biol. Evol. 34, 654–665.

Bingsohn, L., Knorr, E., Vilcinskas, A., 2016. The model beetle *Tribolium castaneum* can be used as an early warning system for transgenerational epigenetic side effects caused by pharmaceuticals. Comp. Biochem. Physiol. C Toxicol. Pharmacol. 185–186, 57–64.

Boffelli, D., Takayama, S., Martin, D.I., 2014. Now you see it: genome methylation makes a comeback in *Drosophila*. Bioessays 36, 1138–1144.
Boija, A., Mannervik, M., 2015. A time of change: dynamics of chromatin and transcriptional regulation during nuclear programming in early *Drosophila* development. Mol. Reprod. Dev. 82, 735–746.
Bonasio, R., 2015. The expanding epigenetic landscape of non-model organisms. J. Exp. Biol. 218, 114–122.
Bonasio, R., Li, Q., Lian, J., Mutti, N.S., Jin, L., Zhao, H., Zhang, P., Wen, P., Xiang, H., Ding, Y., Jin, Z., Shenn, S.S., Wang, Z., Wang, W., Wang, J., Berger, S.L., Liebig, J., Zhang, G., Reinberg, D., 2012. Genome-wide and caste-specific DNA methylomes of the ants *Camponotus floridanus* and *Harpegnathos saltator*. Curr. Biol. 22, 1755–1764.
Borras-Castells, F., Nieva, C., Maestro, J.L., Maestro, O., Belles, X., Martin, D., 2017. Juvenile hormone biosynthesis in adult *Blattella germanica* requires nuclear receptors seven-up and ftz-f1. Sci. Rep. 7, 40234.
Bou Kheir, T., Lund, A.H., 2010. Epigenetic dynamics across the cell cycle. Essays Biochem. 48, 107–120.
Brennecke, J., Malone, C.D., Aravin, A.A., Sachidanandam, R., Stark, A., Hannon, G.J., 2008. An epigenetic role for maternally inherited piRNAs in transposon silencing. Science 322, 1387–1392.
Brune, A., Dietrich, C., 2015. The gut microbiota of termites: digesting the diversity in the light of ecology and evolution. Annu. Rev. Microbiol. 69, 145–166.
Brutscher, L.M., Flenniken, M.L., 2015. RNAi and antiviral defense in the honey bee. J. Immunol. Res. 2015, 941897.
Burggren, W.W., 2014. Epigenetics as a source of variation in comparative animal physiology—or—Lamarck is lookin' pretty good these days. J. Exp. Biol. 217, 682–689.
Burggren, W.W., 2015. Dynamics of epigenetic phenomena: intergenerational and intragenerational phenotype 'washout'. J. Exp. Biol. 218, 80–87.
Burggren, W.W., 2016. Epigenetic inheritance and its role in evolutionary biology: re-evaluation and new perspectives. Biology 5, e24.
Burggren, W.W., Crews, D., 2014. Epigenetics in comparative biology: why we should pay attention. Integr. Comp. Biol. 54, 7–20.
Burrows, M., Rogers, S.M., Ott, S.R., 2011. Epigenetic remodelling of brain, body and behaviour during phase change in locusts. Neural Syst. Circuits 1, 11.
Buttstedt, A., Moritz, R.F., Erler, S., 2014. Origin and function of the major royal jelly proteins of the honeybee (*Apis mellifera*) as members of the yellow gene family. Biol. Rev. Camb. Philos. Soc. 89, 255–269.
Chen, B., Li, S., Ren, Q., Tong, X., Zhang, X., Kang, L., 2015. Paternal epigenetic effects of population density on locust phase-related characteristics associated with heat-shock protein expression. Mol. Ecol. 24, 851–862.
Chujo, T., Yamazaki, T., Hirose, T., 2016. Architectural RNAs (arcRNAs): a class of long noncoding RNAs that function as the scaffold of nuclear bodies. Biochim. Biophys. Acta 1859, 139–146.
Corrales, J., Thornton, C., White, M., Willett, K.L., 2014. Multigenerational effects of benzo[a]pyrene exposure on survival and developmental deformities in zebrafish larvae. Aquat. Toxicol. 148, 16–26.
Cridge, A.G., Leask, M.P., Duncan, E.J., Dearden, P.K., 2015. What do studies of insect polyphenisms tell us about nutritionally-triggered epigenomic changes and their consequences? Forum Nutr. 7, 1787–1797.
Cristino, A.S., Tanaka, E.D., Rubio, M., Piulachs, M.D., Belles, X., 2011. Deep sequencing of organ- and stage-specific microRNAs in the evolutionarily basal insect *Blattella germanica* (l.) (Dictyoptera, Blattellidae). PLoS One 6, e19350.

Cuerda-Gil, D., Slotkin, R.K., 2016. Non-canonical RNA-directed DNA methylation. Nat. Plants 2, 16163.

Dall, S.R., McNamara, J.M., Leimar, O., 2015. Genes as cues: phenotypic integration of genetic and epigenetic information from a Darwinian perspective. Trends Ecol. Evol. 30, 327–333.

Damiani, G., 2007. The Yin and Yang of anti-Darwinian epigenetics and Darwinian genetics. Riv. Biol. 100, 361–402.

de Vanssay, A., Bouge, A.L., Boivin, A., Hermant, C., Teysset, L., Delmarre, V., Antoniewski, C., Ronsseray, S., 2013. piRNAs and epigenetic conversion in *Drosophila*. Fly 7, 237–241.

Dhar, M.K., Vishal, P., Sharma, R., Kaul, S., 2014. Epigenetic dynamics: role of epimarks and underlying machinery in plants exposed to abiotic stress. Int. J. Genomics 2014, 187146.

Dickman, M.J., Kucharski, R., Maleszka, R., Hurd, P.J., 2013. Extensive histone post-translational modification in honey bees. Insect Biochem. Mol. Biol. 43, 125–137.

Dietz, K.N., Di Stefano, L., Maher, R.C., Zhu, H., Macdonald, M.E., Gusella, J.F., Walker, J.A., 2015. The *Drosophila* Huntington's disease gene ortholog *dhtt* influences chromatin regulation during development. Hum. Mol. Genet. 24, 330–345.

Dinan, T.G., Cryan, J.F., 2017. The microbiome-gut-brain axis in health and disease. Gastroenterol. Clin. North Am. 46, 77–89.

Douglas, A.E., 2015. Multiorganismal insects: diversity and function of resident microorganisms. Annu. Rev. Entomol. 60, 17–34.

Drewell, R.A., Bush, E.C., Remnant, E.J., Wong, G.T., Beeler, S.M., Stringham, J.L., Lim, J., Oldroyd, B.P., 2014. The dynamic DNA methylation cycle from egg to sperm in the honey bee *Apis mellifera*. Development 141, 2702–2711.

Engel, P., Moran, N.A., 2013. The gut microbiota of insects—diversity in structure and function. FEMS Microbiol. Rev. 37, 699–735.

Ernst, U.R., Van Hiel, M.B., Depuydt, G., Boerjan, B., De Loof, A., Schoofs, L., 2015. Epigenetics and locust life phase transitions. J. Exp. Biol. 218, 88–99.

Fablet, M., 2014. Host control of insect endogenous retroviruses: small RNA silencing and immune response. Viruses 6, 4447–4464.

Falckenhayn, C., Boerjan, B., Raddatz, G., Frohme, M., Schoofs, L., Lyko, F., 2013. Characterization of genome methylation patterns in the desert locust Schistocerca gregaria. J. Exp. Biol. 216, 1423–1429.

Fedorka, K.M., Mousseau, T.A., 2002. Nuptial gifts and the evolution of male body size. Evolution 56, 590–596.

Feliciello, I., Parazajder, J., Akrap, I., Ugarkovic, D., 2013. First evidence of DNA methylation in insect *Tribolium castaneum*: environmental regulation of DNA methylation within heterochromatin. Epigenetics 8, 534–541.

Field, L.M., Lyko, F., Mandrioli, M., Prantera, G., 2004. DNA methylation in insects. Insect Mol. Biol. 13, 109–115.

Ford, D., 2013. Honeybees and cell lines as models of DNA methylation and aging in response to diet. Exp. Gerontol. 48, 614–619.

Foret, S., Kucharski, R., Pellegrini, M., Feng, S., Jacobsen, S.E., Robinson, G.E., Maleszka, R., 2012. DNA methylation dynamics, metabolic fluxes, gene splicing, and alternative phenotypes in honey bees. Proc. Natl. Acad. Sci. U.S.A. 109, 4968–4973.

Fournier, D., Hellemans, S., Hanus, R., Roisin, Y., 2016. Facultative asexual reproduction and genetic diversity of populations in the humivorous termite *Cavitermes tuberosus*. Proc. Biol. Sci. 283 (1832). http://dx.doi.org/10.1098/rspb.2016.0196, pii: 20160196.

Galbraith, D.A., Yang, X., Nino, E.L., Yi, S., Grozinger, C., 2015. Parallel epigenomic and transcriptomic responses to viral infection in honey bees (*Apis mellifera*). PLoS Pathog. 11, e1004713.

Gibert, J.M., Peronnet, F., Schlotterer, C., 2007. Phenotypic plasticity in *Drosophila* pigmentation caused by temperature sensitivity of a chromatin regulator network. PLoS Genet. 3, e30.

Gill, K.S., 1964. Epigenetics of the promorphology of the egg in *Drosophila melanogaster*. J. Exp. Zool. 155, 91–104.

Giuliani, C., Sazzini, M., Bacalini, M.G., Pirazzini, C., Marasco, E., Fontanesi, E., Franceschi, C., Luiselli, D., Garagnani, P., 2016. Epigenetic variability across human populations: a focus on DNA methylation profiles of the *krtcap3*, *mad1l1* and *brsk2* genes. Genome Biol. Evol. 8, 2760–2773.

Glastad, K.M., Hunt, B.G., Yi, S.V., Goodisman, M.A., 2011. DNA methylation in insects: on the brink of the epigenomic era. Insect Mol. Biol. 20, 553–565.

Glastad, K.M., Goodisman, M.A., Yi, S.V., Hunt, B.G., 2015a. Effects of DNA methylation and chromatin state on rates of molecular evolution in insects. G3 (Bethesda) 6, 357–363.

Glastad, K.M., Hunt, B.G., Goodisman, M.A., 2015b. DNA methylation and chromatin organization in insects: insights from the ant *Camponotus floridanus*. Genome Biol. Evol. 7, 931–942.

Globisch, D., Munzel, M., Muller, M., Michalakis, S., Wagner, M., Koch, S., Bruckl, T., Biel, M., Carell, T., 2010. Tissue distribution of 5-hydroxymethylcytosine and search for active demethylation intermediates. PLoS One 5, e15367.

Gotham, S., Song, H., 2013. Non-swarming grasshoppers exhibit density-dependent phenotypic plasticity reminiscent of swarming locusts. J. Insect Physiol. 59, 1151–1159.

Grentzinger, T., Armenise, C., Brun, C., Mugat, B., Serrano, V., Pelisson, A., Chambeyron, S., 2012. piRNA-mediated transgenerational inheritance of an acquired trait. Genome Res. 22, 1877–1888.

Guo, X.X., Ye, H.Y., Zhang, M., 2011. DNA methylation in *Drosophila*, a review of recent studies. Yi Chuan 33, 713–719.

Guo, X., Su, S., Skogerboe, G., Dai, S., Li, W., Li, Z., Liu, F., Ni, R., Guo, Y., Chen, S., Zhang, S., Chen, R., 2013. Recipe for a busy bee: MicroRNAs in honey bee caste determination. PLoS One 8, e81661.

Guo, S., Jiang, F., Yang, P., Liu, Q., Wang, X., Kang, L., 2016. Characteristics and expression patterns of histone-modifying enzyme systems in the migratory locust. Insect Biochem. Mol. Biol. 76, 18–28.

Hanan, M., Soreq, H., Kadener, S., 2016. CircRNAs in the brain. RNA Biol. 1–7. http://dx.doi.org/10.1080/15476286.2016.1255398 [Epub ahead of print].

Hand, S.C., Denlinger, D.L., Podrabsky, J.E., Roy, R., 2016. Mechanisms of animal diapause: recent developments from nematodes, crustaceans, insects, and fish. Am. J. Physiol. Regul. Integr. Comp. Physiol. 310, R1193–R1211.

He, G., Elling, A.A., Deng, X.W., 2011. The epigenome and plant development. Annu. Rev. Plant Biol. 62, 411–435.

He, X.J., Zhou, L.B., Pan, Q.Z., Barron, A.B., Yan, W.Y., Zeng, Z.J., 2017. Making a queen: an epigenetic analysis of the robustness of the honeybee (*Apis mellifera*) queen developmental pathway. Mol. Ecol. 26, 1598–1607.

Head, J.A., 2014. Patterns of DNA methylation in animals; an ecotoxicological perspective. Integr. Comp. Biol. 54, 77–86.

Holman, L., Trontti, K., Helantera, H., 2016. Queen pheromones modulate DNA methyltransferase activity in bee and ant workers. Biol. Lett. 12, 20151038.

Hu, L., Di, C., Kai, M., Yang, Y.C., Li, Y., Qiu, Y., Hu, X., Yip, K.Y., Zhang, M.Q., Lu, Z.J., 2015. A common set of distinct features that characterize noncoding RNAs across multiple species. Nucleic Acids Res. 43, 104–114.

Huang, C.Y., Chi, L.L., Huang, W.J., Chen, Y.W., Chen, W.J., Kuo, Y.C., Yuan, C.M., Chen, C.N., 2012. Growth stimulating effect on queen bee larvae of histone deacetylase inhibitors. J. Agric. Food Chem. 60, 6139–6149.

Izzo, A., Schneider, R., 2010. Chatting histone modifications in mammals. Brief Funct. Genomics 9, 429–443.
Jablonka, E., Lamb, M.J., Avital, E., 1998. "Lamarckian" mechanisms in Darwinian evolution. Trends Ecol. Evol. 13, 206–210.
Janssen, K.A., Sidoli, S., Garcia, B.A., 2017. Recent achievements in characterizing the histone code and approaches to integrating epigenomics and systems biology. Methods Enzymol. 586, 359–378.
Kamstra, J.H., Alestrom, P., Kooter, J.M., Legler, J., 2015. Zebrafish as a model to study the role of DNA methylation in environmental toxicology. Environ. Sci. Pollut. Res. Int. 22, 16262–16276.
Kay, S., Skowronski, D., Hunt, B.G., 2016. Developmental DNA methyltransferase expression in the fire ant *Solenopsis invicta*. Insect Sci. http://dx.doi.org/10.1111/1744-7917.12413 [Epub ahead of print].
Kim, D., Thairu, M.W., Hansen, A.K., 2016. Novel insights into insect-microbe interactions—role of epigenomics and small RNAs. Front. Plant Sci. 7, 1164.
Kishida, O., Nishimura, K., 2006. Flexible architecture of inducible morphological plasticity. J. Anim. Ecol. 75, 705–712.
Klironomos, F.D., Berg, J., Collins, S., 2013. How epigenetic mutations can affect genetic evolution: model and mechanism. Bioessays 35, 571–578.
Kodai, T., Umebayashi, K., Nakatani, T., Ishiyama, K., Noda, N., 2007. Compositions of royal jelly II. Organic acid glycosides and sterols of the royal jelly of honeybees (*Apis mellifera*). Chem. Pharm. Bull. 55, 1528–1531.
Kodai, T., Nakatani, T., Noda, N., 2011. The absolute configurations of hydroxy fatty acids from the royal jelly of honeybees (*Apis mellifera*). Lipids 46, 263–270.
Kucharski, R., Maleszka, J., Foret, S., Maleszka, R., 2008. Nutritional control of reproductive status in honeybees via DNA methylation. Science 319, 1827–1830.
Kunert, G., Weisser, W.W., 2003. The interplay between density- and trait-mediated effects in predator-prey interactions: a case study in aphid wing polymorphism. Oecologia 135, 304–312.
Kunert, G., Schmoock-Ortlepp, K., Reissmann, U., Creutzburg, S., Weisser, W.W., 2008. The influence of natural enemies on wing induction in *Aphis fabae* and *Megoura viciae* (hemiptera: Aphididae). Bull. Entomol. Res. 98, 59–62.
Laland, K.N., Uller, T., Feldman, M.W., Sterelny, K., Muller, G.B., Moczek, A., Jablonka, E., Odling-Smee, J., 2015. The extended evolutionary synthesis: its structure, assumptions and predictions. Proc. Biol. Sci. 282, 20151019.
Li, Q., Wang, Z., Lian, J., Schiøtt, M., Jin, L., Zhang, P., Zhang, Y., Nygaard, S., Peng, Z., Zhou, Y., Deng, Y., Zhang, W., Boomsma, J.J., Zhang, G., 2014. Caste-specific RNA editomes in the leaf-cutting ant *Acromyrmex echinatior*. Nat. Commun. 5, 4943.
Li, K., Tian, L., Guo, Z., Guo, S., Zhang, J., Gu, S.H., Palli, S.R., Cao, Y., Li, S., 2016. 20-Hydroxyecdysone (20E) primary response gene E75 isoforms mediate steroidogenesis autoregulation and regulate developmental timing in *Bombyx*. J. Biol. Chem. 291, 18163–18175.
Lian, T., Gaur, U., Yang, D., Li, D., Li, Y., Yang, M., 2015. Epigenetic mechanisms of dietary restriction induced aging in *Drosophila*. Exp. Gerontol. 72, 38–44.
Libbrecht, R., Oxley, P.R., Keller, L., Kronauer, D.J., 2016. Robust DNA methylation in the clonal raider ant brain. Curr. Biol. 26, 391–395.
Lockett, G.A., Kucharski, R., Maleszka, R., 2012. DNA methylation changes elicited by social stimuli in the brains of worker honey bees. Genes Brain Behav. 11, 235–242.
Lowdon, R.F., Jang, H.S., Wang, T., 2016. Evolution of epigenetic regulation in vertebrate genomes. Trends Genet. 32, 269–283.
Lu, Y.X., Denlinger, D.L., Xu, W.H., 2013. Polycomb repressive complex 2 (PRC2) protein ESC regulates insect developmental timing by mediating H3K27me3 and activating prothoracicotropic hormone gene expression. J. Biol. Chem. 288, 23554–23564.

Lucas, K.J., Zhao, B., Liu, S., Raikhel, A.S., 2015. Regulation of physiological processes by microRNAs in insects. Curr. Opin. Insect Sci. 11, 1–7.

Lyko, F., Foret, S., Kucharski, R., Wolf, S., Falckenhayn, C., Maleszka, R., 2010. The honey bee epigenomes: differential methylation of brain DNA in queens and workers. PLoS Biol. 8, e1000506.

Ma, Z., Guo, W., Guo, X., Wang, X., Kang, L., 2011. Modulation of behavioral phase changes of the migratory locust by the catecholamine metabolic pathway. Proc. Natl. Acad. Sci. U.S.A. 108, 3882–3887.

Ma, J., Wang, R., Li, X., Gao, B., Chen, S., 2016. Transcriptome and gene expression analysis of *Cylas formicarius* (Coleoptera: Brentidae) during different development stages. J. Insect Sci. 16, 63.

Macedo, L.M., Nunes, F.M., Freitas, F.C., Pires, C.V., Tanaka, E.D., Martins, J.R., Piulachs, M.D., Cristino, A.S., Pinheiro, D.G., Simoes, Z.L., 2016. MicroRNA signatures characterizing caste-independent ovarian activity in queen and worker honeybees (*Apis mellifera* L.). Insect Mol. Biol. 25, 216–226.

Maeno, K., Tanaka, S., 2011. Phase-specific responses to different qualities of food in the desert locust, *Schistocerca gregaria*: developmental, morphological and reproductive characteristics. J. Insect Physiol. 57, 514–520.

Maleszka, R., 2008. Epigenetic integration of environmental and genomic signals in honey bees: the critical interplay of nutritional, brain and reproductive networks. Epigenetics 3, 188–192.

Maleszka, R., 2016. Epigenetic code and insect behavioural plasticity. Curr. Opin. Insect Sci. 15, 45–52.

Matsuura, K., 2017. Evolution of the asexual queen succession system and its underlying mechanisms in termites. J. Exp. Biol. 220, 63–72.

Matzke, M.A., Kanno, T., Matzke, A.J., 2015. RNA-directed DNA methylation: the evolution of a complex epigenetic pathway in flowering plants. Annu. Rev. Plant Biol. 66, 243–267.

Mikolajewski, D.J., Rusen, L., Mauersberger, R., Johansson, F., Rolff, J., 2015. Relaxed predation results in reduced phenotypic integration in a suite of dragonflies. J. Evol. Biol. 28, 1354–1363.

Minelli, A., Brena, C., Deflorian, G., Maruzzo, D., Fusco, G., 2006. From embryo to adult—beyond the conventional periodization of arthropod development. Dev. Genes Evol. 216, 373–383.

Momparler, R.L., Cote, S., 2015. Targeting of cancer stem cells by inhibitors of DNA and histone methylation. Expert Opin. Investig. Drugs 24, 1031–1043.

Mondor, E.B., Rosenheim, J.A., Addicott, J.F., 2005. Predator-induced transgenerational phenotypic plasticity in the cotton aphid. Oecologia 142, 104–108.

Morandin, C., Dhaygude, K., Paviala, J., Trontti, K., Wheat, C., Helantera, H., 2015. Caste-biases in gene expression are specific to developmental stage in the ant *Formica exsecta*. J. Evol. Biol. 28, 1705–1718.

Morris, K.V., 2015. The theory of RNA-mediated gene evolution. Epigenetics 10, 1–5.

Mukherjee, K., Twyman, R.M., Vilcinskas, A., 2015. Insects as models to study the epigenetic basis of disease. Prog. Biophys. Mol. Biol. 118, 69–78.

Nandety, R.S., Kuo, Y.W., Nouri, S., Falk, B.W., 2015. Emerging strategies for RNA interference (RNAi) applications in insects. Bioengineered 6, 8–19.

Ng, K.W., Anderson, C., Marshall, E.A., Minatel, B.C., Enfield, K.S., Saprunoff, H.L., Lam, W.L., Martinez, V.D., 2016. Piwi-interacting RNAs in cancer: emerging functions and clinical utility. Mol. Cancer 15, 5.

Nijhout, H.F., Callier, V., 2015. Developmental mechanisms of body size and wing-body scaling in insects. Annu. Rev. Entomol. 60, 141–156.

Nijhout, H.F., Riddiford, L.M., Mirth, C., Shingleton, A.W., Suzuki, Y., Callier, V., 2014. The developmental control of size in insects. Wiley Interdiscip. Rev. Dev. Biol. 3, 113–134.

Noble, D., 2015. Evolution beyond neo-Darwinism: a new conceptual framework. J. Exp. Biol. 218, 1273.

Noda, N., Umebayashi, K., Nakatani, T., Miyahara, K., Ishiyama, K., 2005. Isolation and characterization of some hydroxy fatty and phosphoric acid esters of 10-hydroxy-2-decenoic acid from the royal jelly of honeybees (*Apis mellifera*). Lipids 40, 833–838.

Onder, O., Sidoli, S., Carroll, M., Garcia, B.A., 2015. Progress in epigenetic histone modification analysis by mass spectrometry for clinical investigations. Expert Rev. Proteomics 12, 499–517.

Oppold, A., Kress, A., Vanden Bussche, J., Diogo, J.B., Kuch, U., Oehlmann, J., Vandegehuchte, M.B., Müller, R., 2015. Epigenetic alterations and decreasing insecticide sensitivity of the Asian tiger mosquito *Aedes albopictus*. Ecotoxicol. Environ. Saf. 122, 45–53.

Orr, T.J., Zuk, M., 2012. Sperm storage. Curr. Biol. 22, R8–10.

Pegoraro, M., Bafna, A., Davies, N.J., Shuker, D.M., Tauber, E., 2016. DNA methylation changes induced by long and short photoperiods in *Nasonia*. Genome Res. 26, 203–210.

Peleg, S., Feller, C., Forne, I., Schiller, E., Sevin, D.C., Schauer, T., Regnard, C., Straub, T., Prestel, M., Klima, C., Schmitt Nogueira, M., Becker, L., Klopstock, T., Sauer, U., Becker, P.B., Imhof, A., Ladurner, A.G., 2016. Life span extension by targeting a link between metabolism and histone acetylation in *Drosophila*. EMBO Rep. 17, 455–469.

Pezer, Z., Ugarkovic, D., 2012. Satellite DNA-associated siRNAs as mediators of heat shock response in insects. RNA Biol. 9, 587–595.

Pigliucci, M., 2009. An extended synthesis for evolutionary biology. Ann. N.Y. Acad. Sci. 1168, 218–228.

Pilpel, Y., Rechavi, O., 2015. The Lamarckian chicken and the Darwinian egg. Biol. Direct 10, 34.

Prendergast, J.G., Chambers, E.V., Semple, C.A., 2014. Sequence-level mechanisms of human epigenome evolution. Genome Biol. Evol. 6, 1758–1771.

Quintana-Murci, L., 2016. Genetic and epigenetic variation of human populations: an adaptive tale. C. R. Biol. 339, 278–283.

Rai, K., Huggins, I.J., James, S.R., Karpf, A.R., Jones, D.A., Cairns, B.R., 2008. DNA demethylation in zebrafish involves the coupling of a deaminase, a glycosylase, and gadd45. Cell 135, 1201–1212.

Redchuk, T.A., Rozhok, A.I., Zhuk, O.W., Kozeretska, I.A., Mousseau, T.A., 2012. DNA methylation in *Drosophila melanogaster* may depend on lineage heterogeneity. Tsitol. Genet. 46, 75–79.

Rehan, S.M., Glastad, K.M., Lawson, S.P., Hunt, B.G., 2016. The genome and methylome of a subsocial small carpenter bee, *Ceratina calcarata*. Genome Biol. Evol. 8, 1401–1410.

Remnant, E.J., Ashe, A., Young, P.E., Buchmann, G., Beekman, M., Allsopp, M.H., Suter, C.M., Drewell, R.A., Oldroyd, B.P., 2016. Parent-of-origin effects on genome-wide DNA methylation in the cape honey bee (*Apis mellifera capensis*) may be confounded by allele-specific methylation. BMC Genomics 17, 226.

Reynolds, J.A., Clark, J., Diakoff, S.J., Denlinger, D.L., 2013. Transcriptional evidence for small RNA regulation of pupal diapause in the flesh fly, *Sarcophaga bullata*. Insect Biochem. Mol. Biol. 43, 982–989.

Reynolds, J.A., Bautista-Jimenez, R., Denlinger, D.L., 2016. Changes in histone acetylation as potential mediators of pupal diapause in the flesh fly, *Sarcophaga bullata*. Insect Biochem. Mol. Biol. 76, 29–37.

Ronsseray, S., 2015. Paramutation phenomena in non-vertebrate animals. Semin. Cell Dev. Biol. 44, 39–46.

Rothbart, S.B., Strahl, B.D., 2014. Interpreting the language of histone and DNA modifications. Biochim. Biophys. Acta 1839, 627–643.

Sasai, M., Kawabata, Y., Makishi, K., Itoh, K., Terada, T.P., 2013. Time scales in epigenetic dynamics and phenotypic heterogeneity of embryonic stem cells. PLoS Comput. Biol. 9, e1003380.

Sas-Chen, A., Srivastava, S., Yarden, Y., 2017. The short and the long: non-coding RNAs and growth factors in cancer progression. Biochem. Soc. Trans. 45, 51–64.

Schiesari, L., O'Connor, M.B., 2013. Diapause: delaying the developmental clock in response to a changing environment. Curr. Top. Dev. Biol. 105, 213–246.

Seymour, D.K., Becker, C., 2017. The causes and consequences of DNA methylome variation in plants. Curr. Opin. Plant Biol. 36, 56–63.

Sharma, A., 2015. Systems genomics analysis centered on epigenetic inheritance supports development of a unified theory of biology. J. Exp. Biol. 218, 3368–3373.

Shi, W., Guo, Y., Xu, C., Tan, S., Miao, J., Feng, Y., Zhao, H., St. Leger, R.J., Fang, W., 2014. Unveiling the mechanism by which microsporidian parasites prevent locust swarm behavior. Proc. Natl. Acad. Sci. U.S.A. 111, 1343–1348.

Simola, D.F., Graham, R.J., Brady, C.M., Enzmann, B.L., Desplan, C., Ray, A., Zwiebel, L.J., Bonasio, R., Reinberg, D., Liebig, J., Berger, S.L., 2016. Epigenetic (re)programming of caste-specific behavior in the ant *Camponotus floridanus*. Science 351, aac6633.

Simon, J.C., Stoeckel, S., Tagu, D., 2010. Evolutionary and functional insights into reproductive strategies of aphids. C. R. Biol. 333, 488–496.

Simpson, S.J., Miller, G.A., 2007. Maternal effects on phase characteristics in the desert locust, *Schistocerca gregaria*: a review of current understanding. J. Insect Physiol. 53, 869–876.

Skinner, M.K., 2015. Environmental epigenetics and a unified theory of the molecular aspects of evolution: a neo-Lamarckian concept that facilitates neo-Darwinian evolution. Genome Biol. Evol. 7, 1296–1302.

Snell-Rood, E.C., Troth, A., Moczek, A.P., 2013. DNA methylation as a mechanism of nutritional plasticity: limited support from horned beetles. J. Exp. Zool. B Mol. Dev. Evol. 320, 22–34.

Spannhoff, A., Kim, Y.K., Raynal, N.J., Gharibyan, V., Su, M.B., Zhou, Y.Y., Li, J., Castellano, S., Sbardella, G., Issa, J.P., Bedfort, M.T., 2011. Histone deacetylase inhibitor activity in royal jelly might facilitate caste switching in bees. EMBO Rep. 12, 238–243.

Springer, N.M., McGinnis, K.M., 2015. Paramutation in evolution, population genetics and breeding. Semin. Cell Dev. Biol. 44, 33–38.

Stilling, R.M., Dinan, T.G., Cryan, J.F., 2016. The brain's geppetto—microbes as puppeteers of neural function and behaviour? J. Neurovirol. 22, 14–21.

Strachecka, A., Olszewski, K., Bajda, M., Demetraki-Paleolog, J., 2015. Natural larval diet differently influences the pattern of developmental changes in DNA 5-methylcytosine levels in *Apis mellifera* queens as compared with workers and drones. Biochemistry 80, 1019–1025.

Stuart, T., Eichten, S.R., Cahn, J., Karpievitch, Y.V., Borevitz, J.O., Lister, R., 2016. Population scale mapping of transposable element diversity reveals links to gene regulation and epigenomic variation. eLife 5, e20777.

Taira, W., Hiyama, A., Nohara, C., Sakauchi, K., Otaki, J.M., 2015. Ingestional and transgenerational effects of the Fukushima nuclear accident on the pale grass blue butterfly. J. Radiat. Res. 56, i2–i18.

Tanaka, S., Maeno, K., 2006. Phase-related body-color polyphenism in hatchlings of the desert locust, *Schistocerca gregaria*: re-examination of the maternal and crowding effects. J. Insect Physiol. 52, 1054–1061.

Tweedie, S., Charlton, J., Clark, V., Bird, A., 1997. Methylation of genomes and genes at the invertebrate-vertebrate boundary. Mol. Cell Biol. 17, 1469–1475.

Urena, E., Chafino, S., Manjon, C., Franch-Marro, X., Martin, D., 2016. The occurrence of the holometabolous pupal stage requires the interaction between e93, kruppel-homolog 1 and *broad-complex*. PLoS Genet. 12, e1006020.

Vaiserman, A., 2014. Developmental epigenetic programming of caste-specific differences in social insects: an impact on longevity. Curr. Aging Sci. 7, 176–186.

Vandegehuchte, M.B., Janssen, C.R., 2014. Epigenetics in an ecotoxicological context. Mutat. Res. Genet. Toxicol. Environ. Mutagen. 764–765, 36–45.

Vidalis, A., Zivkovic, D., Wardenaar, R., Roquis, D., Tellier, A., Johannes, F., 2016. Methylome evolution in plants. Genome Biol. 17, 264.

Vieira, C., Fablet, M., Lerat, E., Boulesteix, M., Rebollo, R., Burlet, N., Akkouche, A., Hubert, B., Mortada, H., Biemont, C., 2012. A comparative analysis of the amounts and dynamics of transposable elements in natural populations of *Drosophila melanogaster* and *Drosophila simulans*. J. Environ. Radioact. 113, 83–86.

Wang, X., Werren, J.H., Clark, A.G., 2015. Genetic and epigenetic architecture of sex-biased expression in the jewel wasps *Nasonia vitripennis* and *giraulti*. Proc. Natl. Acad. Sci. U.S.A. 112, E3545–E3554.

Wendte, J.M., Pikaard, C.S., 2017. The RNAs of RNA-directed DNA methylation. Biochim. Biophys. Acta 1860, 140–148.

Wenseleers, T., Van Oystaeyen, A., 2011. Unusual modes of reproduction in social insects: shedding light on the evolutionary paradox of sex. Bioessays 33, 927–937.

Xiang, H., Zhu, J., Chen, Q., Dai, F., Li, X., Li, M., Zhang, H., Zhang, G., Li, D., Dong, Y., Zhao, L., Lin, Y., Cheng, D., Yu, J., Sun, J., Zhou, X., Ma, K., He, Y., Zhao, Y., Guo, S., Ye, M., Guo, G., Li, Y., Li, R., Zhang, X., Ma, L., Kristiansen, K., Guo, Q., Jiang, J., Beck, S., Xia, Q., Wang, W., Wang, J., 2010. Single base-resolution methylome of the silkworm reveals a sparse epigenomic map. Nat. Biotechnol. 28, 516–520.

Yan, H., Simola, D.F., Bonasio, R., Liebig, J., Berger, S.L., Reinberg, D., 2014. Eusocial insects as emerging models for behavioural epigenetics. Nat. Rev. Genet. 15, 677–688.

Yan, H., Tian, S., Slager, S.L., Sun, Z., Ordog, T., 2016. Genome-wide epigenetic studies in human disease: a primer on -omic technologies. Am. J. Epidemiol. 183, 96–109.

Ye, Y.H., Woolfit, M., Huttley, G.A., Rances, E., Caragata, E.P., Popovici, J., O'Neill, S.L., McGraw, E.A., 2013. Infection with a virulent strain of Wolbachia disrupts genome wide-patterns of cytosine methylation in the mosquito *Aedes aegypti*. PLoS One 8, e66482.

Ye, Y., Gu, L., Chen, X., Shi, J., Zhang, X., Jiang, C., 2016. Chromatin remodeling during the *in vivo* glial differentiation in early *Drosophila embryos*. Sci. Rep. 6, 33422.

Ye, Y., Li, M., Gu, L., Chen, X., Shi, J., Zhang, X., Jiang, C., 2017. Chromatin remodeling during *in vivo* neural stem cells differentiating to neurons in early *Drosophila* embryos. Cell Death Differ. 24, 409–420.

Zemach, A., McDaniel, I.E., Silva, P., Zilberman, D., 2010. Genome-wide evolutionary analysis of eukaryotic DNA methylation. Science 328, 916–919.

Zhang, G., Hussain, M., O'Neill, S.L., Asgari, S., 2013. *Wolbachia* uses a host microRNA to regulate transcripts of a methyltransferase, contributing to dengue virus inhibition in *Aedes aegypti*. Proc. Natl. Acad. Sci. U.S.A. 110, 10276–10281.

Zhang, T., Cooper, S., Brockdorff, N., 2015. The interplay of histone modifications—writers that read. EMBO Rep. 16, 1467–1481.

Zhao, M., Wang, Z., Yung, S., Lu, Q., 2015. Epigenetic dynamics in immunity and autoimmunity. Int. J. Biochem. Cell Biol. 67, 65–74.

Zhou, Q.Z., Fang, S.M., Zhang, Q., Yu, Q.Y., Zhang, Z., 2017. Identification and comparison of long non-coding RNAs in the silk gland between the domestic and wild silkworms. Insect Sci. http://dx.doi.org/10.1111/1744-7917.12443 [Epub ahead of print].

CHAPTER TWO

Nutrition and Epigenetic Change in Insects: Evidence and Implications

Andrew Cridge*[,1], Tom Harrop*[,1], Mackenzie Lovegrove*[,1], Emily Remnant†[,1], Peter Dearden*

*Laboratory for Evolution and Development, Genetics Otago, University of Otago, Dunedin, New Zealand
†Behaviour and Genetics of Social Insects Laboratory, School of Life and Environmental Sciences, University of Sydney, Sydney, NSW, Australia

Contents

Abstract

Insects provide valuable model systems to help us understand the effects of nutrition on body form and life history traits. Insects often show very pronounced, extreme differences in phenotype in response to nutritional cues, and in recent years epigenetic mechanisms have been proposed as key to these responses. Here we review the links between nutrition, epigenetic change, and phenotype in a range of insects, focusing on evidence for epigenetic change linking nutrition and phenotype. While evidence exists that epigenetics plays a key role in responding to nutrition, we suggest that investigating the chain of causality from nutritional change to epigenetic change to phenotype is required before we can understand the role of epigenetics in insect nutrition.

[1] These authors contributed equally to this work.

Advances in Insect Physiology, Volume 53
ISSN 0065-2806
http://dx.doi.org/10.1016/bs.aiip.2017.06.001

1. INTRODUCTION

One critical question in genetics is how environmental influences regulate genes and contribute to phenotype over the lifespan of an organism. This question is particularly pertinent to human health, where environmental challenge and diet are key to the most pernicious disease in the western world: obesity and diabetes (Bray and Popkin, 2014). Understanding how early-life exposure, or diet, leads to life-long health implications has led to the establishment of a new field of research, the Developmental Origins of Health and Disease (DoHaD) paradigm (Gluckman and Hanson, 2006; Gluckman et al., 2007). In this field, understanding how diet, in particular, leads to long-term changes in gene expression (Bateson et al., 2004; Godfrey et al., 2010), and alternative metabolic phenotypes (Campbell et al., 2016), is a critical question. Epigenetics, a branch of gene regulation, has been widely accepted as the key underlying mechanism by which wide-ranging, and long-term, changes in gene regulation are achieved (Waterland and Michels, 2007).

The DoHaD paradigm, with epigenetics at its heart, has always looked for parallels in animal biology that might provide insight into human health phenomenon. This has included predictive adaptive responses (Bateson et al., 2014), as demonstrated by the meadow vole, and polyphenisms, as demonstrated by many insects, especially the honeybee *Apis mellifera*. Polyphenisms and predictive adaptive responses can be reduced to the concept of a signal being received by the organism, which causes a transformation in gene expression, leading to alternative phenotypes (Bateson, 2007; Bateson et al., 2014). In the meadow vole, this is daylength during pregnancy being transmitted to the offspring leading to differences in coat thickness: light for voles born in spring and heavy for those born in autumn (Lee et al., 1987). In the honeybee, the signal, royal jelly, leads to a rearrangement of development trajectories to produce a queen bee, or a worker when royal jelly is not encountered (Cameron et al., 2013). These systems are hypothesised to mirror what may happen in a human pregnancy, where maternal-to-foetal signals about diet, micronutrient, and stress trigger changes in gene expression and metabolic set points that may better match a child with their postnatal nutritional environment, or, if those signals are erroneous or misinterpreted, lead to a mismatch of physiology and environment, and eventually perhaps disease (Gluckman and Hanson, 2004).

The critical contribution of honeybee caste development to the DoHaD field is an experiment in 2008, in which it was reported that knocking out the key de novo methylase during honeybee larval development blocks the development of worker honeybees, i.e., DNA methylation is required to produce castes in honeybees (Kucharski et al., 2008). This key experiment has been taken to demonstrate that alternative phenotypes, triggered by nutritional signals, are produced via epigenetic mechanisms (Kucharski et al., 2008). The insect example here has been taken as strong evidence that epigenetics underpins DoHaD phenomena (Kubota et al., 2015). This translation of results ignores the facts that the DNA methylation in insects follows differing patterns to methylation in vertebrates (Beeler et al., 2014; Bonasio et al., 2012; Drewell et al., 2014; Lyko et al., 2010; Yan et al., 2015), that many insects do not perform CpG DNA methylation at all, and that little functional evidence has emerged, indicating the mechanism by which DNA methylase knockdown has its effects.

Given the importance of polyphenisms to the DoHaD phenomenon, and the ease by which both the genetics and environment of insect model systems can be manipulated, we suggest it is important to review the evidence supporting, or otherwise, the role(s) of epigenetic changes in response to nutrition in insects.

1.1 What Is Epigenetics?

Epigenetics is a confusing term as it has changed its meaning since being coined by Waddington in 1942 (Waddington, 1942). Initially, epigenetics referred to the processes that lie between genetics and phenotype and thus was really a reflection of all aspects of gene regulation. In later years, epigenetics has come to refer to a small range of gene regulation mechanisms, centred around changes in the conformation of chromatin. Here, in line with current definitions, we will refer to modifications in chromatin structure via histone modification (methylation, acetylation, ubiquitination, and phosphorylation), DNA methylation, and regulation by small RNAs as epigenetics (Duncan et al., 2014). It must be noted that this is a very arbitrary definition, narrowly defining a class of chemical changes as 'epigenetics'. It is important to also note that these mechanisms have been implicated in a wide range of activities in the genome, including regulation of transposable elements (Slotkin and Martienssen, 2007), chromosome inactivation (Mohandas et al., 1981), increase in chromosome transcriptional activity

(Conrad and Akhtar, 2012), imprinting (Verona et al., 2003), gene regulation (Jaenisch and Bird, 2003), and more.

In this review, we examine the evidence for nutritional cues affecting epigenetics in nonsocial insects, and examine the role of nutrition in life history traits and then in the development of polyphenisms in social insects. Genome-wide changes in gene expression are a hallmark of epigenetic regulation events; here we try to link the evidence between such changes in response to nutrition and epigenetic change in the genome.

2. EPIGENETIC EFFECTS OF NUTRITION IN INSECTS

Epigenetic modifications play important roles in modulating gene expression and can therefore establish, sustain, and alter organism-level phenotypes, including behaviour and lifespan. The role of nutrition in inducing epigenetic changes has been the target of a number of studies in insects that exhibit environmental influenced phenotypes, including model insect system such as the fruitfly *Drosophila melanogaster*. *Drosophila* provides an excellent model system in which to examine these effects, determine their molecular mechanisms, and perhaps identify a link to epigenetic modification. It must be noted, however, that *Drosophila* does not carry out CpG methylation of its genome (Glastad et al., 2011; Lyko and Maleszka, 2011).

2.1 Molecular Effects of Nutrition in *Drosophila*

Dietary restriction (DR), a reduction in nutrient intake without malnutrition, is the most reproducible way to extend adult lifespan in a wide range of organisms, including insects (Mair et al., 2003). DR exerted during adulthood extends lifespan and reduces fertility in *Drosophila* (Chapman and Partridge, 1996; Nusbaum and Rose, 1999). It has been suggested that this effect is adaptive and results from reallocation of resources from reproduction to somatic maintenance, in order to survive periods of DR in nature (Zajitschek et al., 2016). Indeed, lifetime egg production is a good index of fitness and is maximised when females flies are fed a diet with a protein:carbohydrate ration (P:C) of 1:4, this is at the expense of longevity alone, which is maximised by feeding at P:C 1:16 (Lee et al., 2008). Also, poor nutrition during development results in additional detrimental fitness effects including decreased size, fecundity, and lifespan (Kolss et al., 2009; Zajitschek et al., 2013; Zwaan et al., 1991). It is presumed that poor larval nutrition leads to unfavourable effects in the adult. However, several recent studies imply that the effect of the developmental environment depends on the specific adult environment experienced (May et al., 2015). In *Drosophila,* poor larval food also exerts an adult environment-dependent effect, especially by affecting virgin lifespan and altering adult patterns of reproductive investment (May et al., 2015) indicating that most favourable nutritional conditions are likely to be different for larvae and adults.

A tremendous number of changes in gene expression occur as early as 4 days following exposure to DR and persevere as long as flies are maintained in a reduced nutrient environment (Pletcher et al., 2002). Many of the over 2000 genes that are differentially regulated during life-long DR are regulated similarly by acute starvation in larvae and by yeast starvation in adult flies (Pletcher et al., 2005). Genes that are downregulated specifically upon DR include those involved in DNA repair and cell signalling including protein phosphatase 2 complex formation. Surprisingly, genes involved in muscle formation and function are significantly upregulated upon DR, as are several neuropeptides that have alleged roles in feeding and ageing (Pletcher et al., 2005). This kind of whole-genome response implies epigenetic regulation is at the heart of this phenomenon.

An alternative approach to identifying changes in gene expression comes from analysis of the acute response of transcript levels to a shift in nutrient environment. Initially, this was applied to shifts in nutrient quality such as acute fasting or removing all dietary protein. Zinke et al. (2002) studied second instar *Drosophila* larvae when starved or when fed only sugar with samples taken over several hours. Based on a criterion where a fourfold change occurred in at least one time point, 40 genes were regulated upon starvation, but not in sugar conditions, and 51 genes were regulated in sugar condition, but not under starvation (Zinke et al., 2002). Gershman et al. (2007) studied the global gene expression dynamics at high-temporal resolution in response to nutrition in *Drosophila* but only looked at complete yeast deprivation with acute replacement. Approximately 3500 transcripts responded to nutrition, most notably in the insulin and target of rapamycin (TOR) pathways, purine synthesis, tricarboxylic acid cycle (TCA)-biosynthetic functions and mitochondria biogenesis (Whitaker et al., 2014).

Caloric restriction (CR) without malnutrition delays age-related pathologies and extends survival in flies, but the magnitude of the effect varies with genetic background. CR has multiple effects on fly physiology, such as decreased weight and fecundity, increased physical activity and oxidative stress resistance, and delayed lipid oxidative damage (Bross et al., 2005; Parashar and Rogina, 2009; Zheng et al., 2005).

Caloric excess (CE), an oversupply of calories in the diet, in prenatal insect nutritional environment also affects offspring. In *Drosophila*, a high-sugar maternal diet altered body composition of larval offspring for at least two generations, generating an obese-like phenotype under suboptimal (high-calorie) feeding conditions in adults which was characterised by modification in expression of metabolic genes (Buescher et al., 2013; Li-Byarlay et al., 2013). The highest and earliest activated gene upon sugar ingestion is *sugarbabe*, a zinc finger protein that is induced in the gut and the fat body. Identification of potential targets using microarrays suggests that *sugarbabe* represses genes involved in dietary fat breakdown and absorption (Zinke et al., 2002).

Although DR, CR, and CE all correlated with widespread changes in gene expression in *Drosophila*, establishing the mechanistic relationship between gene expression changes and epigenetic regulation is required to justify the link between diet and phenotype identified in these studies.

2.2 Nutrition and Epigenetic Change in *Drosophila*

In *Drosophila* changes in adult and larval nutritional environment have been shown to exhibit effects on phenotype through changes in gene expression. However, the underlying mechanisms responsible for driving these changes are only now starting to emerge. Studies that have assessed the effect of dietary restriction or excess have been pivotal in identifying the mechanisms responsible. Predominantly these studies have focused on epigenetic signals, including histone modifications, noncoding RNA transcripts (ncRNAs), and transcription factors (TFs), as the mechanism by which nutritional status influences phenotype and allows transgenerational transmission of environmental information.

Modification in methylation patterns of histones is correlated with response to both caloric restriction and excess, indicating a role for diet in regulating the activities of histone acetylases and deacetylases (reviewed in Lian et al., 2015). CR of flies leads to a moderate but significant down-regulation of Rpd3 which possesses histone deacetylase (HDAC) activity (class I lysine deacetylase) and has a major role in epigenetic regulation of chromatin structure (Pletcher et al., 2002). Independent of diet, knockdown of Rpd3 levels leads to a lifespan extension in flies (Rogina et al., 2002). Longevity is increased in control flies fed a low-calorie diet compared to *rpd3* mutants fed a normal diet, whereas CR of the *rpd3* mutants shows no further extension of lifespan (Rogina et al., 2002). Similar effects on

Drosophila lifespan are achieved by feeding fruit flies 4-phenylbutyrate (PBA), an HDAC inhibitor (Kang et al., 2002). Also, reduction of *rpd3* downregulates the expression of members of the insulin/insulin-like growth factor signalling pathway (IIS), which is associated with altered metabolism, increased energy storage, and higher resistance to starvation and oxidative stress (Woods and Rogina, 2016). The role of IIS in *rpd3* longevity is supported by genetic studies that show reduced stress resistance and longevity of flies double mutant for *rpd3* and the transcription factor *dfoxo,* a downstream target of IIS pathway, compared to *rpd3* single-mutant flies. Increased *dfoxo* is a mediator of *rpd3*'s effects on fly longevity and intermediary metabolism, and confers a new link between *rpd3* and IIS longevity pathways (Woods et al., 2016), as well as highlighting a causal link between histone methylation and longevity in *Drosophila.*

The developmental response of larvae to DR may involve similar mechanisms to the effect of DR on the adult. Posteclosion dietary manipulation with a low-protein (LP) diet in *Drosophila* upregulates the protein level of E(z), an H3K27 specific methyltransferase, leading to higher levels of H3K27 trimethylation (H3K27me3) and is correlated with changes in gene expression (Zinke et al., 2002). This diet-mediated change in H3K27me3 corresponds with a shortened longevity of F0 flies as well as their F2 offspring (Xia et al., 2016). Specific RNAi-mediated posteclosion knockdown of E(z) or pharmacological inhibition of its enzymatic function with EPZ-6438 in the F0 parents improved longevity while rendering H3K27me3 low across generations. The addition of EPZ-6438 to the LP diet alleviated the longevity-reducing effect of the poor diet, implicating increased levels of E(z)-dependent H3K27me3 as the primary cause in programmed longevity through epigenetic regulation (Xia et al., 2016). The requirements for H3K9/K27me3-dependent reprogramming of metabolic genes have also been identified in a model for paternal diet-induced intergenerational metabolic reprogramming where paternal sugar acts as a physiological suppressor of variegation, desilencing chromatin-state-defined domains in both mature sperm and offspring embryos (Öst et al., 2014).

Drosophila does not contain any of the canonical DNA methyltransferases (Dnmt1 and Dnmt3) implicated in regulating DNA methylation patterns in response to nutritional status in *A. mellifera* (Kucharski et al., 2008) and other insects discussed below. *Drosophila* DNA is methylated by Dnmt2 (reviewed in Krauss and Reuter, 2011) but is restricted to the early stages of embryonic development (Lyko et al., 2000). To date, Dnmt2 genomic DNA methylation has not been implicated in regulating DNA methylation patterns in response to nutritional status in *Drosophila*. The lack of Dnmt3 in *Drosophila* prevents the investigation of the role of diet on methylation in this model system.

2.3 Nutrition and Epigenetics in Other Insects

While work in *Drosophila* has linked epigenetic regulation with large-scale gene expression, outside *Drosophila* only a few studies have been carried out. Perhaps the best-examined polymorphisms occur in beetles. Mandible size in male broad-horned flour beetles, *Gnatocerus cornutus*, is hyperallometric to body size, meaning that increase in mandible size is disproportionate to increase in body size (Ozawa et al., 2016). In theory, males would develop larger mandibles when nutrient availability is favourable as a mechanism to increase the probability of reproduction, leading to sexual selection for plasticity of mandible size. Injection of an HDAC inhibitor or RNAi of either of two HDAC genes in *G. cornutus* influences mandible size, suggesting that the phenotypic plasticity is linked to epigenetic modifications (Ozawa et al., 2016). However, the effect on histone acetylation caused by changes in *HDAC1* and *HDAC3* expression has not been clarified, and evidence for the link between nutrition and epigenetic marks in beetles is lacking. Correlation between nutritional environment and methylation was detected at only 8%–16% of methylated sites in the scarab beetle *Onthophagus gazella* (Snell-Rood et al., 2013). In contrast, a link between insulin signalling and nutrient-dependent variation in sexually selected beetle weapons has been suggested by RNAi of the insulin receptor in *Trypoxylus dichotomus* (Emlen et al., 2012).

3. INSECT PHENOTYPIC PLASTICITY

Phenotypic plasticity is the ability of an organism to alter its phenotype in response to environmental influences, without altering its genome (Bateson et al., 2004; West-Eberhard, 1989). Plasticity is an interesting field of study as it provides a way to observe the interaction between genomes and their environments. Further, adding to the appeal of plasticity is the fact that it is a feature of all major braches of plant and animal life, making it fundamental to our understanding of biology. Phenotypically plastic events can be classified into two groups: reaction norms and polyphenisms. A reaction norm is the product of an environmental stimuli acting on gene expression leading to phenotypic variation. This variation is continuous over the reaction norm curve, meaning that a multitude of slightly differing phenotypes can be produced (Stearns, 1989). The second type of plasticity is polyphenisms. Polyphenisms are defined as discrete alternative phenotypes. They can be viewed as one phenotype for values of an environmental variable below a threshold and one phenotype for values above the threshold (Stearns, 1989). In this way, there are limited distinct phenotypes that can be produced (Simpson et al., 2011). Polyphenisms are astounding when one considers that an individual genome encodes two or more completely different developmental trajectories that can be accessed and their expression regulated based on the environmental cues they receive.

Polyphenisms are widely prevalent within the insects, in particular among the social insects, where the production of differing castes aids in the division of tasks within a community. For eusocial insects, this division includes reproductive division of labour, whereby the queen reproduces and her daughter workers aid in rearing her offspring (Oster and Wilson, 1978). It is defined by three characteristics: cooperative care of brood, overlapping generations of adults living together, and reproductive skew towards certain females (Michener, 1974). It is a fascinating life history strategy that exhibits one of the greatest levels of biological complexity yet observed (Maynard Smith and Szathmary, 1995). It has obvious benefits, with eusocial insects dominating the insect fauna of ecosystems in which they live (Wilson, 1990).

Eusociality has evolved independently in several animal lineages (Wilson and Holldobler, 2005), and estimated 20 separate times in the insects (Grimaldi and Engel, 2005). In particular, it has often evolved within the group known as the Hymenoptera (which contains the bees, ants, and wasps), with eight independent transitions (Crozier, 2008; Hughes et al., 2008). Eusociality relies on a reproductive dominance system, where one female caste produces offspring and the other is constrained. In many cases, reproductive constraint relies on both behavioural and physiological mechanisms, such as aggression, whereby worker-laid eggs are killed by the queen or other workers, or the use of pheromones (Hefetz et al., 2016).

3.1 Nutrition and Polyphenism in Social Insects

Insects in general, and social insects in particular, provide elegant examples of the extreme polyphenisms that can be generated by different environmental stimuli during key developmental time points. Polyphenisms can be triggered by a range of environmental conditions such as seasonality and density (Simpson et al., 2011). However, some of the best characterised are those that are nutritionally mediated. One particularly well-known example is the honeybee *A. mellifera*. Female honeybees share the same genetic complement, yet develop into two distinct castes—that of queen or workers—depending on the environmental factor of early-life nutrition, namely, feeding of royal jelly by nurse bees (Buttstedt et al., 2014; Cameron et al., 2013; Huang et al., 2012; Kamakura, 2011; Kucharski et al., 2008; Townsend and Lucas, 1940). Royal jelly contains a variety of components, including vitamins, lipids, amino acids, and a family of proteins—major royal jelly proteins—that are believed to be vital for reproductive maturation (Maleszka, 2008). Female larvae are bipotent at less than 3 days of age and can still develop into either a queen or a worker (Shuel and Dixon, 1959, 1960). Their developmental fate relies on the quality and quantity of food provided for them by nurse bees throughout the next two instars (Haydak, 1943; Jay, 1964). Those larvae destined to become queens are fed a surplus of royal jelly and will develop more rapidly. By 4 days into the larval stage, they will weigh 50% more than their worker counterparts (Wang, 1965; Weaver, 1957). Queens require only 16 days to develop (Bertholf, 1925), which is 5 days faster than the development of the worker bee.

After development, there are various behavioural and biological differences between these two castes. Queens are larger, possess jagged mandibles, produce queen mandibular pheromone, have a greater fecundity with a spermatheca and active ovaries (120–200 ovarioles on average), and have an extended lifespan of 1–3 years (Seeley, 1978). Workers are reproductively inactive in the presence of a queen and possess inactive ovaries with only 2–12 ovarioles and no spermatheca (Cridge et al., 2016). They have corbicula (pollen carrying 'baskets'), undertake nursing and foraging roles within the hive, and have a far shorter lifespan than their sister queens, 3–5 weeks in summer and up to 20 weeks in winter (Fukuda and Kiichi, 1966; Winston et al., 1981). The queen bee is usually the only reproductive female within the hive, with worker bees tending to the young and caring for the queen.

Although eusociality has evolved independently in different hymenopteran lineages, much like honeybees, ants and wasps display physiologically and reproductively distinct castes, with underlying regulatory networks derived from similar origins (Corona et al., 2016). In particular, ants show a diverse range of extreme caste morphologies including queens, ergatoid (wingless) queens, workers, and soldiers (Trible and Kronauer, 2017). Caste development in various ant species can be triggered by a spectrum of factors ranging from species that exhibit solely environmental caste determination, to other species where caste is genetically influenced (Schwander et al., 2010). In honeybees, the environmental variable of nutrition is necessary and sufficient to initiate a queen-like developmental pathway regardless of genetic background, whereas in some ant species, a genetic component can predispose individuals to particular caste trajectories, or caste can be strictly genetically determined (Schwander et al., 2010). Nevertheless, nutritional signals such as food quality and quantity influence caste development in many species of ants (Smith et al., 2008; Trible and Kronauer, 2017) and wasps (Berens et al., 2014).

3.2 Nutrition and Epigenetics in Social Insect Polyphenisms

It remains unclear exactly how the genome communicates with environmental and nutritional cues to produce different insect morphologies from bipotent larvae. There is considerable evidence to suggest that epigenetics may provide the missing link. Epigenetic mechanisms are critical for determining tissue differentiation and stabilising cell fate in multicellular organisms (Jaenisch and Bird, 2003). Different castes of social insect colonies have been likened to the various cell types and tissues that make up an organism, and therefore the same epigenetic processes have been implicated as candidate mechanisms for caste differentiation in social insects (Bonasio, 2014; Kucharski et al., 2008; Lyko and Maleszka, 2011). Now, molecular investigations into the link between nutritionally mediated polyphenism and epigenetic regulation in insects are beginning to reveal aspects of how the environment and epigenome interact (Lyko et al., 2010). However, it is not clear whether epigenetic factors cause, facilitate, or simply maintain caste differentiation once transcriptional differences have been triggered by nutritional stimuli.

3.3 Histones and Small RNAs in Caste Development

Interestingly, royal jelly contains an HDAC (phenyl butyrate), generally involved in the removal of acetyl groups from histones. These allow chromatin to be compacted and prevent access by transcription machinery, silencing gene expression (Kouzarides, 2007). These enzymes often act together with DNA methyltransferases (Jaenisch and Bird, 2003), making direct epigenetic regulation of gene expression possible in caste development in honeybees. Furthermore, a variety of small RNAs are found in royal jelly and worker jelly, the alternate diet fed to developing worker bees (Guo et al., 2013). Small RNAs such as microRNAs (miRNAs), piwi-interacting

RNAs (piRNAs), and short interfering RNAs (siRNAs) are involved in posttranscriptional gene silencing, providing an added level of epigenetic regulation to transcribed genes (Carthew and Sontheimer, 2009). There is a greater abundance of miRNAs in worker jelly compared to royal jelly (Guo et al., 2013), as well as differences in miRNA expression profiles between honeybee castes (Ashby et al., 2016). Reproductive queens also possess higher levels of piRNAs compared to sterile workers, possibly correlated with germline suppression of transposable elements, but may also be involved in the development of caste-specific phenotypes (Wang et al., 2017). Further, miRNA expression differs between castes in the ant species *Camponotus floridanus* and *Harpegnathos saltator* (Bonasio et al., 2010), and histone acetylation varies between castes in *C. floridanus* (Simola et al., 2013, 2016). Thus, histones and small noncoding RNAs such as miRNAs and piRNAs could well be involved in the nutritional component of caste determination.

3.4 DNA Methylation in Caste Development

Since the initial discovery of an active DNA methylation system in social insects (Wang et al., 2006), a number of studies have attempted to demonstrate a link between nutrition, epigenetics, and caste fate (Bonasio et al., 2012; Foret et al., 2012; Kucharski et al., 2008; Lyko et al., 2010; Weiner et al., 2013). Genome-wide analysis of DNA methylation in insects has revealed that methylation patterns differ from that observed in mammals and plants. Vertebrate and plant genomes are densely methylated across the genome with between 60% and 90% of methylation at CpG sites, compared to less than 1% of methylated CpG sites in honeybees, wasps, and ants (Beeler et al., 2014; Bonasio et al., 2012; Drewell et al., 2014; Lyko et al., 2010; Yan et al., 2015). Further, while vertebrate and plant methylation occurs in genes, transposable elements, and upstream sequences, insect methylation is largely restricted to gene bodies—the exons and introns of genes (Sarda et al., 2012; Yan et al., 2015). Methylation in gene bodies is associated with constitutively expressed, highly conserved genes, as opposed to transcriptional repression observed when methylation occurs in promoters (Bird, 2002; Foret et al., 2009; Sarda et al., 2012). How these patterns of methylation correlate to caste, whether methylation is regulated by environmental variables such as nutrition and the mechanisms by which DNA methylation is directed to particular sites and genes, is subject to wide investigation (Lyko and Maleszka, 2011; Maleszka, 2016).

The first functional evidence that DNA methylation was involved in caste differentiation was shown in honeybees (Kucharski et al., 2008). Silencing the expression of the de novo DNA methyltransferase enzyme, DNMT3, in newly hatched lab-reared larvae led to a skew in the development of adults with queen-like morphology. Queens emerging from the DNMT3-silenced group had significantly more ovarioles than control treatments (Kucharski et al., 2008). To correlate nutrition and methylation levels, the gene *dynactin p62* was used as a measure of altered methylation status, as it has been previously shown to be differentially methylated during honeybee development (Wang et al., 2006). Samples were taken from hive-reared whole larvae, larval heads, and heads of the Dnmt3-silenced larvae. In all comparisons, methylation of *dynactin p62* was reduced in queen larvae compared to worker larvae by approximately 10%. This shows that early-life nutrition is capable of leading to a difference in methylation levels. However, further work is required before a causal link can be established between diet-induced methylation changes, altered gene expression, and reproductive phenotype (Kucharski et al., 2008).

Diet is clearly the primary factor influencing the development of a queen caste in honeybees. However, the size of the brood cell in which a larva is reared may also contribute to the process. Larvae destined to be queens are laid in queen cells, which are larger than those designated for workers. Both nutrition and cell size were shown to affect caste determination by altering methyltransferase activity, suggesting that nutrition is not the only environmental cue that can influence DNA methylation levels. Larvae that were reared in queen cells and larvae with increased exposure to royal jelly showed decreased expression levels and activity of DNMT3 and decreased methylation at specific CpG sites in *dynactin p62* (Shi et al., 2011). This led to an increase in the number of queens produced and a decrease in workers and intercastes. Environmental cues of diet or cell size were each associated with differential methylation at particular CpG sites, suggesting that methylation is targeted and specific in nature.

Comparison of genome-wide methylation between social insect castes has shown differences in honeybee queen and worker brain methylomes (Lyko et al., 2010), developing larval methylomes (Foret et al., 2012), and methylomes of the ants *C. floridanus* and *H. saltator* (Bonasio et al., 2012). However, methylation in gene bodies is associated with maintenance of housekeeping gene expression (Glastad et al., 2011; Sarda et al., 2012). Furthermore, genes predicted to have low methylation based on low levels of CpG depletion were more likely to be overexpressed in different castes

(Elango et al., 2009). The presence of methylation in housekeeping genes rather than caste-specific genes is not consistent with a role in caste development.

One proposed mechanism for how DNA methylation might be altering caste development is through alternative splicing. DNA methylation has a role in regulating alternative splicing of mRNA precursors, with exons having higher levels of DNA methylation in comparison to flanking introns (Maor et al., 2015). Whole-genome high-resolution bisulphite sequencing was carried out on brains from queens and workers (Lyko et al., 2010), which showed methylation was not as widespread as in the human genome. Instead, it was limited to smaller, specific regions, such as those with critical cellular functions. Approximately 550 genes showed altered methylation patterns between the castes, which may account for some of the differences observed in behaviour between queen and worker honeybees. There was a correlation between methylation clustering and regions of genes with the potential to generate alternative transcripts (Lyko et al., 2010). This led to the hypothesis that the polyphenisms produced when female larvae develop into either queens or workers are due to DNA methylation altering the accessibility of certain regions of genes to the cellular machinery that controls their regulation.

A second study looking at whole-genome DNA methylation differences in larval and adult heads also found that DNA methylation is reduced in queens. In particular, differential methylation was enriched in regions that showed conserved domains (such as helicases and zinc fingers), and metabolic networks, providing a link between nutrition and altered metabolic functioning (Foret et al., 2012). Among these pathways of interest were the TCA cycle (related to altered insulin signalling in the brain of adult honeybees and links cellular metabolism to epigenetic regulation via histone acetylation); inositol phosphate/TOR pathway (linking nutrient sensing to the regulation of cell growth and metabolism); the ubiquitin-proteasome pathway (regulating protein degradation, implicated in epigenetic regulation and histone methylation); and spliceosome, splicing factors, and DNA methylation (methylation being more common around splice sites and alternatively expressed exons). The gene *alk* was found to be differentially methylated in intron 25, leading to alternative splicing of exon 25. This is proposed to work by unmethylated CTCF sites (bound by CCCTC-binding factor) allowing binding of nearby insulator elements, leading to the inclusion of weak upstream exons as RNA polymerase II pauses leading to the less methylated exon 25 being included more frequently

(Foret et al., 2012; Oberdoerffer, 2012). *Alk* was differentially methylated and alternately spliced between queens and workers, as well as in developmental stages, making it an interesting, dynamic candidate for linking nutritional environmental cues and cellular responses (Foret et al., 2012).

There have been other studies of the link between caste determination in social insects and alternative splicing through DNA methylation. Some of these have included species, such as the jewel wasp *Nasonia vitripennis*, which once again corroborated this theory. Methylated genes for both honeybee and *Nasonia* were enriched for alternative splicing. These genes were longer, had more exons, and appeared to be subjected to slower rates of evolution (Flores et al., 2012).

Kamakura (2011) proposed a mechanism through which nutrition and altered DNA methylation might be linked. The aim of the study was to isolate the component of royal jelly that induces caste differentiation in honeybees. Monomeric MRJP1, dubbed 'royalactin', was identified as the key factor which induced increased body size, ovary development, and shortened developmental time (Kamakura, 2011). It was also found that a similar effect could be observed in *Drosophila*. Furthermore, knockdown of epidermal growth factor receptor (*Egfr*) in the fat bodies of both *A. mellifera* and *D. melanogaster* blocked the effects of royalactin, and hence was proposed as a mechanism through which it acts (Kamakura, 2011). DNA methylation was thought to play a vital role in the regulation of the *Apime-egfr* gene. Larvae that develop into queens were shown to have ~75% less methylation at the loci for the *Apime-egfr* gene in comparison to those larvae destined to become workers. However, this theory could not be corroborated in 2015, where it was shown that in honeybees the *egfr* locus is never methylated (Kucharski et al., 2015). Subsequently, many have questioned the validity of the claims made by Kamakura. The findings were publically called into question in the journal *Nature*, when it was shown experimentally that they were unable to be replicated, and that none of the MRJP (1, 2, 3, and 5) were sufficient to be the key factor for queen caste development (Buttstedt et al., 2016). In response Kamakura has questioned the stringency of their experimental conditions, citing differences in the protocol (Kamakura, 2016). More research is required to determine the role of royalactin and *egfr* methylation in controlling the queen phenotype.

Nutrition in honeybees not only alters their developmental trajectory with regard to caste development, but there is also evidence to imply that larvae use early nutritional cues to predict the environment in which they will emerge (Wang et al., 2016). During development an organism has the ability to plastically alter its phenotype in response to environmental signals it receives. In some ways, these adaptations may appear detrimental. Developmental nutritional stress has been known to decrease body size (Searcy et al., 2004), reduce immune function (Butler and McGraw, 2011), repress reproduction (Lindström, 1999), and impair cognition (Fisher et al., 2006). However, it has been proposed that this stress during development may lead to adaptations that are useful when under stress in future (Gluckman et al., 2007). For example, type II diabetes in humans can be the result of nutritive stress in utero, producing a phenotype of insulin resistance and central fat stores. Under conditions of plentiful resources this is detrimental, yet under starvation conditions (like those anticipated based on environmental cues during development) this adaptation is beneficial, a proposed predictive adaptive response (Bateson et al., 2014). To investigate if honeybees are capable of showing predictive adaptations, larvae were starved during the fifth instar of development and then later subjected to starvation stress in adulthood. It was found that under normal conditions there was no difference in metabolic rates or haemolymph sugar levels in adult bees. Adaptation was seen, however, when adults were starved. Those adults who had previously experienced starvation as a larvae were able to shift from carbohydrate metabolism, reduce their metabolic rate, and maintain stable haemolymph levels (Wang et al., 2016). Experiments such as this, whereby this altered developmental programming in response to environmental cues leads to an adult phenotype, may show another way in which nutrition is affecting epigenetics in insects.

3.5 Limitations and Required Evidence

Investigations into the link between DNA methylation and caste determination in social insects have deepened our understanding of how nutrition and epigenetics are involved in insect polyphenism, yet they are not free of limitations. Comparisons of DNA methylation often use samples that pool different tissues and cell types, are sensitive to lack of biological replication, and could be confounded by genetically influenced, *cis*-mediated allele-specific methylation (Libbrecht et al., 2016; Maleszka, 2016; Remnant et al., 2016).

Methylation is inherently tissue specific; all cell types will have a distinct methylation pattern that arises during cellular differentiation (Bird, 2002). Thus, when many cell types are mixed, it becomes possible to see only an amalgamation of multiple signals. If one is interested in ovaries, for example, a whole individual being sampled will cause the pattern of interest to be lost among the numerous others from differing tissues. Even within tissues, there are multiple cell types. If one is interested in ovarian terminal filament cells, the majority of a whole ovary sample will be oocytes or nurse cells, and any terminal filament-specific pattern will be overwhelmed by these. In this way, careful isolation of specific cell types is required to accurately compare between different treatment conditions and phenotypes.

Lack of biological replication could result in sample-specific rather than caste-specific differences, leading to incorrect conclusions arising from genome-wide methylation patterns (Libbrecht et al., 2016). These limitations were highlighted in genome-wide methylation comparisons of the ant *Cerapachys biroi* that indicated between-sample variation could account for high levels of methylation differences, calling into question previous claims in ants and honeybees (Bonasio, 2014; Lyko et al., 2010). Methylation differences reported between honeybee queen and worker brains could not be replicated in subsequent studies (Flores et al., 2012; Lyko et al., 2010), highlighting the need for appropriate biological replication in genome-wide methylation analyses.

Between-replicate differences in methylation may be due to stochastic variation between sampling conditions or differences in tissue composition, or they could even have a deeper genetic basis. While previous genome-wide studies in insects have overlooked genetic background as a potential factor influencing methylation state, sequence-dependent or *cis*-mediated allele-specific methylation is widespread in the human genome, with many genomic regions associated with variation in DNA methylation (Kerkel et al., 2008; Mendizabal et al., 2014). There is increasing evidence of *cis*-mediated allele-specific methylation in honeybees where differing methylation patterns are strongly correlated with genotype, suggesting there could be a genetic basis to methylated epialleles (Maleszka, 2016; Remnant et al., 2016; Wedd et al., 2016). This could be a way for the genome and methylome to interact and creates another level of regulation where the genotype and epigenotype are both involved in generating the phenotype.

A final issue is in the interpretation of differences in methylation levels. Obviously, methylation levels were greater at some CpG sites than at others. But what does this mean? What conclusions can one make about these data when we study only a small number of genes and look at methylation as a percentage across a gene, as opposed to having a specific, targeted effect? When so many CpG sites are showing altered methylation, is it likely all of these changes are having an effect? Are there many small alterations acting in a cumulative fashion? Or is it a case of a few target genes having a major role? Is DNA methylation the causative alteration in caste development? Or is it a genetic 'placeholder' of changes made by other mechanisms? These are questions we still have not answered, yet data are frequently interpreted as if we have. There is a need within this field to take a step back from the hasty interpretation of results and to investigate how this system is truly functioning.

4. SUMMARY AND CONCLUSIONS

Understanding the links between nutrition and epigenetic change and phenotype is challenging in any system. Insects provide perhaps a more manipulable system than most to investigate these relationships, but difficulties still arise. Perhaps the most problematic are the background epigenetic changes associated with the differentiation of cells during development that makes each cell type, and possibly each cell, a unique epigenetic modification of the genome. To understand how nutrition leads to epigenetic change and how that affects phenotype, it may be required to understand what is happening in individual cell types and how those cell types affect the phenotype of the organism. Alongside this, our current inability, in animals, to inducibly direct an epigenetic change to a region of the genome makes it very difficult to interpret causality in epigenetic experiments.

With current tools, much exciting data point to a key role in linking nutrition, epigenetics, and final phenotype in insects. Epigenetic mechanisms are plausible for controlling polyphenisms, predictive adaptive responses, and controlling the genome-wide changes in gene expression in response to nutrition. What is not complete is the chain of causality from (1) nutritional stimulus, through (2) epigenetic mark in a specific cell type, to (3) changes in gene expression/splicing, to a (4) final nutrition-triggered phenotype. We believe that research should focus on this chain of causality as the only way to clearly demonstrate what is tantalisingly close; a clear understanding of how epigenetics may link environmental effects and final phenotype, with all the implications for insect life history and human health, that that would imply.

REFERENCES

Ashby, R., Forêt, S., Searle, I., Maleszka, R., 2016. MicroRNAs in honey bee caste determination. Sci. Rep. 6, 18794.

Bateson, P., 2007. Developmental plasticity and evolutionary biology. J. Nutr. 137, 1060–1062.

Bateson, P., Barker, D., Clutton-Brock, T., Deb, D., D'Udine, B., Foley, R.A., Gluckman, P., Godfrey, K., Kirkwood, T., Lahr, M.M., et al., 2004. Developmental plasticity and human health. Nature 430, 419–421.

Bateson, P., Gluckman, P., Hanson, M., 2014. The biology of developmental plasticity and the predictive adaptive response hypothesis. J. Physiol. 592, 2357–2368.

Beeler, S.M., Wong, G.T., Zheng, J.M., Bush, E.C., Remnant, E.J., Oldroyd, B.P., Drewell, R.A., 2014. Whole-genome DNA methylation profile of the jewel wasp (*Nasonia vitripennis*). G3 4, 383–388.

Berens, A.J., Hunt, J.H., Toth, A.L., 2014. Nourishment level affects caste-related gene expression in Polistes wasps. BMC Genomics 16, 1–12.

Bertholf, L.M., 1925. The moults of the honeybee. J. Econ. Entomol. 18, 380–384.

Bird, A., 2002. DNA methylation patterns and epigenetic memory. Genes Dev. 16, 6–21.

Bonasio, R., 2014. The role of chromatin and epigenetics in the polyphenisms of ant castes. Brief. Funct. Genomics 13, 235–245.

Bonasio, R., Zhang, G., Ye, C., Mutti, N.S., Fang, X., Qin, N., Donahue, G., Yang, P., Li, Q., Li, C., et al., 2010. Genomic comparison of the ants *Camponotus floridanus* and *Harpegnathos saltator*. Science 329, 1068–1071.

Bonasio, R., Li, Q., Lian, J., Wen, P., Xiang, H., Ding, Y., Jin, Z., 2012. Genome-wide and caste-specific DNA methylomes of the ants *Camponotus floridanus* and *Harpegnathos saltator*. Curr. Biol. 22, 1755–1764.

Bray, G.A., Popkin, B.M., 2014. Dietary sugar and body weight: have we reached a crisis in the epidemic of obesity and diabetes? Diabetes Care 37, 950–956.

Bross, T.G., Rogina, B., Helfand, S.L., 2005. Behavioral, physical, and demographic changes in Drosophila populations through dietary restriction. Aging Cell 4, 309–317.

Buescher, J.L., Musselman, L.P., Wilson, C.A., Lang, T., Keleher, M., Baranski, T.J., Duncan, J.G., 2013. Evidence for transgenerational metabolic programming in *Drosophila*. Dis. Model. Mech. 6, 1123–1132.

Butler, M.W., McGraw, K.J., 2011. Past or present? Relative contributions of developmental and adult conditions to adult immune function and coloration in mallard ducks (*Anas platyrhynchos*). J. Comp. Physiol. B Biochem. Syst. Environ. Physiol. 181, 551–563.

Buttstedt, A., Moritz, R.F.A., Erler, S., 2014. Origin and function of the major royal jelly proteins of the honeybee (*Apis mellifera*) as members of the yellow gene family. Biol. Rev. 89, 255–269.

Buttstedt, A., Ihling, C.H., Pietzsch, M., Moritz, R.F.A., 2016. Royalactin is not a royal making of a queen. Nature 537, E10–E12.

Cameron, R.C., Duncan, E.J., Dearden, P.K., 2013. Biased gene expression in early honeybee larval development. BMC Genomics 14, 903.

Campbell, C.P., Raubenheimer, D., Badaloo, A.V., Gluckman, P.D., Martinez, C., Gosby, A., Simpson, S.J., Osmond, C., Boyne, M.S., Forrester, T.E., 2016. Developmental contributions to macronutrient selection: a randomized controlled trial in adult survivors of malnutrition. Evol. Med. Public Health 2016 (1), 158–169.

Carthew, R.W., Sontheimer, E.J., 2009. Origins and mechanisms of miRNAs and siRNAs. Cell 136, 642–655.

Chapman, T., Partridge, L., 1996. Female fitness in *Drosophila melanogaster*: an interaction between the effect of nutrition and of encounter rate with males. Proc. R. Soc. Lond. Ser. B Biol. Sci. 263, 755–759.

Conrad, T., Akhtar, A., 2012. Dosage compensation in *Drosophila melanogaster*: epigenetic fine-tuning of chromosome-wide transcription. Nat. Rev. Genet. 13, 123–134.

Corona, M., Libbrecht, R., Wheeler, D.E., 2016. Molecular mechanisms of phenotypic plasticity in social insects. Curr. Opin. Insect Sci. 13, 55–60.

Cridge, A.G., Leask, M.P., Duncan, E.J., Dearden, P.K., 2016. What do studies of insect polyphenisms tell us about nutritionally-triggered epigenomic changes and their consequences? Forum Nutr. 7, 1787–1797.

Crozier, R.H., 2008. Advanced eusociality, kin selection and male haploidy. Aust. J. Entomol. 47, 2–8.

Drewell, R.A., Bush, E.C., Remnant, E.J., Wong, G.T., Beeler, S.M., Stringham, J.L., Lim, J., Oldroyd, B.P., 2014. The dynamic DNA methylation cycle from egg to sperm in the honey bee *Apis mellifera*. Development 141, 2702–2711.

Duncan, E.J., Gluckman, P.D., Dearden, P.K., 2014. Epigenetics, plasticity, and evolution: how do we link epigenetic change to phenotype? J. Exp. Zool. B Mol. Dev. Evol. 322, 208–220.

Elango, N., Hunt, B.G., Goodisman, M.A.D., Yi, S.V., 2009. DNA methylation is widespread and associated with differential gene expression in castes of the honeybee, *Apis mellifera*. Proc. Natl. Acad. Sci. U. S. A. 106, 11206–11211.

Emlen, D.J., Warren, I.A., Johns, A., Dworkin, I., Lavine, L.C., 2012. A mechanism of extreme growth and reliable signaling in sexually selected ornaments and weapons. Science 337, 860–864.

Fisher, M.O., Nager, R.G., Monaghan, P., 2006. Compensatory growth impairs adult cognitive performance. PLoS Biol. 4, 4–8.

Flores, K., Wolschin, F., Corneveaux, J.J., Allen, A.N., Huentelman, M.J., Amdam, G.V., 2012. Genome-wide association between DNA methylation and alternative splicing in an invertebrate. BMC Genomics 13, 461–465. http://dx.doi.org/10.1186/1471-2164-13-480.

Foret, S., Kucharski, R., Pittelkow, Y., Lockett, G.A., Maleszka, R., 2009. Epigenetic regulation of the honey bee transcriptome: unravelling the nature of methylated genes. BMC Genomics 10, 1–11.

Foret, S., Kucharski, R., Pellegrini, M., Feng, S., Jacobsen, S.E., Robinson, G.E., 2012. DNA methylation dynamics, metabolic fluxes, gene splicing, and alternative phenotypes in honey bees. Proc. Natl. Acad. Sci. U. S. A. 109, 4968–4973.

Fukuda, H., Kiichi, S., 1966. Seasonal change of the honeybee worker longevity in Sapporo, North Japan, with notes on some factors affecting the life-span. Jap. J. Ecol. 16, 206–212.

Gershman, B., Puig, O., Hang, L., Peitzsch, R.M., Tatar, M., Garofalo, R.S., 2007. High-resolution dynamics of the transcriptional response to nutrition in *Drosophila*: a key role for dFOXO. Physiol. Genomics 29, 24–34.

Glastad, K.M., Hunt, B.G., Yi, S.V., Goodisman, M.A.D., 2011. DNA methylation in insects: on the brink of the epigenomic era. Insect Mol. Biol. 20, 553–565.

Gluckman, P.D., Hanson, M.A., 2004. The Fetal Matrix: Evolution, Development and Disease. Cambridge University Press, Cambridge.

Gluckman, P.D., Hanson, M.A., 2006. The developmental origins of health and disease. In: Marelyn Wintour, E., Owens, J.A. (Eds.), Early Life Origins of Health and Disease Springer, Boston, MA, pp. 1–7.

Gluckman, P., Hanson, M., Beedle, A., 2007. Early life events and their consequences for later disease: a life history and evolutionary perspective. Am. J. Hum. Biol. 19, 1–19.

Godfrey, K.M., Gluckman, P.D., Hanson, M.A., 2010. Developmental origins of metabolic disease: life course and intergenerational perspectives. Trends Endocrinol. Metab. 21, 199–205.

Grimaldi, D., Engel, M.S., 2005. Evolution of the Insects. Cambridge University Press, Cambridge, UK.

Guo, X., Su, S., Skogerboe, G., Dai, S., Li, W., Li, Z., Liu, F., Ni, R., Guo, Y., Chen, S., et al., 2013. Recipe for a busy bee: microRNAs in honey bee caste determination. PLoS One 8, e81661.

Haydak, M.H., 1943. Larval food and development of castes in the honeybee. J. Econ. Entomol. 36, 778–792.

Hefetz, A., Grozinger, C.M., Padilla, M., Amsalem, E., Altman, N., 2016. Chemical communication is not sufficient to explain reproductive inhibition in the bumblebee *Bombus impatiens*. R. Soc. Open Sci. 3, 160576.

Huang, C.Y., Chi, L.L., Huang, W.J., Chen, Y.W., Chen, W.J., Kuo, Y.C., Yuan, C.M., Chen, C.N., 2012. Growth stimulating effect on queen bee larvae of histone deacetylase inhibitors. J. Agric. Food Chem. 60, 6139–6149.

Hughes, W.O.H., Oldroyd, B.P., Beekman, M., Ratinieks, F.L., 2008. Ancestral monogamy shows kin selection is key to the evolution of eusociality. Science 320, 1213–1217.

Jaenisch, R., Bird, A., 2003. Epigenetic regulation of gene expression: how the genome integrates intrinsic and environmental signals. Nat. Genet. 33, 245–254.

Jay, S.C., 1964. Rearing honeybee brood outside the hive. J. Apic. Res. 3, 51–60.

Kamakura, M., 2011. Royalactin induces queen differentiation in honeybees. Nature 473, 478–483.

Kamakura, M., 2016. Kamakura replies. Nature 537, E13.

Kang, H.-L., Benzer, S., Min, K.-T., 2002. Life extension in *Drosophila* by feeding a drug. Proc. Natl. Acad. Sci. U. S. A. 99, 838–843.

Kerkel, K., Spadola, A., Yuan, E., Kosek, J., Jiang, L., Hod, E., Li, K., Murty, V.V., Schupf, N., Vilain, E., et al., 2008. Genomic surveys by methylation-sensitive SNP analysis identify sequence-dependent allele-specific DNA methylation. Nat. Genet. 40, 904–908.

Kolss, M., Vijendravarma, R.K., Schwaller, G., Kawecki, T.J., 2009. Life-history consequences of adaptation to larval nutritional stress in *Drosophila*. Evolution 63, 2389–2401.

Kouzarides, T., 2007. Chromatin modifications and their function. Cell 128, 693–705.

Krauss, V., Reuter, G., 2011. DNA methylation in *Drosophila*—a critical evaluation. Prog. Mol. Biol. Transl. Sci. 101, 177–191.

Kubota, T., Miyake, K., Hariya, N., Mochizuki, K., 2015. Understanding the epigenetics of neurodevelopmental disorders and DOHaD. J. Dev. Orig. Health Dis. 6, 96–104.

Kucharski, R., Maleszka, J., Foret, S., Maleszka, R., 2008. Nutritional control of reproductive status in honeybees via DNA methylation. Science 319, 1827–1830.

Kucharski, R., Foret, S., Maleszka, R., 2015. EGFR gene methylation is not involved in Royalactin controlled phenotypic polymorphism in honey bees. Sci. Rep. 5, 14070.

Lee, T.M., Smale, L., Zucker, I., Dark, J., 1987. Influence of daylength experienced by dams on post-natal development of young meadow voles (*Microtus pennsylvanicus*). J. Reprod. Fertil. 81, 337–342.

Lee, K.P., Simpson, S.J., Clissold, F.J., Brooks, R., Ballard, J.W.O., Taylor, P.W., Soran, N., Raubenheimer, D., 2008. Lifespan and reproduction in *Drosophila*: new insights from nutritional geometry. Proc. Natl. Acad. Sci. U. S. A. 105, 2498–2503.

Lian, T., Gaur, U., Yang, D., Li, D., Li, Y., Yang, M., 2015. Epigenetic mechanisms of dietary restriction induced aging in *Drosophila*. Exp. Gerontol. 72, 38–44.

Libbrecht, R., Oxley, P.R., Keller, L., Jan, D., Kronauer, C., Libbrecht, R., Oxley, P.R., Keller, L., Jan, D., Kronauer, C., 2016. Robust DNA methylation in the clonal raider ant brain. Curr. Biol. 26, 391–395.

Li-Byarlay, H., Li, Y., Stroud, H., Feng, S., Newman, T.C., Kaneda, M., Hou, K.K., Worley, K.C., Elsik, C.G., Wickline, S.A., et al., 2013. RNA interference knockdown of DNA methyl-transferase 3 affects gene alternative splicing in the honey bee. Proc. Natl. Acad. Sci. U. S. A. 110, 12750–12755.

Lindström, J., 1999. Early development and fitness in birds and mammals. Trends Ecol. Evol. 14, 343–348.

Lyko, F., Maleszka, R., 2011. Insects as innovative models for functional studies of DNA methylation. Trends Genet. 27, 127–131.

Lyko, F., Ramsahoye, B.H., Jaenisch, R., 2000. Development: DNA methylation in *Drosophila melanogaster*. Nature 408, 538–540.

Lyko, F., Foret, S., Kucharski, R., Wolf, S., Falckenhayn, C., 2010. The honey bee epigenome: differential methylation of brain DNA in queens and workers. PLoS Biol. 8, e1000506.

Mair, W., Goymer, P., Pletcher, S.D., Partridge, L., 2003. Demography of dietary restriction and death in *Drosophila*. Science 301, 1731–1733.

Maleszka, R., 2008. Epigenetic integration of environmental and genomic signals in honey bees: the critical interplay of nutritional, brain and reproductive networks. Epigenetics 3, 188–192.

Maleszka, R., 2016. Epigenetic code and insect behavioural plasticity. Curr. Opin. Insect Sci. 15, 45–52.

Maor, G.L., Yearim, A., Ast, G., 2015. The alternative role of DNA methylation in splicing regulation. Trends Genet. 31, 274–280.

May, C.M., Doroszuk, A., Zwaan, B.J., 2015. The effect of developmental nutrition on life span and fecundity depends on the adult reproductive environment in *Drosophila melanogaster*. Ecol. Evol. 5, 1156–1168.

Maynard Smith, J., Szathmary, E., 1995. The Major Transitions in Evolution. Oxford University Press, Oxford, UK.

Mendizabal, I., Keller, T.E., Zeng, J., Yi, S.V., 2014. Epigenetics and evolution. Integr. Comp. Biol. 54, 31–42.

Michener, D.C., 1974. The Social Behaviour of the Bees: A Comparative Study. Harvard University Press, Cambridge.

Mohandas, T., Sparkes, R.S., Shapiro, L.J., 1981. Reactivation of an inactive human X chromosome: evidence for X inactivation by DNA methylation. Science 211, 393–396.

Nusbaum, T.J., Rose, M.R., 1999. The effects of nutritional manipulation and laboratory selection on lifespan in *Drosophila melanogaster*. J. Gerontol. A Biol. Sci. Med. Sci. 54, B192–B198.

Oberdoerffer, S., 2012. A conserved role for intragenic DNA methylation in alternative pre-mRNA splicing. Transcription 3, 106–109.

Öst, A., Lempradl, A., Casas, E., Weigert, M., Tiko, T., Deniz, M., Pantano, L., Boenisch, U., Itskov, P.M., Stoeckius, M., et al., 2014. Paternal diet defines offspring chromatin state and intergenerational obesity. Cell 159, 1352–1364.

Oster, G.F., Wilson, E.O., 1978. Caste and Ecology in the Social Insects. Princeton University Press, Princeton, NJ.

Ozawa, T., Mizuhara, T., Arata, M., Shimada, M., Niimi, T., Okada, K., Okada, Y., Ohta, K., 2016. Histone deacetylases control module-specific phenotypic plasticity in beetle weapons. Proc. Natl. Acad. Sci. U. S. A. 113, 15042–15047.

Parashar, V., Rogina, B., 2009. dSir2 mediates the increased spontaneous physical activity in flies on calorie restriction. Aging 1, 529–541.

Pletcher, S.D., Macdonald, S.J., Marguerie, R., Certa, U., Stearns, S.C., Goldstein, D.B., Partridge, L., 2002. Genome-wide transcript profiles in aging and calorically restricted *Drosophila melanogaster*. Curr. Biol. 12, 712–723.

Pletcher, S.D., Libert, S., Skorupa, D., 2005. Flies and their golden apples: the effect of dietary restriction on *Drosophila* aging and age-dependent gene expression. Ageing Res. Rev. 4, 451–480.

Remnant, E.J., Ashe, A., Young, P.E., Buchmann, G., Beekman, M., Allsopp, M.H., Suter, C.M., Drewell, R.A., Oldroyd, B.P., 2016. Parent-of-origin effects on genome-wide DNA methylation in the Cape honey bee (*Apis mellifera capensis*) may be confounded by allele-specific methylation. BMC Genomics 17, 1–14.

Rogina, B., Helfand, S.L., Frankel, S., 2002. Longevity regulation by *Drosophila* Rpd3 deacetylase and caloric restriction. Science 298, 1745.

Sarda, S., Zeng, J., Hunt, B.G., Yi, S.V., 2012. The evolution of invertebrate gene body methylation. Mol. Biol. Evol. 29, 1907–1916.

Schwander, T., Lo, N., Beekman, M., Oldroyd, B.P., Keller, L., 2010. Nature versus nurture in social insect caste differentiation. Trends Ecol. Evol. 25, 275–282.

Searcy, W.A., Peters, S., Nowicki, S., 2004. Effects of early nutrition on growth rate and adult size in song sparrows *Melospiza melodia*. J. Avian Biol. 35, 269–279.
Seeley, T.D., 1978. Life history strategy of the honey bee, *Apis mellifera*. Oecologia 32, 109–118.
Shi, Y.Y., Huang, Z.Y., Zeng, Z.J., Wang, Z.L., Wu, X.B., Yan, W.Y., 2011. Diet and cell size both affect queen-worker differentiation through DNA methylation in honey bees (*Apis mellifera, Apidae*). PLoS One 6, 2–7.
Shuel, R.W., Dixon, S.E., 1959. Studies in the mode of action of royal jelly in honeybee development. II. Respiration of newly emerged larvae on various substrates. Can. J. Zool. 37, 803–813.
Shuel, R.W., Dixon, S.E., 1960. The early establishment of dimorphism in the female honeybee, *Apis mellifera* L. Insectes Soc. 7, 265–282.
Simola, D.F., Ye, C., Mutti, N.S., Dolezal, K., Bonasio, R., Reinberg, D., Berger, S.L., 2013. A chromatin link to caste identity in the carpenter ant *Camponotus floridanus*. Genome Res. 23, 486–496.
Simola, D.F., Graham, R.J., Brady, C.M., Enzmann, B.L., Desplan, C., Ray, A., Zwiebel, L.J., Bonasio, R., Reinberg, D., Liebig, J., et al., 2016. Epigenetic (re)programming of caste-specific behavior in the ant *Camponotus floridanus*. Science 351, aac6633.
Simpson, S.J., Sword, G.A., Lo, N., 2011. Polyphenism in insects. Curr. Biol. 21, R738–R749.
Slotkin, R.K., Martienssen, R., 2007. Transposable elements and the epigenetic regulation of the genome. Nat. Rev. Genet. 8, 272–285.
Smith, C.R., Anderson, K.E., Tillberg, C.V., Gadau, J., Suarez, A.V., 2008. Caste determination in a polymorphic social insect: nutritional, social, and genetic factors. Am. Nat. 172, 497–507.
Snell-Rood, E.C., Troth, A., Moczek, A.P., 2013. DNA methylation as a mechanism of nutritional plasticity: limited support from horned beetles. J. Exp. Zool. B Mol. Dev. Evol. 320, 22–34.
Stearns, S.C., 1989. The evolutionary significance of phenotypic plasticity. Am. Inst. Biol. Sci. 39, 436–445.
Townsend, G.F., Lucas, C.C., 1940. The chemical nature of royal jelly. Biochem. J. 34, 1155–1162.
Trible, W., Kronauer, D.J.C., 2017. Caste development and evolution in ants: it's all about size. J. Exp. Biol. 220, 53–62.
Verona, R.I., Mann, M.R.W., Bartolomei, M.S., 2003. Genomic imprinting: intricacies of epigenetic regulation in clusters. Annu. Rev. Cell Dev. Biol. 19, 237–259.
Waddington, C.H., 1942. The epigenotype. Endeavour 1, 18–20.
Wang, D.-I., 1965. Growth rates of young queen and worker honeybee larvae. J. Apic. Res. 4, 3–5.
Wang, Y., Jorda, M., Jones, P.L., Maleszka, R., Ling, X., Robertson, H.M., Mizen, C.A., Peinado, M.A., Robinson, G.E., 2006. Functional CpG methylation system in a social insect. Science 314, 645–647.
Wang, Y., Campbell, J.B., Kaftanoglu, O., Page, R.E., Amdam, G.V., Harrison, J.F., 2016. Larval starvation improves metabolic response to adult starvation in honey bees (*Apis mellifera L.*). J. Exp. Biol. 219, 960–968.
Wang, W.W., Ashby, R., Ying, H., Maleszka, R., Forêt, S., 2017. Contrasting sex-and caste-dependent piRNA profiles in the transposon depleted haplodiploid honeybee *Apis mellifera*. Genome Biol. Evol. 9, 1341–1356. http://dx.doi.org/10.1093/gbe/evx087.
Waterland, R.A., Michels, K.B., 2007. Epigenetic epidemiology of the developmental origins hypothesis. Annu. Rev. Nutr. 27, 363–388.

Weaver, N., 1957. Effects of larval age on dimorphic differentiation of the female honey bee. Ann. Entomol. Soc. Am. 50, 283–294.
Wedd, L., Kucharski, R., Maleszka, R., 2016. Differentially methylated obligatory epialleles modulate context-dependent LAM gene expression in the honeybee. Epigenetics 11, 1–10.
Weiner, S.A., Galbraith, D.A., Adams, D.C., Valenzuela, N., Noll, F.B., Grozinger, C.M., Toth, A.L., 2013. A survey of DNA methylation across social insect species, life stages, and castes reveals abundant and caste-associated methylation in a primitively social wasp. Naturwissenschaften 100, 795–799.
West-Eberhard, M.J., 1989. Phenotypic plasticity and the origins of diversity. Annu. Rev. Ecol. Syst. 20, 249–278.
Whitaker, R., Gil, M.P., Ding, F., Tatar, M., Helfand, S.L., Neretti, N., 2014. Dietary switch reveals fast coordinated gene expression changes in *Drosophila melanogaster*. Aging 6, 355–368.
Wilson, E.O., 1990. Success and Dominance in Ecosystems: The Case of the Social Insects. Federal Repulic of Germany: Ecology Institute, Oldendorf/Luhe.
Wilson, E.O., Holldobler, B., 2005. Eusociality: origin and consequences. Proc. Natl. Acad. Sci. U. S. A. 102, 13367–13371.
Winston, M.L., Dropkin, J.A., Taylor, O.R., 1981. Demography and life-history characteristics of two honey bee races (*Apis mellifera*). Oecologia 48, 407–413.
Woods, J.K., Rogina, B., 2016. The effects of Rpd3 on fly metabolism, health, and longevity. Exp. Gerontol. 86, 124–128.
Woods, J.K., Ziafazeli, T., Rogina, B., 2016. Rpd3 interacts with insulin signaling in *Drosophila* longevity extension. Aging (Albany NY) 8, 3028–3040.
Xia, B., Gerstin, E., Schones, D.E., Huang, W., de Belle, J.S., 2016. Transgenerational programming of longevity through E(z)-mediated histone H3K27 trimethylation in *Drosophila*. Aging 8, 2988–3002.
Yan, H., Bonasio, R., Simola, D.F., Liebig, J., Berger, S.L., Reinbeg, D., 2015. DNA methylation in social insects: how epigenetics can conteol behavior and longevity. Annu. Rev. Entomol. 60, 435–452.
Zajitschek, F., Zajitschek, S.R.K., Friberg, U., Maklakov, A.A., 2013. Interactive effects of sex, social environment, dietary restriction, and methionine on survival and reproduction in fruit flies. Age (Dordr.) 35, 1193–1204.
Zajitschek, F., Zajitschek, S.R.K., Canton, C., Georgolopoulos, G., Friberg, U., Maklakov, A.A., 2016. Evolution under dietary restriction increases male reproductive performance without survival cost. Proc. R. Soc. B Biol. Sci. 283, 20152726.
Zheng, J., Mutcherson 2nd, R., Helfand, S.L., 2005. Calorie restriction delays lipid oxidative damage in *Drosophila melanogaster*. Aging Cell 4, 209–216.
Zinke, I., Schütz, C.S., Katzenberger, J.D., Bauer, M., Pankratz, M.J., 2002. Nutrient control of gene expression in *Drosophila* microarray analysis of starvation and sugar-dependent response. EMBO J. 21, 6162–6173.
Zwaan, B.J., Bijlsma, R., Hoekstra, R.F., 1991. On the developmental theory of ageing. I. Starvation resistance and longevity in *Drosophila melanogaster* in relation to pre-adult breeding conditions. Heredity 66 (Pt. 1), 29–39.

CHAPTER THREE

The Role of MicroRNAs in *Drosophila* Regulation of Insulin-Like Peptides and Ecdysteroid Signalling: Where Are We Now?

Jian Q. Cao*[,1], Wai S. Tong*[,1], Hiu Y. Yu*[,1], Stephen S. Tobe[†], William G. Bendena[‡], Jerome H.L. Hui*

*School of Life Sciences, Simon F.S. Li Marine Science Laboratory and Centre of Soybean Research, Partner State Key Laboratory of Agrobiotechnology, The Chinese University of Hong Kong, Shatin, Hong Kong
[†]University of Toronto, Toronto, ON, Canada
[‡]Queen's University, Kingston, ON, Canada
[1]Cofirst authors.

Contents

Abstract

Arthropods are the most successful group of terrestrial animals in terms of their diverse habitats and species numbers. Understanding their evolution has conventionally been linked to comparison of their developmental modes and morphologies. Within the field

Advances in Insect Physiology, Volume 53
ISSN 0065-2806
http://dx.doi.org/10.1016/bs.aiip.2017.02.002

of evolutionary developmental biology, the actions of hormones are usually ignored as hormones are thought to function more as regulators of metabolic activity associated with late developmental stages, such as ecdysis and metamorphosis. In recent years, the field of evolutionary endocrinology has emerged with the advances in sequencing techniques. Here, we consolidate the current findings of microRNA actions on insulin-like peptides and ecdysteroids in the major insect genetic model, *Drosophila melanogaster*. MicroRNA regulation of hormone activity remains a neglected area that needs to be developed using other insect and arthropod models to understand the diverse adaptations exhibited by this most successful group of animals.

1. INTRODUCTION

Evolutionary developmental biology is a field that merges the study of evolution and developmental biology, aiming to unravel the ancestral relationships between organisms, the origin and evolution of embryonic development and the evolution of developmental processes. In the last three decades, developmental genes, especially those involved in embryonic development, have been intensively studied. It is becoming widely accepted that many of the developmental genes are highly conserved in terms of their sequences and expression domains across the animal kingdom. The latest consensus is that most animals have inherited a basic developmental genetic toolkit or regulatory network from a common animal ancestor (e.g. De Robertis, 2008). Endocrine systems, which probably developed after the embryonic stage, have been generally neglected for their roles during animal evolution.

Neglecting how hormonal systems contribute to animal evolution is unwise, as hormones are crucial signalling molecules that regulate a diverse range of physiological and behavioural activities. With the advancement of sequencing technologies, our understanding of endocrine systems in different groups of animals, have improved (e.g. Cheong et al., 2015; Hui et al., 2013; Kenny et al., 2013). Evolutionary endocrinology, a field that involves the comparison of hormonal systems, in particular the hormonal pathway gene content in different animal genomes, has subsequently arisen (Zera et al., 2007). Similar to the existence of an established developmental genetic and programming system in the animal common ancestor, some hormonal systems also have a common ancestor followed by modifications in different lineages (e.g. the different sesquiterpenoids and ecdysteroids in arthropods; Qu et al., 2015).

Insect hormones can be divided into several major groups, including peptides, steroids and sesquiterpenoids. In recent years, much attention has been paid to regulatory microRNAs (miRNAs) which are small non-coding RNAs that may regulate one or more key protein-coding genes. Considering that there are several extensive reviews on sesquiterpenoids (e.g. Cheong et al., 2015; Hui et al., 2013; Kenny et al., 2013; Tobe and Bendena, 1999), here, we will only focus on the contribution of miRNAs to the timing and expression of genes that are regulated by action of the best characterized insect peptides and steroid hormones, namely, insulin-like peptides (ILPs) and ecdysteroids.

2. ILPs AND ECDYSTEROIDS

The ILPs control growth and metabolism in insects whereas ecdysteroids (specifically 20-hydroxyecdysone; 20E) initiate the transition between developmental stages. ILPs and ecdysteroids have antagonistic actions that coordinate the growth and maturation (i.e. size and shape) of insects (Colombani et al., 2005). In this section, we provide a brief overview of their biogenesis and actions linking their roles in animal evolution.

2.1 Insulin-Like Molecules

In most animals, insulin regulates the metabolism of carbohydrates and fats in response to the animals' nutritional states. In vertebrates, absence or reduced release of insulin leads to diabetes mellitus, a condition whereby blood sugar levels are continuously elevated. The discovery of ILPs in unicellular eukaryotes (Le Roith et al., 1980) suggested that similar peptide regulators of metabolic activity would be conserved throughout the metazoans (de Souza and López, 2004). The latest consensus is that the metazoan ancestor had acquired the biosynthetic pathway for ILP and physiological actions involved in the response to varying nutrient conditions, and this system continues to be used in extant animals (Teleman, 2010).

The first invertebrate ILP, named bombyxin, was accidentally identified in the silkworm *Bombyx mori* during the purification of prothoracicotropic hormone (PTTH). Whereas PTTH stimulated the prothoracic glands (PGs) to release ecdysone (Nagasawa et al., 1984, 1986), the N-terminal sequence of bombyxin is homologous to vertebrate insulin (Akira et al., 1987; Ishizaki et al., 1983; Nagasawa et al., 1984). Bombyxin is now viewed as an important hormone in controlling nutrient-dependent growth and metabolism (for details, see reviews Gu et al., 2015; Mizoguchi and Okamoto, 2013).

ILPs have been identified in numerous insects and have been extensively studied in the genetic model organism, *Drosophila melanogaster*. In *Drosophila*, eight ILPs (ILP1–8) share high sequence identity to mammalian insulin (Brogiolo et al., 2001; Colombani et al., 2012; Fernandez et al., 1995; Garelli et al., 2012; Meneses and de los Angeles Ortíz, 1975; Yamaguchi et al., 1995). In general, each *Drosophila* ILP exhibits different expression patterns and functions in various biological processes such as metabolism, growth, development, reproduction, aging and behaviour (Erion and Sehgal, 2013; Wu and Brown, 2006). Insulin signalling pathways have been extensively studied in the last two decades, and there are many in-depth reviews (e.g. Nässel and Vanden Broeck, 2016; Sell, 2015; these actions are summarized in Figs. 1 and 3).

2.2 Ecdysteroids

Ecdysone is a steroidal prohormone of 20E. Primarily produced in the PGs, it is then released into the haemolymph and converted to the active form, 20E, at the target tissues (Figs. 2 and 3). 20E will then bind to the nuclear receptor and initiate various gene expression cascades to regulate insect developmental transitions, such as molting and metamorphosis. The production of ecdysone correlates well with the developmental timing (for a detailed review, see Huang et al., 2008; summarized in Figs. 2 and 3). Ecdysone and insulin have been extensively studied for their ability to control animal growth and development. Ecdysone functions in orchestrating developmental transitions such as larval molting and metamorphosis, whereas insulin mainly controls metabolism and growth (Géminard et al., 2006; Kozlova and Thummel, 2003). It was reported that insulin stimulated ecdysteroid production in vitro by ovaries of the mosquito *Aedes aegypti*, and *Drosophila* insulin receptor mutants showed a phenotype of reduced ecdysone synthesis (Riehle and Brown, 1999; Tu et al., 2002). Ecdysone and insulin were found to have antagonistic actions in controlling the final body size of adult *Drosophila*, and the crosstalk between ecdysone and insulin was crucial for coupling the control of growth and developmental timing (Colombani et al., 2005).

3. miRNA BIOGENESIS PATHWAY AND ACTION

miRNAs are a class of noncoding RNAs, about 21–23 nucleotides in length, that can target messenger RNAs (mRNAs) for posttranscriptional sequence-specific repression or degradation by binding to complementary

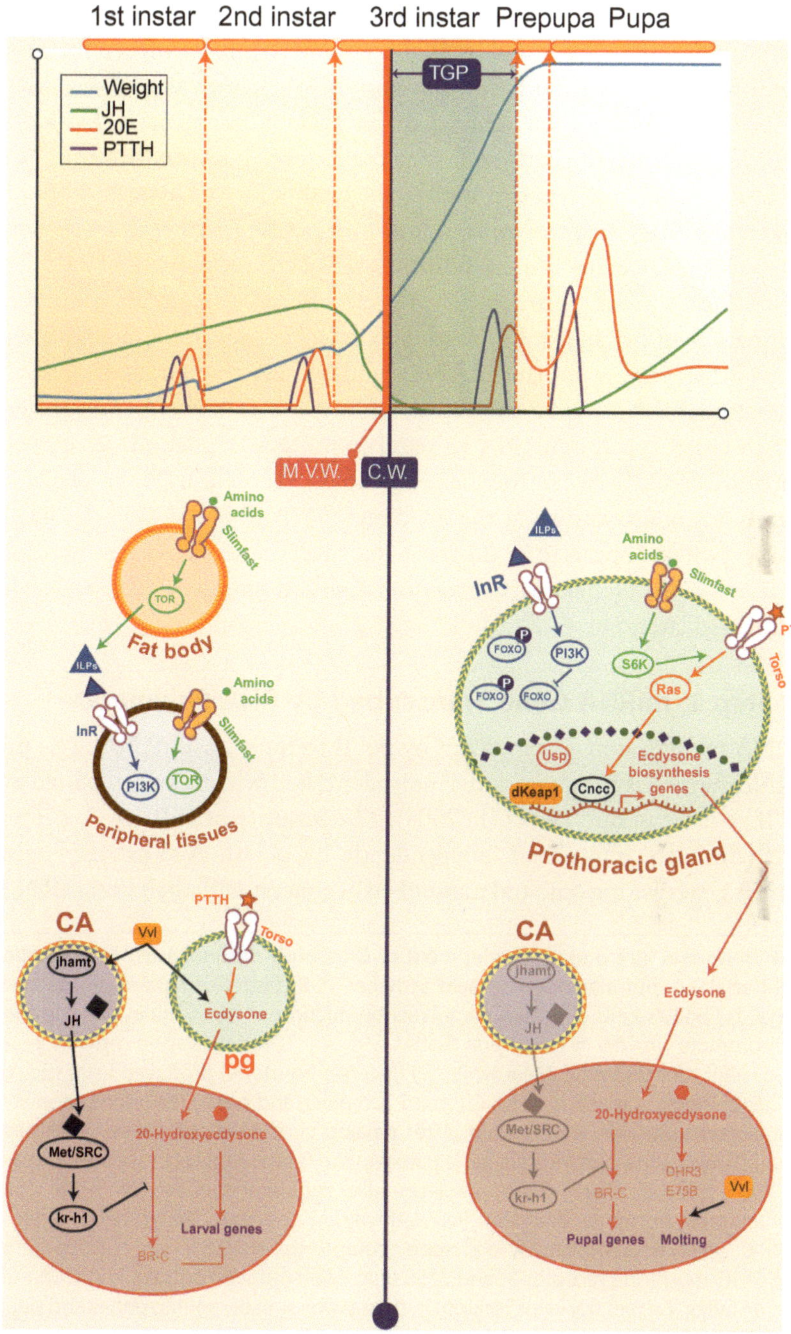

Fig. 1 See legend on next page.

sites in target mRNAs. In the canonical miRNA biogenesis pathway, miRNA genes are transcribed by RNA polymerase II (pol II) to form the long primary miRNA transcripts (pri-miRNAs) (Lee et al., 2004; Fig. 4). The pri-mRNAs are then cleaved by an RNase III, termed Drosha, in the nucleus forming the hairpin-shaped preliminary miRNA (pre-miRNA) (Lee et al., 2003), and then exported to the cytoplasm by Exportin-5 (Exp5) (Lund et al., 2004). In the cytoplasm, pre-miRNAs are further cleaved by the endonuclease Dicer to form a miRNA–miRNA* duplex. The HSC70–HSP90 (Heat shock cognate protein 70–Heat Shock Protein 90) chaperone machinery then aids in loading the duplex into the Argonaute (Ago) protein that will then form the mature RNA-induced silencing complex (RISC). The mature RISC unwinds the duplex RNA and uses miRNA complementarity to target specific mRNA to regulate protein translation. In animals, complementarity between the messenger and a segment in the 5′ region of the miRNA is also known as the ‘seed region’ (2nd–8th nucleotides of the miRNA) which appears to be the most crucial element of target recognition. The details of the biogenesis pathway and their actions (Steps 1–6, Fig. 4) will be discussed later.

3.1 Step 1: miRNA Gene Transcribed by RNA Polymerase

miRNA transcription is carried out by pol II to form pri-miRNA. The pri-miRNA formed is 5′ capped and 3′ polyadenylated which can be defined as a class II gene transcript (Lee et al., 2004). The majority of miRNAs are intergenic which are transcribed independently by pol II. Alternatively, some miRNA genes are intronic and transcribed in concert with their host mRNAs

Fig. 1 Hormonal titre during development of *Drosophila melanogaster*. Before critical weight attained, nutritional amino acid activates TOR (target of rapamycin) pathway in larval fat body's cells and then stimulates production of ILPs (insulin-like peptides) which promote growth in peripheral tissues with the TOR pathway, in order to reach critical weight for pupation. Meanwhile, JH (juvenile hormone) produced from the *corpora allata* inhibit the action of BR-C (broad complex), and hence, ecdysone-mediated larval genes expression is permitted. After passing critical weight, inside prothoracic gland (PG), ecdysone biosynthetic gene transcription is favoured by the protein dKeap1 and Cncc (Cap-n-collar). Cncc mediates the regulatory effect of PTTH (prothoracicotropic hormone)-activated Ras pathway on ecdysone biosynthetic genes in PG. BR-C which is controlled by ecdysone promote the transcription of pupa genes while DHR3 (hormone receptor 3) and E75B (ecdysone-induced protein 75B) with transcription factor Vvl (ventral veins lacking) initiate expression of molting genes and pupation. *TGP*, terminal growth period. *Adapted from Edgar, B.A., 2006. How flies get their size: genetics meets physiology. Nat. Rev. Genet. 7 (12), 907–916.*

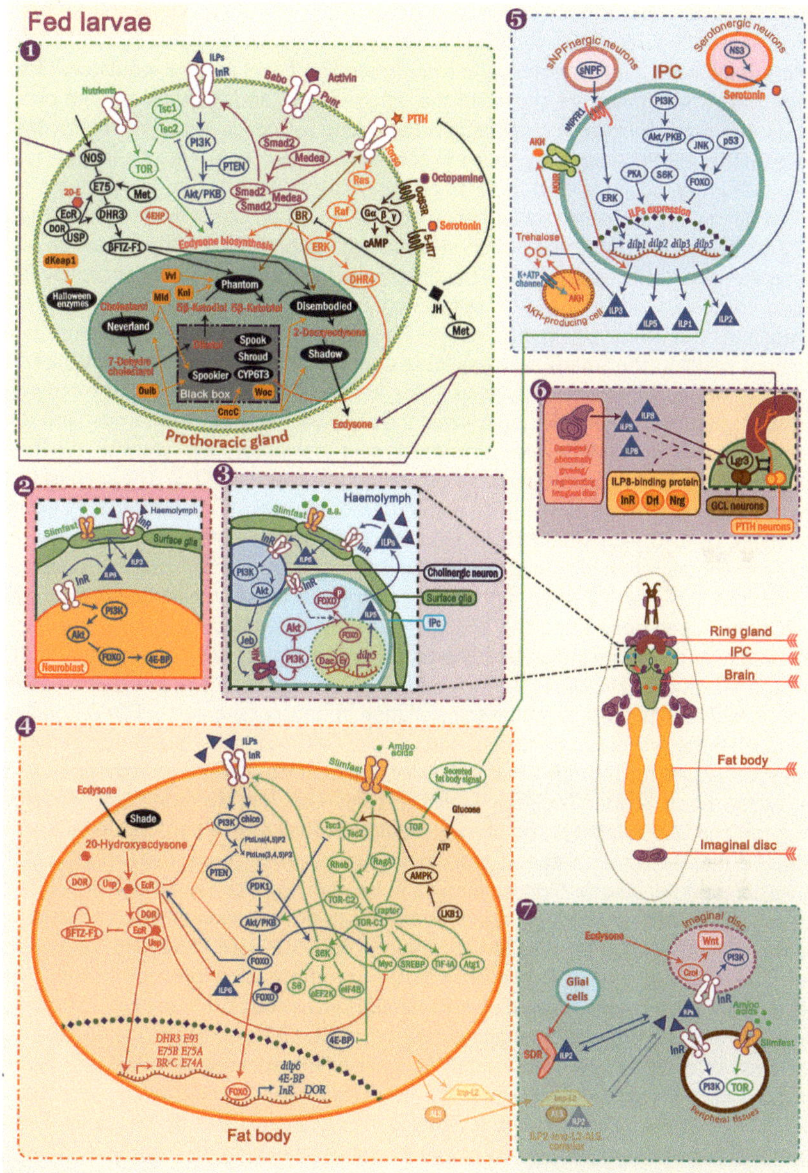

Fig. 2 The IIS (insulin/insulin growth factors signalling) pathway and ecdysteriod signalling pathway in the fed larvae of *Drosophila melanogaster*. *Blue arrows* and components depict the InR/PI3K (insulin receptor/phosphoinositide 3-kinase) pathway; *green* refers to the TOR pathway and *orange* represents the fat body-mediated signals. (1) Ecdysone biosynthesis within cells of the prothoracic gland. The inner *blue-green* ellipse details the biosynthetic pathway of ecdysone in the cell. Ecdysone is synthesized from cholesterol *(Continued)*

Fig. 2—Cont'd that is obtained from the diet. Ecdysone undergoes a series of reactions catalysed by the Halloween (CYP430) enzymes that are regulated by transcription factors. Halloween genes are shown in *black* in the *grey boxes* and the regulatory factors of the Halloween genes are shown in the *orange boxes*, and their action with *orange arrows*. The outer *light green circle* depicts regulating factors acting on the biosynthetic process. The InR/PI3K pathway (*blue*) is activated by circulating insulin-like peptides (ILPs) in the haemolymph and the TOR pathway (*green*) is activated by nutrients consumed by the larvae. These two pathways interact and promote ecdysone biosynthesis. PTTH activates the ecdysone production-favouring Ras pathway (*orange*) through binding to membrane receptor Torso. The action of PTTH on Torso is enhanced by BR-C (broad complex) and inhibited by circulating juvenile hormone (JH). BR also directly upregulates the Halloween genes. Activin interaction with receptor complex of Babo and Punt activates the Smad2 pathway (*purple*) which also promotes signalling via Torso as well as InR. NOS (nitric oxide synthase) interaction with components of the ecdysone pathway (*grey square*) directly upregulates the halloween genes. Octopamine and serotonin both raise the cAMP level in prothoracic cells to promote the synthesis. (2) The IIS pathway in glia and neuroblasts of the central nervous system (CNS). The secreted ILPs from glia activate the InR in neuroblasts and promote growth. (3) Regulated ILP5 expression in IPC by ILP6 secreted from glia neurons. The activation of InR in cholinergic cells promotes the secretion of ligand Jeb (jelly belly) which binds to receptor Alk (anaplastic lymphoma kinase) on IPC (insulin-producing cell) and activates the Alk pathway (*purple*) in IPC. (4) InR, ecdysone signalling and TOR pathway in larval fat body cells. Circulating ILPs bind to and activate InR on fat body cells, which activates downstream kinases PI3K (phosphoinositide 3-kinase), PDK1 (phosphoinositide-dependent kinase 1) and Akt/PKB (protein kinase B) that act to inhibit translocation of transcription factor FOXO (forkhead box O) from the cytoplasm to nucleus and inhibit FOXO-mediated translation. Ecdysone (*red*) synthesized from prothoracic cells is converted into 20-hydroxyecdysone by the Halloween gene Shade. Together with receptors Usp (ultraspiracle), EcR (ecdysone receptor) and DOR (diabetes and obesity regulated), they form a transcriptionally active complex that activates a series of ecdysone-responsive genes dependent on the corresponding life stages. Glucose (*Brown*) in haemolymph promotes the signal transduction of TOR pathway, which regulates a series of downstream growth and developmental proteins. (5) Expression and secretion of ILP1, ILP2, ILP3 and ILP5 from IPCs in CNS. PI3K pathway and PKA (protein kinase A) promote ILP expression in cells while JNK (c-Jun N-terminal kinase) and p53-activated FOXO inhibit the expression. Short neuropeptide F (sNPF) from sNPFnergic neurons promote ILP1 and ILP2 expression through activation of receptors sNPFR1 and ERK (extracellular-signal-regulated kinase). Serotonin from serotonergic neurons which are very close to IPCs promotes the secretion of ILP2 of IPCs. Presence of circulating trehalose stimulates release of AKH (adipokinetic hormone) from AKH-secreting neurons which sense haemolymph sugar levels through ATP-sensitive potassium (K ATP) channels; AKH further stimulates receptor AKHR on IPC and promotes downstream ILP3 secretion from IPC. AKH also positively regulates circulating sugar level, through a positive feedback loop. The loop is limited by secreted ILP3 which downregulates circulating sugar level. (6) Regulation of ecdysone production by imaginal disc-secreted ILP8. To ensure proper developmental timing, damaged or abnormally growing or regenerating imaginal disc(s) secrete ILP8-activating Lgr3 (G protein-coupled receptor 3) on GCL (glutamate-cysteine ligase)

(Continued)

(Rodriguez et al., 2004). The transcription of miRNAs can either be positively or negatively regulated dependent on the binding of different transcription factors and epigenetic regulators (Cai et al., 2004; Davis-Dusenbery and Hata, 2010; Kim et al., 2009; Krol et al., 2010; Lee et al., 2004).

3.2 Step 2: Pri-miRNA Transcript Processed by Microprocessor

Pri-miRNA transcript is the long miRNA precursor which can contain one or more miRNA precursors (each hairpin loop structure contains ~70 nucleotides) (Gregory et al., 2004). The flanking sequence is generally thought to be functionally constrained and essential for efficient processing (Griffiths-Jones et al., 2011; Kenny et al., 2015). The cleavage of the pri-miRNA is first initiated by an RNase III enzyme Drosha and its double-stranded RNA-binding partner, Pasha, forming a multiprotein complex of ~500 kDa termed the 'microprocessor complex' (Denli et al., 2004; Fig. 4). Suppression of Drosha or Pasha in *Drosophila* cells will lead to an accumulation of pri-miRNA and a decrease in mature miRNAs, suggesting that both components have individual important roles in pri-miRNA processing (Denli et al., 2004; Lee et al., 2003). The cleavage site of the microprocessor on the pri-miRNA is mainly located on the 11th nucleotide of the stem-ssRNA junction (Han et al., 2006). Pasha specifically interacts with the target pri-miRNA and binds to the flanking ssRNA, which positions the centre of Drosha to the right place for cleaving the pri-miRNA into pre-miRNA (Han et al., 2006).

3.3 Step 3: Pre-miRNA Leaving the Nucleus by Exp5

Export of pre-miRNA from nucleus to cytoplasm is one of the key steps for miRNA maturation. After the cleavage by microprocessor, the pre-miRNA contains a 2-nucleotide overhang at the 3′ end and will be exported to the cytoplasm through the nuclear pore mediated by Exp5 for further processing (Wan et al., 2013; Fig. 4). Exp5 directly binds to the properly processed pre-miRNA in the nucleus which is dependent on RanGTP, and depletion of

Fig. 2—Cont'd neurons directly or indirectly (through a relay of ILP8-binding protein), leading to inhibition of PTTH secretion in PTTH neurons. Thus, ecdysone production is limited. Furthermore, ILP8 activates nitric oxide synthase (NOS) in the prothoracic gland, which inhibits ecdysone biosynthesis. (7) Regulation of IIS of peripheral tissues by circulating SDR. Secreted decoy of InR (SDR) secreted from glial cells is a constitutive regulators of circulating ILPs, which inhibit IIS in peripheral tissues, including imaginal discs, through forming a complex with ILPs (mainly ILP3). Meanwhile, ecdysone activates Wnt (Wingless) pathway in imaginal discs through Crol (Crooked legs).

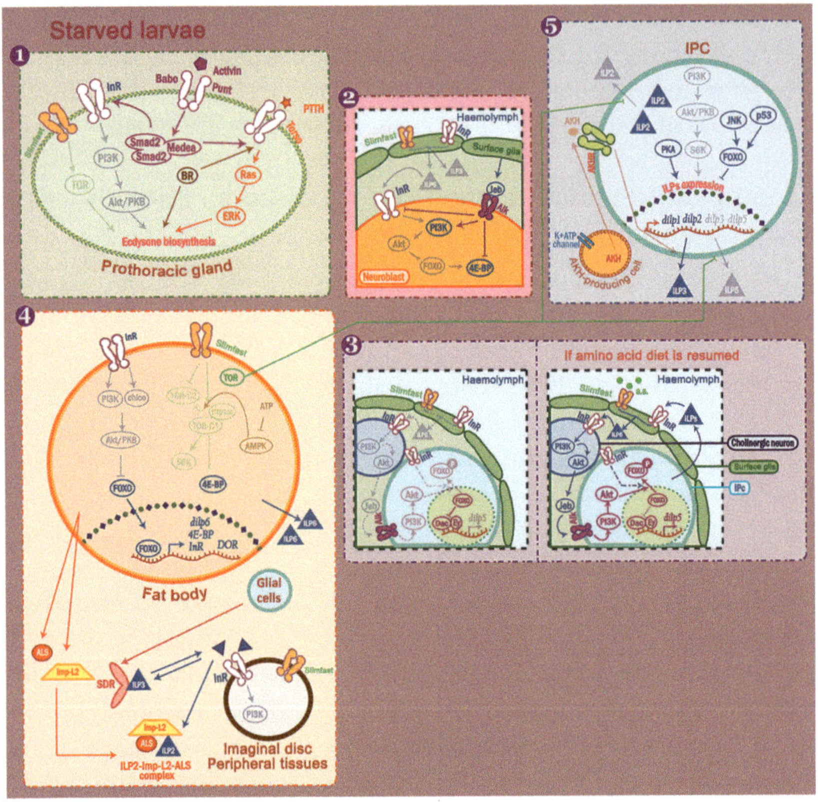

Fig. 3 The IIS pathway and ecdysteroid signalling pathway in starved larvae of *Drosophila melanogaster*. In comparison to fed larvae, a number of differences, especially on ILP expression and secretion, have been observed in starved larvae. Note that pathways that are identical to the fed larvae are not shown here. *Blue arrows* and components depict the InR/PI3K pathway; *Green* refers to the TOR pathway and *orange* represents the fat body-mediated signals. (1) Reduced IIS and TOR pathway signalling lower ecdysone production by prothoracic gland, which affects the baseline ecdysone level and larval developmental timing, respectively. (2) Activated Jeb/Alk pathway in neuroblasts of the central nervous system (CNS). During starvation, circulating ILPs drop, while the IIS and TOR pathways remain inactivated in the glia cells. The glia cells will then secrete Jeb which activates the Jeb/Alk pathway in neuroblasts. (3) ILP5 expression under restricted diet. In adverse conditions, when both sugar and amino acids are highly limited, the IIS pathway in surface glia cells and cholinergic cells is restricted, and thus, the Alk pathway is inactivated in IPC (insulin-producing cell), and the nonphosphorylated FOXO will translocate to the nucleus and inhibit the formation of the Dac/Ey (Dachshund/Eyeless) complex and dilp5 expression. Resumption of an amino acid diet results in the reactivation of expression of dilp5 to maintain larval growth during

(Continued)

Fig. 3—Cont'd starvation. (4) Reduced activation of IIS and TOR in the fat body and Imp-L2 negatively regulates IIS in the peripheral tissues. FOXO is nonphosphorylated as InR/PI3K is activated and FOXO translocates to the nucleus, promoting transcription of a series of developmental genes. ILP6 is induced. Furthermore, TOR signalling in the fat body downregulates the secretion of ILP2 and ILP5. Imp-L2 and ALS, which are secreted from fat body, form a growth suppressed complex with ILP 2, 3, 5 and 6 (but mainly ILP2) as the Imp-L2/ALS/ILPs complex lower the number of circulating free ILPs in haemolymph, hence, binding of ILPs to InR and larval growth is also reduced. Thus, the complex limits IIS in the larva as a whole, allowing the larva to survive starvation. (5) Restricted expression and secretion of ILPs in IPC. Expression levels of dilp3 and dilp5 are reduced, but not dilp2 which remain unchanged. Increased accumulation of ILP2 and ILP5 in IPC is also observed. Due to low haemolymph sugar level, AKH-induced ILP3 secretion is also limited. Abbreviations: *AKH*, adipokinetic hormone; *AKHR*, adipokinetic hormone receptor; *Alk*, anaplastic lymphoma kinase; *AMPK*, AMP-activated protein kinase; *Atg1*, autophagy-specific gene 1; *babo*, baboon; βFTZ-F1, ftz transcription factor 1; *BR/BR-C*, broad complex; *CA*, corpus allatum; *CC*, corpus cardiacum; *Cncc*, cap-n-collar; *CREB*, cAMP-response-element-binding protein; *CW*, critical weight; *Dac*, Dachshund; *dALS/ALS*, *Drosophila* acid-labile subunit; *DHR3*, hormone receptor 3; *DHR4*, hormone receptor 4; *DOR*, diabetes and obesity regulated; *Drl*, derailed; *E74A*, ecdysone-induced protein 74EF; *E75/E75A/E75B*, ecdysone-induced protein 75B; *E93*, ecdysone-induced protein 93F; *4E-BP*, 4E-binding protein; *EcR*, ecdysone receptor; *eIF4B*, eukaryotic initiation factor 4B; *eIF4E*, eukaryotic initiation factor 4E; *ERK*, extracellular-signal-regulated kinase; *Ey*, eyeless; *FOXO*, forkhead box O; *GAP*, GTPase-activating protein; *GEF*, guanine-nucleotide-exchange factor; *GLUT4*, glucose transport protein 4; *GRB2*, growth factor receptor-bound protein 2; *5-HT7*, 5-hydroxytryptamine (serotonin) receptor 7; *IGF*, insulin-like growth factor; *ILP*, insulin-like peptide; *Imp-L2*, imaginal morphogenesis protein-late 2; *InR*, insulin-like receptor; *IPC*, insulin-producing cell; *IRS*, insulin receptor substrate; *Jeb*, jelly belly; *JH*, juvenile hormone; *jhamt*, juvenile hormone acid methyltransferase; *JNK*, c-Jun N-terminal kinase; *kr-h1*, Kruppel homolog 1; *Lgr3*, G protein-coupled receptor 3; *LKB1*, Lkb1 kinase; *MAP4K3*, MAPK kinase kinase kinase-3; *MAPK*, mitogen-activated protein kinase; *Met*, methoprene-tolerant; *Mld*, molting defective; *MVW*, minimal viable weight; *NLaz*, neural lazarillo; *Nos*, nitric oxide synthase; *Nrg*, neuroglian; *NS3*, nucleostemin 3; *NSC*, neurosecretory cell; *Path*, pathetic; *Octβ3R*, octopamine β3 receptor; *Ouib*, ouija board; *PDK1*, phosphoinositide-dependent kinase 1; *PG*, prothoracicotropic hormone; *PH*, pleckstrin homology; *PI3K*, phosphoinositide 3-kinase; *PKA*, protein kinase A; *PTEN*, phosphatase and tensin homologue deleted on chromosome 10; *PTTH*, prothoracicotropic hormone; *Raf*, Raf oncogene; *RagA*, Ras-related GTP-binding A/B; S6K, RPS6 (ribosomal protein S6)-p70-protein kinase; *SDR*, Secreted decoy of InR; *Slif*, slimfast; *Smad2*, Smad on X; *sNPF*, short neuropeptide F; *NPFR1*, sNPF receptor; *SREBP*, sterol regulatory element-binding protein; *Step*, steppke; *TIF-IA*, transcriptional intermediary factor-IA; *Tobi*, target of brain insulin; *TOP*, terminal oligopyrimidine tract; *TOR*, target of rapamycin; *TOR-C1*, rapamycin-sensitive TORC complex; *TOR-C2*, rapamycin-resistant TORC complex; *Tsc1*, tuberous sclerosis complex 1; *Tsc2*, tuberous sclerosis complex 2; *Usp*, ultraspiracle; *Vvl*, ventral veins lacking; *Woc*, without children.

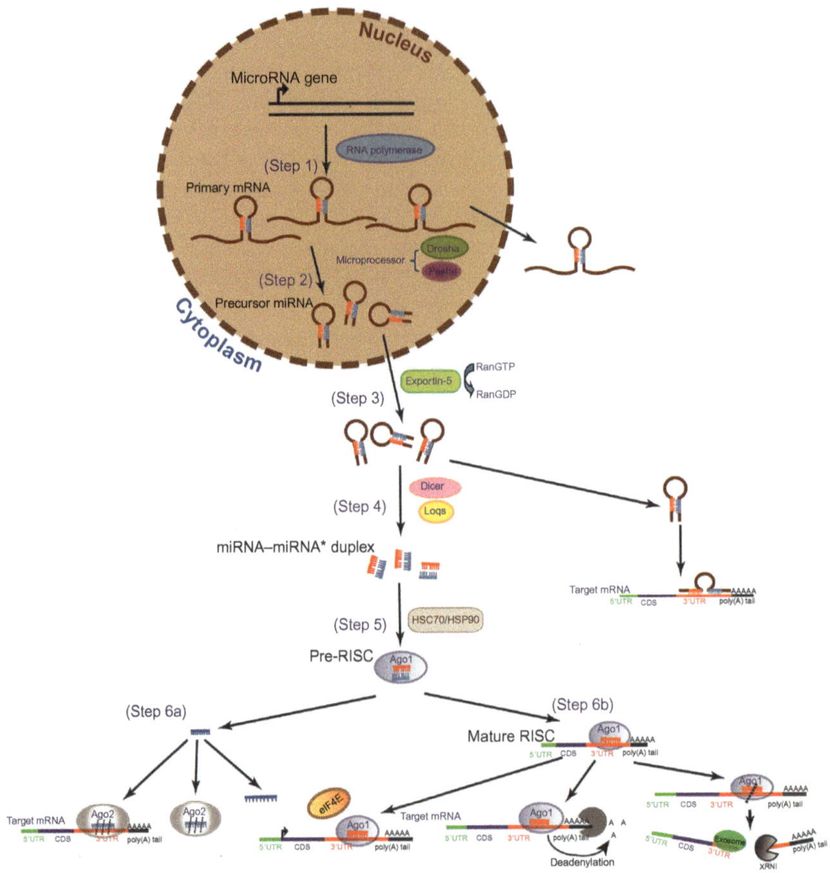

Fig. 4 Schematic diagram showing the canonical biogenesis pathway and actions of miRNAs in *Drosophila*. In the canonical miRNA biogenesis pathway, the primary miRNA (pri-mRNA) transcript is first cleaved by Drosha in the nucleus to form preliminary miRNA (pre-miRNA). The pre-miRNA will then be exported from the nucleus to cytoplasm by Exportin, which is further cleaved by endonuclease Dicer to form a miRNA–miRNA* duplex. With the help of HSC70–HSP90 (Heat shock cognate protein 70–Heat Shock Protein 90) chaperone machinery, the duplex will be loaded into an Argonaute (Ago) protein to produce a mature RNA-induced silencing complex (RISC). The mature RISC will then bind to its target mRNA to regulate the protein translation. For details, please refer to text.

Exp5 leads to a reduction in mature miRNA levels which may be related to the loss of protection from nuclease digestion (Bohnsack et al., 2004; Lund et al., 2004). Export of pre-miRNA is accelerated upon DNA damage through phosphorylation of the nucleopore protein Nup153 by an Ataxia Telangiectasia Mutated (ATM)-activated Akt kinase (a protein kinase

B family member) or Akt/PKB, which enhances the interaction of Nup153 with Exp5 (Wan et al., 2013). The hydrolysis of RanGTP to RanGDP will then release the pre-miRNA for further cleavage.

3.4 Step 4: Pre-miRNA Cleavage by Dicer

In the cytoplasm, an RNase III enzyme named as Dicer cleaves the pre-RNA forming a miRNA–miRNA* duplex (Ketting et al., 2001; Fig. 4). In flies, there are two Dicer proteins that are required for the different types of small RNA biogenesis—Dicer-1 is involved in the miRNA pathway which generates the mature miRNA–miRNA* duplex from pre-miRNA, and Dicer-2 liberates the siRNA duplex forming long dsRNA (Lee et al., 2004). In the presence of 5′-terminal phosphate, the PAZ domain of Dicer binds to the pre-miRNA termini and cleaves at a fixed position on the stem loop structure (Macrae et al., 2006; Park et al., 2011; Zhang et al., 2004). The length of each strand in the miRNA–miRNA* duplex was usually 21–25 nucleotides depending on the type of Dicer being used. In flies, Dicer-binding partner proteins have the ability to change the cleavage site of Dicer, which allows the same sequence to produce various miRNAs with different target specificities (Fukunaga et al., 2012). Moreover, depletion of R3D1-L, a dsRNA-binding protein that forms a stable complex with Dicer-1, leads to the accumulation of pre-miRNA, which alters the miRNA production by Dicer-1 and reproductive development (Jiang et al., 2005).

3.5 Step 5: Ago Loading and RISC Formation

The miRNA–miRNA* duplex generated by Dicer is then loaded into the Ago effector protein to form an RISC, leading to subsequent gene silencing. Ago proteins comprise an RNA-binding PAZ domain, an RNA endonuclease PIWI domain and a middle (MID) domain (Song et al., 2004; Yan et al., 2003); both Ago-1 and Ago-2 in *Drosophila* contain PIWI domains that retain their RNA endonuclease activities for forming miRNA and siRNA duplexes, respectively (Okamura et al., 2004; Saito et al., 2005; Tomari et al., 2007). The 5′ phosphate of the miRNA binds tightly to the binding pocket between the PIWI and MID domains (Parker et al., 2005), and the PIWI domain contains an RNase-like active site that allows it to cleave specifically between position 10 and 11 at 5′ end of the miRNA (Parker et al., 2005; Song et al., 2004).

There are generally two models that depict the Ago loading and RISC formation. In the case of central sequence mismatches within the miRNA duplex, the complex is then directed to Ago-1 for RISC assembly (Czech et al., 2009; Fig. 4, Step 6b). With the help of HSC70/HSP90 chaperone machinery, conformational change occurs, allowing the opening of the binding site and Ago proteins can then bind to the miRNA duplexes (Iwasaki et al., 2010). Another model (Fig. 4, Step 6a) proposes that miRNA* with more double-stranded character will be loaded into Ago-2 and silenced mRNA with fully complementary miRNA-binding sites in flies (Förstemann et al., 2007).

The types of proteins that bind to the Ago protein will also affect the stability of the miRNA-mediated gene silencing action. For example, Ago-2 interacts with a protein termed 'GW182' originally identified in the sera of patients suffering from motor and sensory neuropathy (named after its molecular size and the presence of glycine and tryptophan repeats). The GW repeats in the N-terminus of GW182 forms an Ago-binding domain for Ago–GW182 interaction (Pfaff et al., 2013), which protects the Ago-bound miRNA from degradation by exoribonuclease (Yao et al., 2012). On the other hand, in Mei-P26, a Trim-NHL protein (a potential E3 ubiquitin ligase), binds to the Ago-1 protein (Neumüller et al., 2008). Overexpression of *mei-p26* will broadly reduce miRNA levels (Neumüller et al., 2008), but in undifferentiated germ cells, Mei-P26 physically associates with Ago-1 and GW182 and promotes miRNA action instead (Li et al., 2012).

3.6 Step 6: miRNA Actions

The majority of miRNAs that bind to their targets depend on complementarity of the 5′ 2nd–8th nucleotides, which is also known as the 'seed region' of the miRNA and the 3′ UTR of the mRNA (Lee et al., 1993; Lewis et al., 2005; Wightman et al., 1993). In some cases, the promoter and 5′ UTR (Lytle et al., 2007; Ørom et al., 2008; Place et al., 2008) and open reading frame (Forman et al., 2008; Qian et al., 2016; Shen et al., 2008) of mRNAs have also been reported as functional binding sites of the miRNAs. The three major actions of the RISC complex regulating protein-coding genes in *Drosophila* are translational repression, mRNA deadenylation or translation upregulation.

In the case of translational repression, most miRNAs inhibit translational initiation of mRNA targets in a 7-methyl guanosine (m^7G) cap-dependent manner which reduces protein production (Kiriakidou et al., 2007). In brief, the cap-binding/MC motif in *Drosophila* Ago-1 can also be found in the human Ago-2 MID domain, which allows these Ago proteins to bind to the m^7G cap of mRNAs and thus displace the cap-binding initiation factor eIF4E that further represses the initiation of mRNA translation. In some cases, the miRNA works through the poly(A)-binding proteins (PABPs) and target mRNA poly(A) tails, resulting in mRNA deadenylation and degradation. The association of PABPs with target mRNAs facilitates the binding of miRISC to its target mRNA, and the RISC then displaces PABPs and promote deadenylation (Moretti et al., 2012). The efficiency of such miRNA-mediated repression also correlates with the length of the poly(A) tail (Moretti et al., 2012). In some cases, the GW182 protein may also interact with the Ago-1 protein to recruit deadenylases and decapping enzyme resulting in mRNA degradation, so that some miRNA targets will be degraded whereas some stay in a translationally silent state for further regulation (Behm-Ansmant et al., 2006). In *Drosophila*, the miRNA-mediated silencing mechanism usually starts with effects on translational inhibition, and subsequently leads to the deadenylation and degradation of mRNA (Djuranovic et al., 2012). In addition, some recent studies also show that miRNA and their associated microribonucleoproteins have the ability to stimulate gene expression posttranscriptionally through an unknown mechanism to directly activate or relieve the mRNA repression (Vasudevan, 2012).

In addition to the canonical biogenesis pathway, miRNAs-like small RNAs and miRNAs can also be generated via alternative pathways (Fig. 5; modified from Ha and Kim, 2014). In *Drosophila*, a group of miRNAs are found to be m^7G-capped and to be both Drosha and Pasha independent. miR-320 is one of these miRNAs, whereas its m^7G-cap pre-miRNA can bypass Drosha processing and are exported to the cytoplasm via Exportin-1, cleaved by Dicer and loaded into Ago-1 for mRNA target regulation (Babiarz et al., 2008; Okamura et al., 2013). Mirtrons, TUTase codependent and Dicer-independent miRNAs also enter noncanonical biogenesis pathway in *Drosophila*. Mirtron refers to miRNAs generated from the splicing and debranching of a short intron by spliceosomes without undergoing Drosha-mediated cleavage (Ruby et al., 2007). In brief, nonlinear pre-miRNA mimics are generated during the splicing of introns, followed by transportation to cytoplasm by Exp5, cleavage by Dicer and Ago loading to perform the actions. On the other hand, TUTase-dependent miRNAs carry a shorter 3′ overhang after Drosha cleavage, and thus its production requires an additional mono-uridylation to create a 2 nt 3′ overhang for Dicer recognition (Heo et al., 2012). Lastly, Dicer-independent

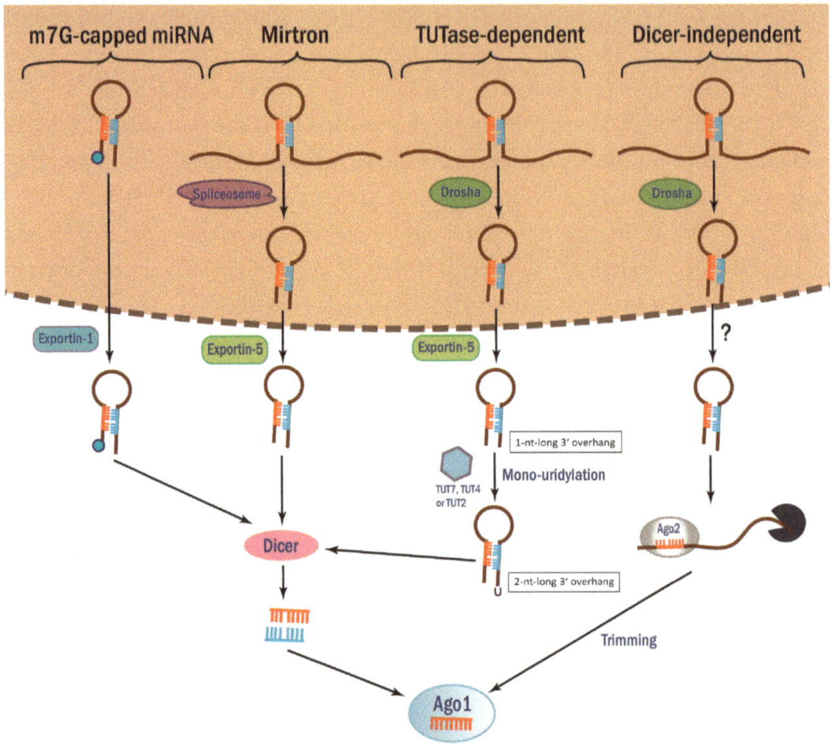

Fig. 5 Noncanonical pathways of miRNA biogenesis.

pre-miRNAs are cleaved by Ago-2 rather than by Dicer to generate functional miRNAs; and in some cases, the miRNA hairpin might undergo 3′ end trimming to generate a ~20–26 nt functional miRNA (Cheloufi et al., 2010; Cifuentes et al., 2010).

4. miRNA REGULATION OF ILPs AND ECDYSTEROIDS

Despite the identification of the first miRNA in nematodes more than 20 years ago (Lee et al., 1993), the importance of these small RNAs in regulation were only recognized with the identification of highly conserved miRNAs in both invertebrates and vertebrates a decade later (e.g. Lagos-Quintana et al., 2001). To date, understanding how miRNA functions is one of the most intensely studied areas in biology and medicine. In animals, there are already a large number of miRNAs identified (28,645 entries in the miRBase Release 21), and different miRNA target prediction tools have also predicted numerous genes involved in different hormonal signalling pathways.

As one miRNA can regulate multiple targets, they form a complex regulatory network in the control of hormonal production and signalling. For example, combinatory miRNAs can repress target genes in an additive fashion, thus fine tuning gene expression rather than switching gene actions in a dramatic on–off manner (Baek et al., 2008; Selbach et al., 2008). Moreover, miRNAs are often involved in feedback or feedforward loops to confer robustness (Herranz and Cohen, 2010; Tsang et al., 2007). Furthermore, miRNAs inhibit gene expression posttranscriptionally, thus avoiding the energy-intensive process of protein synthesis while still enabling rapid translation of miRNA targets in response to various stimuli. Although our understanding of miRNA biogenesis and function remains limited, numerous studies have demonstrated the important roles of miRNAs in the ecdysone and insulin signalling pathways of insects. Here, we will utilize these two well-studied signalling pathways in *Drosophila* to illustrate how these two kinds of intracellular signal molecules can coordinate the responses to stimuli and to maintain homeostasis. In the following section, two types of miRNAs related to ecdysone or insulin signalling pathways will be discussed: (I) direct regulation of either of genes of the hormonal signalling pathways or induced in the presence of either hormones and (II) regulation of genes of both hormonal pathways (Table 1).

4.1 (I) miRNAs Related to Either Insulin or Ecdysteroid Signalling Pathways

4.1.1 let-7

let-7 was the second miRNA discovered in animals and was found to regulate late developmental events in the nematode *Caenorhabditis elegans* (Pasquinelli et al., 2000; Reinhart et al., 2000). The sequences of let-7 are highly conserved in bilaterians, indicating a strong selective pressure

Table 1 Table Summarizing the Current Knowledge of the miRNAs Involved in the Regulation of Ecdysone and Insulin Signalling Pathways

miRNAs that are regulated or induced in the presence of either insulin or ecdysteroid	let-7, miR-34-miR-277-miR-317 cluster, miR-278, mir-9a, miR-275-miR-305 cluster
miRNAs that cross-regulate insulin and ecdysone pathways	bantam, miR-14, miR-8

in maintaining its function (Griffiths-Jones et al., 2011; Pasquinelli et al., 2000). In arthropods, let-7 is usually genomically clustered with miR-100 and miR-125, and is transcribed as a single polycistronic transcript (Bashirullah et al., 2003; Chawla et al., 2016; Griffiths-Jones et al., 2011; Hertel et al., 2012; Sempere et al., 2003; Sokol et al., 2008).

In *Drosophila*, the broad complex (BR-C) transcription factor is expressed in last instar larvae and prepupae before the triggering of pupal morphogenesis (Kiss et al., 1988). Ecdysone was first thought to regulate the temporal expression of let-7 during metamorphosis mediated by BR-C (Sempere et al., 2002, 2003). However, three functional Ecdysone response elements have also been identified in the first intron of the let-7 locus which can be activated by the heterodimer composed of Usp (ultraspiracle) and EcR (ecdysone receptor) (Chawla and Sokol, 2012). Nevertheless, comparing the expression profiles and the self-transcribing active regulatory region sequencing enhancer-screening assay data suggest the possibility of some currently unknown potential regulators of the let-7 in addition to the EcR–Usp heterodimer (Ryazansky et al., 2011; Shlyueva et al., 2014).

Other targets of let-7 have also been identified in *Drosophila*. In the remodelling of the abdominal neuromusculature during the larval-to-adult transition, let-7 regulates the transcription factors Abrupt and Chinmo which determine the cell fate of mushroom body neuroblast progeny and neurogenesis (Caygill and Johnston, 2008; Chawla et al., 2016; Kucherenko et al., 2012; Sokol et al., 2008; Wu et al., 2012). In the testis, let-7 has been found to target IGF-II mRNA-binding protein (Imp) to regulate aging (Toledano et al., 2012); whereas in the eye disc, let-7 targets the *lin-41* homologue Dappled to regulate eye development (O'Farrell et al., 2008). In S2 cells, let-7 has been shown to regulate innate immunity by targeting the antimicrobial peptide Diptericin (Garbuzov and Tatar, 2010). In the presence of ecdysteroids, let-7 can also regulate the circadian rhythm via Core Clock Component CWO or Sexual identity (Chen et al., 2014; Fagegaltier et al., 2014). These studies on ecdysone-induced let-7 demonstrate how a hormone can achieve tissue- or cell-specific functions via the same miRNA.

4.1.2 miR-34–miR-277–miR-317 Cluster

In *Drosophila*, miR-34 and miR-277 of the genomic cluster of miR-34-miR-277-miR-317 are related to either insulin or ecdysteroid regulation. miR-34 is a conserved miRNA with identical seed sequences in the fly, nematode, mouse and human (Liu et al., 2012). In *Drosophila*, despite its relatively low abundance, miR-34 was found to be maternally deposited in the unfertilized oocytes, with its expression dependent on the maternal Dicer-1 (Marco, 2015; Soni et al., 2013). During metamorphosis, its expression is downregulated by ecdysone and the activity of the ecdysone-inducible gene BR-C (Aravin et al., 2003; Sempere et al., 2003). Several ecdysone signalling pathway genes have been found to be regulated by miR-34, including Eip71CD and E74A; E74A is also known to modulate aging and neurodegeneration in *Drosophila* (Liu et al., 2012). miR-277 was first found as the modulator of rCGG repeat-mediated neurodegeneration in the *Drosophila* brain (Tan et al., 2012). Constitutive expression of miR-277 under reduced insulin levels as a consequence of the mutation in the insulin receptor substrate Chico results in lethality in *Drosophila*, indicating a potential relationship between miR-277 and the insulin signalling system (Esslinger et al., 2013). For the other genomic cluster member miR-317, it has been found to be involved in the epistatic network that determines startle-induced locomotion (Yamamoto et al., 2008) or influences the reproductive responses of females in receiving the sex peptide from males (Fricke et al., 2014). No function related to either insulin or ecdysteroid has been proposed for miR-317 to date.

4.1.3 miR-278

miR-278 appears to be insect-specific, and misexpression of miR-278 in *Drosophila* results in massive overgrowth in the developing eye, in part as a result of inhibition of apoptosis; while reduction of miR-278 function does not attenuate growth or cause any other detectable phenotypes (Nairz et al., 2006). However, complete loss of miR-278 mutants fails to repress translation of the gene *expanded* which results in flies that are insulin resistant and lean, storing less triglyceride despite the elevation of insulin and circulating sugar levels mobilized from the adipose-tissue glycogen storage (Teleman et al., 2006). miR-278 has also been found to directly regulate the cyclin-dependent kinase inhibitor *Decapo* that regulates germ-line cell division and it was further demonstrated that this miRNA-mediated regulation may be controlled by insulin signalling (Yu et al., 2009).

4.1.4 miR-9a

miR-9 is a highly conserved miRNA family in animals (Yuva-Aydemir et al., 2011), and miR-9a in *Drosophila* was found to be important for various developmental processes from early embryonic development to formation of nervous system, sensory organ and wing formation (Bejarano et al., 2010; Biemar et al., 2006; Biryukova et al., 2009; Cassidy et al., 2013, 2016; Fu et al., 2014; Leaman et al., 2005; Li et al., 2006, 2013; Wang et al., 2016; Yatsenko and Shcherbata, 2014). Genome-wide miRNA screening has linked the function of miR-9a to regulation of body growth and insulin production by controlling the insulin-producing cells through interaction with the short neuropeptide F receptor (Suh et al., 2015).

4.1.5 miR-275–miR-305 Cluster

In the intestinal stem cells of *Drosophila*, the expression of miR-275–miR-305 cluster is upregulated in the presence of insulin, and miR-305 in turn targets the insulin signalling pathway genes InR and PI3K92E (Foronda et al., 2014). In the fat body, miR-305 regulates Dp53 of the nutritional-sensing-TOR signalling pathway in maintaining metabolic homeostasis (Barrio et al., 2014).

4.2 (II) miRNAs That Links Up Ecdysone and Insulin Signalling

4.2.1 Bantam

In 2002, the *bantam* gene product was first reported as a novel regulator of *Drosophila* tissue growth and cell proliferation. Loss-of-function *bantam* mutants have phenotypes that range from reduced size to pupal lethality with the absence of imaginal discs; and overexpression of *bantam* results in the overgrowth of wing and eye tissue. The altered phenotypes support *bantam* acting to regulate tissue growth by altering cell number rather than cell size (Hipfner et al., 2002). Although the *bantam* locus encodes a miRNA that was originally found to promote cell proliferation and prevent apoptosis during *Drosophila* development (Brennecke et al., 2003), it is now known to be the mediator or regulator of various signalling pathways, namely Hippo (Nolo et al., 2006; Thompson and Cohen, 2006), Notch (Becam et al., 2011; Herranz et al., 2008), Decapentaplegic (Oh and Irvine, 2011), epidermal growth factor receptor (Herranz et al., 2012), insulin (Boulan et al., 2013) and ecdysone biosynthesis (Boulan et al., 2013).

In *Drosophila* late prepupae, the salivary glands undergo programmed cell death. Ecdysone pulses during the prepupal period activate the EcR which in turn represses a transcription factor (β-FTZ-1) that activates transcription of the *Drosophila* inhibitor of apoptosis *diap2*. The activated EcR then directly activates the death activator reaper (*rpr*) and two transcription factors, BR-C and E74A, which in turn regulate the transcription and level of expression of the proapoptotic gene head involution defective (*hid*), which was identified as having three to six binding sites for *bantam* miRNA in its 3′ UTR (Brennecke et al., 2003). *Bantam* gene expression would thus interfere with *hid* mRNA translation to favour inhibition of apoptosis. In addition to bantam, multiple other miRNAs can also posttranscriptionally regulate *hid*, indicating the interconnectivity between *hid*-inhibiting miRNAs and *hid*-activating ecdysone signalling in regulating apoptosis in *Drosophila* (Hilgers et al., 2010; Leaman et al., 2005).

Bantam expression specifically targeted ecdysone-producing cells in the PG and promoted systemic body overgrowth by inhibiting ecdysone biosynthesis (Boulan et al., 2013). The role of bantam in promoting growth also relies on both its systemic and cell-autonomous activities, as the targeted expression of *bantam* in the PG could not fully rescue the wing imaginal discs defects in *bantam* mutant larvae (Boulan et al., 2013). This is consistent with a previous finding that bantam functions in the *Drosophila* wing to regulate G1–S transition and boundary formation (Becam et al., 2011; Herranz et al., 2008). Other than its relationship to the ecdysone pathway, insulin represses bantam activity to regulate ecdysone production in *Drosophila*, revealing that bantam can buffer the antagonistic action of insulin and ecdysone (Boulan et al., 2013).

4.2.2 miR-14

miR-14 was initially identified as a regulator of autophagy in *Drosophila*, by regulating Rpr-, Hid-, Grim- and Dronc-dependent cell death through Drice (Death related ICE-like caspase) (Xu et al., 2003). In the last decade, its role in regulating cell death including regulation of inositol 1,4,5-trisphosphate kinase 2 (ip3k2) to modulate inositol 1,4,5-trisphosphate (IP3) signalling during salivary gland cell death which is better known (Nelson et al., 2014). In addition, miR-14 also plays a role in hormone signalling and metabolism. In the insulin-producing neurosecretory cells in the adult brain of *Drosophila*, miR-14 targets the genes sugarbabe and EcR, and thus insulin and ecdysone production, to control metabolism (Varghese and Cohen, 2007; Varghese et al., 2010). An autoregulatory loop mechanism has also been proposed between miR-14 and ecdysone, whereby miR-14 expression would be reduced in the fat body in the presence of 20E, in turn modulating ecdysone signalling by targeting the EcR (Varghese and Cohen, 2007).

4.2.3 miR-8

miR-8 is a miRNA conserved throughout bilaterians and is expressed in all developmental stages of *Drosophila*. Its expression is strongly upregulated in the larval stage and sustained at a high level until adulthood (Lagos-Quintana et al., 2001), suggesting that it may play a variety of roles in development. For example, miR-8 regulates atrophin levels by directly targeting *atrophin* in the regulation of apoptosis in the brain (Karres et al., 2007). miR-8 targets *enabled* in neuromuscular junction formation (Karres et al., 2007; Loya et al., 2009, 2014), and also regulates other developmental genes involved in signalling pathways such as the Notch ligand Serrate, Wnt and the escargot (Antonello et al., 2015; Kennell et al., 2008; Vallejo et al., 2011). In the fat body, miR-8 targeted u-shaped (USH) to regulate insulin production, and this regulation is also conserved in humans (Hyun et al., 2009). Multiple growth factor hormones including Dilp6 (*Drosophila* ILP6) was found to be regulated by miR-8 (Lee et al., 2015). In the presence of ecdysone, miR-8 level is reduced and its target USH that regulates insulin signalling is reciprocally upregulated (Jin et al., 2012). Moreover, miR-8 is transcriptionally repressed by ecdysone early response genes, BR-C and E74A, which bind the promoter region of miR-8 (Jin et al., 2012). These findings together imply a variable interaction between ecdysone and insulin signalling through miR-8.

5. SUMMARY

The understanding of the role of noncoding RNAs in the regulation of hormone production and action, irrespective of the emergence of evolutionary endocrinology becoming a new area in current comparative endocrinology, reveals the complexity of interactions between hormones, target genes and miRNAs provides much area for further study. Studies of regulation by miRNAs in nondrosophilid insects are also rather limited. For example, there are 190 miRNAs reported with experimentally validated target interactions in *D. melanogaster*, and yet there are only three validated cases in the silkworm *B. mori* (miRtarbase v.6.1). Using high-throughput sequencing, miRNAs are rapidly being identified in a variety of insects. For example, 833 miRNAs have been uncovered in the migratory locust *Locusta migratoria* and 382 in the planthopper *Sogatella furcifera* (Chang et al., 2016; Wang et al., 2015), yet, the majority of their roles in the regulation of endocrine processes remain unknown. Furthermore, the relationships of miRNAs in sesquiterpenoid production and action are also limited.

To date, only a 20E-activated miRNA, miR-1890, has been identified that regulates the expression of a juvenile hormone-induced serine protease in the midgut of the mosquito *Aedes aeypti* (Lucas et al., 2015). Likewise, the miR-2 family has been identified in a cockroach *Blattella germanica* that downregulates a juvenile hormone-dependent transcription factor, Krüppel, that is involved in repression of metamorphosis (Lozano et al., 2015). In the present work, we reviewed the current understanding of how miRNAs mechanistically contribute to the regulation of production and action of two key hormones in the insect model *D. melanogaster*. Recent studies are looking into how miRNAs may contribute to controlling insect pests, such as the artificial miRNA was constructed to target the EcR in cotton bollworm *Helicoverpa armigera* (Yogindran and Rajam, 2016). Given the lack of comparative data on the conserved (or even lineage-specific) miRNAs in the nondrosophilid insects and other arthropods in the action of hormones, it remains unclear how their regulatory circuits and target interactions could have evolved and be useful for promoting beneficial insects or combating pests.

ACKNOWLEDGEMENTS

This work was funded by the Hong Kong Research Grant Council General Research Fund (24100015) (J.H.L.H.).

REFERENCES

Akira, M., Hironori, I., Hiromichi, N., Hiroshi, K., Akira, I., et al., 1987. A monoclonal antibody against a synthetic fragment of bombyxin (4K-prothoracicotropic hormone) from the silkmoth, *Bombyx mori*: characterization and immunohistochemistry. Mol. Cell. Endocrinol. 51 (3), 227–235.

Antonello, Z.A., Reiff, T., Ballesta-Illan, E., Dominguez, M., 2015. Robust intestinal homeostasis relies on cellular plasticity in enteroblasts mediated by miR-8–Escargot switch. EMBO J. 34 (15), 2025–2041.

Aravin, A.A., Lagos-Quintana, M., Yalcin, A., et al., 2003. The small RNA profile during *Drosophila melanogaster* development. Dev. Cell 5 (2), 337–350.

Babiarz, J.E., Ruby, J.G., Wang, Y., et al., 2008. Mouse ES cells express endogenous shRNAs, siRNAs, and other microprocessor-independent, Dicer-dependent small RNAs. Genes Dev. 22 (20), 2773–2785.

Baek, D., Villén, J., Shin, C., et al., 2008. The impact of microRNAs on protein output. Nature 455 (7209), 64–71.

Barrio, L., Dekanty, A., Milán, M., 2014. MicroRNA-mediated regulation of Dp53 in the *Drosophila* fat body contributes to metabolic adaptation to nutrient deprivation. Cell Rep. 8 (2), 528–541.

Bashirullah, A., Pasquinelli, A.E., Kiger, A.A., et al., 2003. Coordinate regulation of small temporal RNAs at the onset of *Drosophila* metamorphosis. Dev. Biol. 259 (1), 1–8.

Becam, I., Rafel, N., Hong, X., et al., 2011. Notch-mediated repression of bantam miRNA contributes to boundary formation in the *Drosophila* wing. Development 138 (17), 3781–3789.

Behm-Ansmant, I., Rehwinkel, J., Doerks, T., et al., 2006. mRNA degradation by miRNAs and GW182 requires both CCR4:NOT deadenylase and DCP1:DCP2 decapping complexes. Genes Dev. 20 (14), 1885–1898.

Bejarano, F., Smibert, P., Lai, E.C., 2010. miR-9a prevents apoptosis during wing development by repressing *Drosophila LIM-only*. Dev. Biol. 338 (1), 63–73.

Biemar, F., Nix, D.A., Piel, J., et al., 2006. Comprehensive identification of *Drosophila* dorsal-ventral patterning genes using a whole-genome tiling array. Proc. Natl. Acad. Sci. U.S.A. 103 (34), 12763–12768.

Biryukova, I., Asmar, J., Abdesselem, H., Heitzler, P., 2009. *Drosophila* mir-9a regulates wing development via fine-tuning expression of the LIM only factor, dLMO. Dev. Biol. 327 (2), 487–496.

Bohnsack, M.T., Czaplinski, K., Gorlich, D., 2004. Exportin 5 is a RanGTP-dependent dsRNA-binding protein that mediates nuclear export of pre-miRNAs. RNA 10 (2), 185–191.

Boulan, L., Martín, D., Milán, M., 2013. *bantam* miRNA promotes systemic growth by connecting insulin signaling and ecdysone production. Curr. Biol. 23 (6), 473–478.

Brennecke, J., Hipfner, D.R., Stark, A., et al., 2003. *bantam* encodes a developmentally regulated microRNA that controls cell proliferation and regulates the proapoptotic gene *hid* in *Drosophila*. Cell 113 (1), 25–36.

Brogiolo, W., Stocker, H., Ikeya, T., et al., 2001. An evolutionarily conserved function of the *Drosophila* insulin receptor and insulin-like peptides in growth control. Curr. Biol. 11 (4), 213–221.

Cai, X., Hagedorn, C.H., Cullen, B.R., 2004. Human microRNAs are processed from capped, polyadenylated transcripts that can also function as mRNAs. RNA 10 (12), 1957–1966.

Cassidy, J.J., Jha, A.R., Posadas, D.M., et al., 2013. miR-9a minimizes the phenotypic impact of genomic diversity by buffering a transcription factor. Cell 155 (7), 1556–1567.

Cassidy, J.J., Straughan, A.J., Carthew, R.W., 2016. Differential masking of natural genetic variation by miR-9a in *Drosophila*. Genetics 202 (2), 675–687.

Caygill, E.E., Johnston, L.A., 2008. Temporal regulation of metamorphic processes in *Drosophila* by the *let-7* and *miR-125* heterochronic microRNAs. Curr. Biol. 18 (13), 943–950.

Chang, Z.-X., Tang, N., Wang, L., et al., 2016. Identification and characterization of microRNAs in the white-backed planthopper, *Sogatella furcifera*. Insect Sci. 23 (3), 452–468.

Chawla, G., Sokol, N.S., 2012. Hormonal activation of *let-7-C* microRNAs via EcR is required for adult *Drosophila melanogaster* morphology and function. Development 139 (10), 1788–1797.

Chawla, G., Deosthale, P., Childress, S., et al., 2016. A *let-7-to-miR-125* MicroRNA switch regulates neuronal integrity and lifespan in *Drosophila*. PLoS Genet. 12 (8), e1006247.

Cheloufi, S., Dos Santos, C.O., Chong, M.M., Hannon, G.J., 2010. A dicer-independent miRNA biogenesis pathway that requires Ago catalysis. Nature 465 (7298), 584–589.

Chen, W., Liu, Z., Li, T., et al., 2014. Regulation of *Drosophila* circadian rhythms by miRNA let-7 is mediated by a regulatory cycle. Nat. Commun. 5, 5549.

Cheong, S.P.S., Huang, J., Bendena, W.G., et al., 2015. Evolution of ecdysis and metamorphosis in arthropods: the rise of regulation of juvenile hormone. Integr. Comp. Biol. 55 (5), 878–890.

Cifuentes, D., Xue, H., Taylor, D.W., et al., 2010. A novel miRNA processing pathway independent of Dicer requires Argonaute2 catalytic activity. Science 328 (5986), 1694–1698.

Colombani, J., Bianchini, L., Layalle, S., et al., 2005. Antagonistic actions of ecdysone and insulins determine final size in *Drosophila*. Science 310 (5748), 667–670.

Colombani, J., Andersen, D.S., Léopold, P., 2012. Secreted peptide Dilp8 coordinates *Drosophila* tissue growth with developmental timing. Science 336 (6081), 582–585.

Czech, B., Zhou, R., Erlich, Y., et al., 2009. Hierarchical rules for Argonaute loading in *Drosophila*. Mol. Cell 36 (3), 445–456.

Davis-Dusenbery, B.N., Hata, A., 2010. Mechanisms of control of microRNA biogenesis. J. Biochem. 148 (4), 381–392.

De Robertis, E.M., 2008. Evo-devo: variations on ancestral themes. Cell 132 (2), 185–195.

de Souza, A.M.F., López, J.A., 2004. Insulin or insulin-like studies on unicellular organisms: a review. Braz. Arch. Biol. Technol. 47 (6), 973–981.

Denli, A.M., Tops, B.B., Plasterk, R.H., et al., 2004. Processing of primary microRNAs by the microprocessor complex. Nature 432 (7014), 231–235.

Djuranovic, S., Nahvi, A., Green, R., 2012. miRNA-mediated gene silencing by translational repression followed by mRNA deadenylation and decay. Science 336 (6078), 237–240.

Erion, R., Sehgal, A., 2013. Regulation of insect behavior via the insulin-signaling pathway. Front. Physiol. 4, 353.

Esslinger, S.M., Schwalb, B., Helfer, S., et al., 2013. *Drosophila* miR-277 controls branched-chain amino acid catabolism and affects lifespan. RNA Biol. 10 (6), 1042–1056.

Fagegaltier, D., König, A., Gordon, A., et al., 2014. A genome-wide survey of sexually dimorphic expression of *Drosophila* miRNAs identifies the steroid hormone-induced miRNA let-7 as a regulator of sexual identity. Genetics 198 (2), 647–668.

Fernandez, R., Tabarini, D., Azpiazu, N., et al., 1995. The *Drosophila* insulin receptor homolog: a gene essential for embryonic development encodes two receptor isoforms with different signaling potential. EMBO J. 14 (14), 3373–3384.

Forman, J.J., Legesse-Miller, A., Coller, H.A., 2008. A search for conserved sequences in coding regions reveals that the *let-7* microRNA targets Dicer within its coding sequence. Proc. Natl. Acad. Sci. U.S.A. 105 (39), 14879–14884.

Foronda, D., Weng, R., Verma, P., et al., 2014. Coordination of insulin and notch pathway activities by microRNA miR-305 mediates adaptive homeostasis in the intestinal stem cells of the *Drosophila* gut. Genes Dev. 28 (21), 2421–2431.

Förstemann, K., Horwich, M.D., Wee, L., et al., 2007. *Drosophila* microRNAs are sorted into functionally distinct Argonaute complexes after production by Dicer-1. Cell 130 (2), 287–297.

Fricke, C., Green, D., Smith, D., et al., 2014. MicroRNAs influence reproductive responses by females to male sex peptide in *Drosophila melanogaster*. Genetics 198 (4), 1603–1619.

Fu, S., Nien, C.Y., Liang, H.L., Rushlow, C., 2014. Co-activation of microRNAs by Zelda is essential for early *Drosophila* development. Development 141 (10), 2108–2118.

Fukunaga, R., Han, B.W., Hung, J.H., et al., 2012. Dicer partner proteins tune the length of mature miRNAs in flies and mammals. Cell 151 (3), 533–546.

Garbuzov, A., Tatar, M., 2010. Hormonal regulation of *Drosophila* microRNA let-7 and miR-125 that target innate immunity. Fly 4 (4), 306–311.

Garelli, A., Gontijo, A.M., Miguela, V., et al., 2012. Imaginal discs secrete insulin-like peptide 8 to mediate plasticity of growth and maturation. Science 336 (6081), 579–582.

Géminard, C., Arquier, N., Layalle, S., Bourouis, M., et al., 2006. Control of metabolism and growth through insulin-like peptides in Drosophila. Diabetes 55 (Suppl. 2), S5–S8.

Gregory, R.I., Yan, K.P., Amuthan, G., et al., 2004. The microprocessor complex mediates the genesis of microRNAs. Nature 432 (7014), 235–240.

Griffiths-Jones, S., Hui, J.H., Marco, A., Ronshaugen, M., 2011. MicroRNA evolution by arm switching. EMBO Rep. 12 (2), 172–177.

Gu, S.-H., Chen, C.H., Hsieh, Y.C., et al., 2015. Modulatory effects of bombyxin on ecdysteroidogenesis in *Bombyx mori* prothoracic glands. J. Insect Physiol. 72, 61–69.

Ha, M., Kim, V.N., 2014. Regulation of microRNA biogenesis. Nat. Rev. Mol. Cell Biol. 15 (8), 509–524.

Han, J., Lee, Y., Yeom, K.H., et al., 2006. Molecular basis for the recognition of primary microRNAs by the Drosha-DGCR8 complex. Cell 125 (5), 887–901.

Heo, I., Ha, M., Lim, J., et al., 2012. Mono-uridylation of pre-microRNA as a key step in the biogenesis of group II let-7 microRNAs. Cell 151 (3), 521–532.

Herranz, H., Cohen, S.M., 2010. MicroRNAs and gene regulatory networks: managing the impact of noise in biological systems. Genes Dev. 24 (13), 1339–1344.

Herranz, H., Pérez, L., Martín, F.A., Milán, M., 2008. A Wingless and Notch double-repression mechanism regulates G1-S transition in the *Drosophila* wing. EMBO J. 27 (11), 1633–1645.

Herranz, H., Hong, X., Cohen, S.M., 2012. Mutual repression by bantam miRNA and Capicua links the EGFR/MAPK and Hippo pathways in growth control. Curr. Biol. 22 (8), 651–657.

Hertel, J., Bartschat, S., Wintsche, A., et al., 2012. Evolution of the let-7 microRNA family. RNA Biol. 9 (3), 231–241.

Hilgers, V., Bushati, N., Cohen, S.M., 2010. *Drosophila* microRNAs 263a/b confer robustness during development by protecting nascent sense organs from apoptosis. PLoS Biol. 8 (6), e1000396.

Hipfner, D.R., Weigmann, K., Cohen, S.M., 2002. The *bantam* gene regulates *Drosophila* growth. Genetics 161 (4), 1527–1537.

Huang, X., Warren, J.T., Gilbert, L.I., 2008. New players in the regulation of ecdysone biosynthesis. J. Genet. Genomics 35 (1), 1–10.

Hui, J., Bendena, W., Tobe, S., 2013. Future perspectives for research on the biosynthesis of juvenile hormones and related sesquiterpenoids in Arthropod endocrinology and ecotoxicology. In: Devillers, J. (Ed.), Juvenile Hormones and Juvenoids: Modeling Biological Effects and Environmental Fate. CRC Press, Florida, USA, pp. 15–30.

Hyun, S., Lee, J.H., Jin, H., et al., 2009. Conserved microRNA miR-8/miR-200 and its target USH/FOG2 control growth by regulating PI3K. Cell 139 (6), 1096–1108.

Ishizaki, H., Mizoguchi, A., Fujishita, M., Suzuki, A., et al., 1983. Species specificity of the insect prothoracicotropic hormone (PITH): the presence of Bombyx- and Samia-specific PTTHs in the brain of Bombyx mori. Dev. Growth Differ. 25 (6), 593–600.

Iwasaki, S., Kobayashi, M., Yoda, M., et al., 2010. Hsc70/Hsp90 chaperone machinery mediates ATP-dependent RISC loading of small RNA duplexes. Mol. Cell 39 (2), 292–299.

Jiang, F., Ye, X., Liu, X., et al., 2005. Dicer-1 and R3D1-L catalyze microRNA maturation in *Drosophila*. Genes Dev. 19 (14), 1674–1679.

Jin, H., Kim, V.N., Hyun, S., 2012. Conserved microRNA miR-8 controls body size in response to steroid signaling in *Drosophila*. Genes Dev. 26 (13), 1427–1432.

Karres, J.S., Hilgers, V., Carrera, I., et al., 2007. The conserved microRNA miR-8 tunes atrophin levels to prevent neurodegeneration in *Drosophila*. Cell 131 (1), 136–145.

Kennell, J.A., Gerin, I., MacDougald, O.A., Cadigan, K.M., 2008. The microRNA miR-8 is a conserved negative regulator of Wnt signaling. Proc. Natl. Acad. Sci. U.S.A. 105 (40), 15417–15422.

Kenny, N.J., Quah, S., Holland, P.W., et al., 2013. How are comparative genomics and the study of microRNAs changing our views on arthropod endocrinology and adaptations to the environment? Gen. Comp. Endocrinol. 188, 16–22.

Kenny, N.J., Sin, Y.W., Hayward, A., et al., 2015. The phylogenetic utility and functional constraint of microRNA flanking sequences. Proc. Biol. Sci. 282 (1803), 20142983.

Ketting, R.F., Fischer, S.E., Bernstein, E., et al., 2001. Dicer functions in RNA interference and in synthesis of small RNA involved in developmental timing in *C. elegans*. Genes Dev. 15 (20), 2654–2659.

Kim, V.N., Han, J., Siomi, M.C., 2009. Biogenesis of small RNAs in animals. Nat. Rev. Mol. Cell Biol. 10 (2), 126–139.

Kiriakidou, M., Tan, G.S., Lamprinaki, S., et al., 2007. An mRNA m7G cap binding-like motif within human Ago2 represses translation. Cell 129 (6), 1141–1151.

Kiss, I., Beaton, A.H., Tardiff, J., et al., 1988. Interactions and developmental effects of mutations in the broad-complex of *Drosophila melanogaster*. Genetics 118 (2), 247–259.

Kozlova, T., Thummel, C.S., 2003. Essential roles for ecdysone signaling during *Drosophila* mid-embryonic development. Science 301 (5641), 1911–1914.

Krol, J., Loedige, I., Filipowicz, W., 2010. The widespread regulation of microRNA biogenesis, function and decay. Nat. Rev. Genet. 11 (9), 597–610.

Kucherenko, M.M., Barth, J., Fiala, A., Shcherbata, H.R., 2012. Steroid-induced microRNA *let-7* acts as a spatio-temporal code for neuronal cell fate in the developing *Drosophila* brain. EMBO J. 31 (24), 4511–4523.

Lagos-Quintana, M., Rauhut, R., Lendeckel, W., Tuschl, T., 2001. Identification of novel genes coding for small expressed RNAs. Science 294 (5543), 853–858.

Le Roith, D., Shiloach, J., Roth, J., Lesniak, M.A., 1980. Evolutionary origins of vertebrate hormones: substances similar to mammalian insulins are native to unicellular eukaryotes. Proc. Natl. Acad. Sci. U.S.A. 77 (10), 6184–6188.

Leaman, D., Chen, P.Y., Fak, J., et al., 2005. Antisense-mediated depletion reveals essential and specific functions of microRNAs in *Drosophila* development. Cell 121 (7), 1097–1108.

Lee, R.C., Feinbaum, R.L., Ambros, V., 1993. The *C. elegans* heterochronic gene *lin-4* encodes small RNAs with antisense complementarity to *lin-14*. Cell 75 (5), 843–854.

Lee, Y., Ahn, C., Han, J., et al., 2003. The nuclear RNase III Drosha initiates microRNA processing. Nature 425 (6956), 415–419.

Lee, Y.S., Nakahara, K., Pham, J.W., et al., 2004. Distinct roles for *Drosophila* Dicer-1 and Dicer-2 in the siRNA/miRNA silencing pathways. Cell 117 (1), 69–81.

Lee, G.J., Jun, J.W., Hyun, S., 2015. MicroRNA miR-8 regulates multiple growth factor hormones produced from *Drosophila* fat cells. Insect Mol. Biol. 24 (3), 311–318.

Lewis, B.P., Burge, C.B., Bartel, D.P., 2005. Conserved seed pairing, often flanked by adenosines, indicates that thousands of human genes are microRNA targets. Cell 120 (1), 15–20.

Li, Y., Wang, F., Lee, J.A., Gao, F.B., 2006. *MicroRNA-9a* ensures the precise specification of sensory organ precursors in *Drosophila*. Genes Dev. 20 (20), 2793–2805.

Li, Y., Maines, J.Z., Tastan, Ö.Y., et al., 2012. Mei-P26 regulates the maintenance of ovarian germline stem cells by promoting BMP signaling. Development 139 (9), 1547–1556.

Li, Z., Lu, Y., Xu, X.L., Gao, F.B., 2013. The FTD/ALS-associated RNA-binding protein TDP-43 regulates the robustness of neuronal specification through microRNA-9a in *Drosophila*. Hum. Mol. Genet. 22 (2), 218–225.

Liu, N., Landreh, M., Cao, K., et al., 2012. The microRNA miR-34 modulates ageing and neurodegeneration in *Drosophila*. Nature 482 (7386), 519–523.

Loya, C.M., Lu, C.S., Van Vactor, D., Fulga, T.A., 2009. Transgenic microRNA inhibition with spatiotemporal specificity in intact organisms. Nat. Methods 6 (12), 897–903.

Loya, C.M., McNeill, E.M., Bao, H., et al., 2014. miR-8 controls synapse structure by repression of the actin regulator enabled. Development 141 (9), 1864–1874.

Lozano, J., Montañez, R., Belles, X., 2015. MiR-2 family regulates insect metamorphosis by controlling the juvenile hormone signaling pathway. Proc. Natl. Acad. Sci. U.S.A. 112 (12), 3740–3745.

Lucas, K.J., Zhao, B., Roy, S., et al., 2015. Mosquito-specific microRNA-1890 targets the juvenile hormone-regulated serine protease JHA15 in the female mosquito gut. RNA Biol. 12 (12), 1383–1390.

Lund, E., Güttinger, S., Calado, A., et al., 2004. Nuclear export of microRNA precursors. Science 303 (5654), 95–98.

Lytle, J.R., Yario, T.A., Steitz, J.A., 2007. Target mRNAs are repressed as efficiently by microRNA-binding sites in the 5' UTR as in the 3' UTR. Proc. Natl. Acad. Sci. U.S.A. 104 (23), 9667–9672.

Macrae, I.J., Zhou, K., Li, F., et al., 2006. Structural basis for double-stranded RNA processing by Dicer. Science 311 (5758), 195–198.

Marco, A., 2015. Selection against maternal microRNA target sites in maternal transcripts. G3 (Bethesda) 5 (10), 2199–2207.

Meneses, P., de los Angeles Ortíz, M., 1975. A protein extract from Drosophila melanogaster with insulin-like activity. Comp. Biochem. Physiol. A Physiol. 51 (2), 483–485.

Mizoguchi, A., Okamoto, N., 2013. Insulin-like and IGF-like peptides in the silkmoth *Bombyx mori*: discovery, structure, secretion, and function. Front. Physiol. 4, 217.

Moretti, F., Kaiser, C., Zdanowicz-Specht, A., Hentze, M.W., 2012. PABP and the poly(A) tail augment microRNA repression by facilitated miRISC binding. Nat. Struct. Mol. Biol. 19 (6), 603–608.

Nagasawa, H., Kataoka, H., Isogai, A., et al., 1984. Amino-terminal amino acid sequence of the silkworm prothoracicotropic hormone: homology with insulin. Science 226 (4680), 1344–1345.

Nagasawa, H., Kataoka, H., Isogai, A., et al., 1986. Amino acid sequence of a prothoracicotropic hormone of the silkworm *Bombyx mori*. Proc. Natl. Acad. Sci. U.S.A. 83 (16), 5840–5843.

Nairz, K., Rottig, C., Rintelen, F., et al., 2006. Overgrowth caused by misexpression of a microRNA with dispensable wild-type function. Dev. Biol. 291 (2), 314–324.

Nässel, D.R., Vanden Broeck, J., 2016. Insulin/IGF signaling in *Drosophila* and other insects: factors that regulate production, release and post-release action of the insulin-like peptides. Cell. Mol. Life Sci. 73 (2), 271–290.

Nelson, C., Ambros, V., Baehrecke, E.H., 2014. *miR-14* regulates autophagy during developmental cell death by targeting *ip3-kinase 2*. Mol. Cell 56 (3), 376–388.

Neumüller, R.A., Betschinger, J., Fischer, A., et al., 2008. Mei-P26 regulates microRNAs and cell growth in the *Drosophila* ovarian stem cell lineage. Nature 454 (7201), 241–245.

Nolo, R., Morrison, C.M., Tao, C., et al., 2006. The *bantam* microRNA is a target of the hippo tumor-suppressor pathway. Curr. Biol. 16 (19), 1895–1904.

O'Farrell, F., Esfahani, S.S., Engström, Y., Kylsten, P., 2008. Regulation of the *Drosophila lin-41* homologue *dappled* by *let-7* reveals conservation of a regulatory mechanism within the LIN-41 subclade. Dev. Dyn. 237 (1), 196–208.

Oh, H., Irvine, K.D., 2011. Cooperative regulation of growth by Yorkie and Mad through *bantam*. Dev. Cell 20 (1), 109–122.

Okamura, K., Ishizuka, A., Siomi, H., Siomi, M.C., 2004. Distinct roles for Argonaute proteins in small RNA-directed RNA cleavage pathways. Genes Dev. 18 (14), 1655–1666.

Okamura, K., Ladewig, E., Zhou, L., Lai, E.C., 2013. Functional small RNAs are generated from select miRNA hairpin loops in flies and mammals. Genes Dev. 27 (7), 778–792.

Ørom, U.A., Nielsen, F.C., Lund, A.H., 2008. MicroRNA-10a binds the 5'UTR of ribosomal protein mRNAs and enhances their translation. Mol. Cell 30 (4), 460–471.

Park, J.-E., Heo, I., Tian, Y., et al., 2011. Dicer recognizes the 5' end of RNA for efficient and accurate processing. Nature 475 (7355), 201–205.

Parker, J.S., Roe, S.M., Barford, D., 2005. Structural insights into mRNA recognition from a PIWI domain-siRNA guide complex. Nature 434 (7033), 663–666.

Pasquinelli, A.E., Reinhart, B.J., Slack, F., et al., 2000. Conservation of the sequence and temporal expression of *let-7* heterochronic regulatory RNA. Nature 408 (6808), 86–89.

Pfaff, J., Hennig, J., Herzog, F., et al., 2013. Structural features of Argonaute-GW182 protein interactions. Proc. Natl. Acad. Sci. U.S.A. 110 (40), E3770–E3779.

Place, R.F., Li, L.C., Pookot, D., et al., 2008. MicroRNA-373 induces expression of genes with complementary promoter sequences. Proc. Natl. Acad. Sci. U.S.A. 105 (5), 1608–1613.

Qian, J., Tu, R., Yuan, L., Xie, W., 2016. Intronic miR-932 targets the coding region of its host gene, *Drosophila neuroligin2*. Exp. Cell Res. 344 (2), 183–193.

Qu, Z., Kenny, N.J., Lam, H.M., et al., 2015. How did arthropod sesquiterpenoids and ecdysteroids arise? Comparison of hormonal pathway genes in non-insect arthropod genomes. Genome Biol. Evol. 7 (7), 1951–1959.

Reinhart, B.J., Slack, F.J., Basson, M., et al., 2000. The 21-nucleotide let-7 RNA regulates developmental timing in *Caenorhabditis elegans*. Nature 403 (6772), 901–906.

Riehle, M.A., Brown, M.R., 1999. Insulin stimulates ecdysteroid production through a conserved signaling cascade in the mosquito *Aedes aegypti*. Insect Biochem. Mol. Biol. 29 (10), 855–860.

Rodriguez, A., Griffiths-Jones, S., Ashurst, J.L., Bradley, A., 2004. Identification of mammalian microRNA host genes and transcription units. Genome Res. 14 (10A), 1902–1910.

Ruby, J.G., Jan, C.H., Bartel, D.P., 2007. Intronic microRNA precursors that bypass Drosha processing. Nature 448 (7149), 83–86.

Ryazansky, S.S., Gvozdev, V.A., Berezikov, E., et al., 2011. Evidence for post-transcriptional regulation of clustered microRNAs in *Drosophila*. BMC Genomics 12 (1), 371.

Saito, K., Ishizuka, A., Siomi, H., Siomi, M.C., 2005. Processing of pre-microRNAs by the Dicer-1-Loquacious complex in *Drosophila* cells. PLoS Biol. 3 (7), e235.

Selbach, M., Schwanhäusser, B., Thierfelder, N., et al., 2008. Widespread changes in protein synthesis induced by microRNAs. Nature 455 (7209), 58–63.

Sell, C., 2015. Minireview: the complexities of IGF/insulin signaling in aging: why flies and worms are not humans. Mol. Endocrinol. 29 (8), 1107–1113.

Sempere, L.F., Dubrovsky, E.B., Dubrovskaya, V.A., et al., 2002. The expression of the *let-7* small regulatory RNA is controlled by ecdysone during metamorphosis in *Drosophila melanogaster*. Dev. Biol. 244 (1), 170–179.

Sempere, L.F., Sokol, N.S., Dubrovsky, E.B., et al., 2003. Temporal regulation of microRNA expression in *Drosophila melanogaster* mediated by hormonal signals and broad-complex gene activity. Dev. Biol. 259 (1), 9–18.

Shen, W.-F., Hu, Y.L., Uttarwar, L., et al., 2008. MicroRNA-126 regulates HOXA9 by binding to the homeobox. Mol. Cell. Biol. 28 (14), 4609–4619.

Shlyueva, D., Stelzer, C., Gerlach, D., et al., 2014. Hormone-responsive enhancer-activity maps reveal predictive motifs, indirect repression, and targeting of closed chromatin. Mol. Cell 54 (1), 180–192.

Sokol, N.S., Xu, P., Jan, Y.N., Ambros, V., 2008. *Drosophila let-7* microRNA is required for remodeling of the neuromusculature during metamorphosis. Genes Dev. 22 (12), 1591–1596.

Song, J.-J., Smith, S.K., Hannon, G.J., Joshua-Tor, L., 2004. Crystal structure of Argonaute and its implications for RISC slicer activity. Science 305 (5689), 1434–1437.

Soni, K., Choudhary, A., Patowary, A., et al., 2013. miR-34 is maternally inherited in *Drosophila melanogaster* and *Danio rerio*. Nucleic Acids Res. 41 (8), 4470–4480.

Suh, Y.S., Bhat, S., Hong, S.H., et al., 2015. Genome-wide microRNA screening reveals that the evolutionary conserved miR-9a regulates body growth by targeting sNPFR1/NPYR. Nat. Commun. 6, 7693.

Tan, H., Poidevin, M., Li, H., et al., 2012. MicroRNA-277 modulates the neurodegeneration caused by Fragile X premutation rCGG repeats. PLoS Genet. 8 (5), e1002681.

Teleman, A.A., 2010. Molecular mechanisms of metabolic regulation by insulin in *Drosophila*. Biochem. J. 425 (1), 13–26.

Teleman, A.A., Maitra, S., Cohen, S.M., 2006. *Drosophila* lacking microRNA miR-278 are defective in energy homeostasis. Genes Dev. 20 (4), 417–422.

Thompson, B.J., Cohen, S.M., 2006. The Hippo pathway regulates the *bantam* microRNA to control cell proliferation and apoptosis in *Drosophila*. Cell 126 (4), 767–774.

Tobe, S.S., Bendena, W.G., 1999. The regulation ofjuvenile hormone production in arthropods: functional and evolutionary perspectives. Ann. N. Y. Acad. Sci. 897 (1), 300–310.

Toledano, H., D'Alterio, C., Czech, B., et al., 2012. The *let*-7–Imp axis regulates ageing of the *Drosophila* testis stem-cell niche. Nature 485 (7400), 605–610.

Tomari, Y., Du, T., Zamore, P.D., 2007. Sorting of *Drosophila* small silencing RNAs. Cell 130 (2), 299–308.

Tsang, J., Zhu, J., van Oudenaarden, A., 2007. MicroRNA-mediated feedback and feedforward loops are recurrent network motifs in mammals. Mol. Cell 26 (5), 753–767.

Tu, M.-P., Yin, C.-M., Tatar, M., 2002. Impaired ovarian ecdysone synthesis of *Drosophila melanogaster* insulin receptor mutants. Aging Cell 1 (2), 158–160.

Vallejo, D.M., Caparros, E., Dominguez, M., 2011. Targeting notch signalling by the conserved miR-8/200 microRNA family in development and cancer cells. EMBO J. 30 (4), 756–769.

Varghese, J., Cohen, S.M., 2007. microRNA miR-14 acts to modulate a positive autoregulatory loop controlling steroid hormone signaling in *Drosophila*. Genes Dev. 21 (18), 2277–2282.

Varghese, J., Lim, S.F., Cohen, S.M., 2010. *Drosophila* miR-14 regulates insulin production and metabolism through its target, *sugarbabe*. Genes Dev. 24 (24), 2748–2753.

Vasudevan, S., 2012. Posttranscriptional upregulation by microRNAs. Wiley Interdiscip. Rev. RNA 3 (3), 311–330.

Wan, G., Zhang, X., Langley, R.R., et al., 2013. DNA-damage-induced nuclear export of precursor microRNAs is regulated by the ATM-AKT pathway. Cell Rep. 3 (6), 2100–2112.

Wang, Y., Jiang, F., Wang, H., et al., 2015. Evidence for the expression of abundant microRNAs in the locust genome. Sci. Rep. 5, 13608.

Wang, Y., Wang, H., Li, X., Li, Y., 2016. Epithelial microRNA-9a regulates dendrite growth through Fmi-Gq signaling in *Drosophila* sensory neurons. Dev. Neurobiol. 76 (2), 225–237.

Wightman, B., Ha, I., Ruvkun, G., 1993. Posttranscriptional regulation of the heterochronic gene *lin-14* by *lin-4* mediates temporal pattern formation in *C. elegans*. Cell 75 (5), 855–862.

Wu, Q., Brown, M.R., 2006. Signaling and function of insulin-like peptides in insects. Annu. Rev. Entomol. 51, 1–24.

Wu, Y.-C., Chen, C.H., Mercer, A., Sokol, N.S., 2012. *let-7-Complex* microRNAs regulate the temporal identity of *Drosophila* mushroom body neurons via *chinmo*. Dev. Cell 23 (1), 202–209.

Xu, P., Vernooy, S.Y., Guo, M., Hay, B.A., 2003. The *Drosophila* microRNA Mir-14 suppresses cell death and is required for normal fat metabolism. Curr. Biol. 13 (9), 790–795.

Yamaguchi, T., Fernandez, R., Roth, R.A., 1995. Comparison of the signaling abilities of the *Drosophila* and human insulin receptors in mammalian cells. Biochemistry 34 (15), 4962–4968.

Yamamoto, A., Zwarts, L., Callaerts, P., et al., 2008. Neurogenetic networks for startle-induced locomotion in *Drosophila melanogaster*. Proc. Natl. Acad. Sci. U.S.A. 105 (34), 12393–12398.

Yan, K.S., Yan, S., Farooq, A., et al., 2003. Structure and conserved RNA binding of the PAZ domain. Nature 426 (6965), 468–474.
Yao, B., La, L.B., Chen, Y.C., et al., 2012. Defining a new role of GW182 in maintaining miRNA stability. EMBO Rep. 13 (12), 1102–1108.
Yatsenko, A.S., Shcherbata, H.R., 2014. *Drosophila miR-9a* targets the ECM receptor dystroglycan to canalize myotendinous junction formation. Dev. Cell 28 (3), 335–348.
Yogindran, S., Rajam, M.V., 2016. Artificial miRNA-mediated silencing of ecdysone receptor (EcR) affects larval development and oogenesis in *Helicoverpa armigera*. Insect Biochem. Mol. Biol. 77, 21–30.
Yu, J.-Y., Reynolds, S.H., Hatfield, S.D., et al., 2009. Dicer-1-dependent Dacapo suppression acts downstream of insulin receptor in regulating cell division of *Drosophila* germline stem cells. Development 136 (9), 1497–1507.
Yuva-Aydemir, Y., Simkin, A., Gascon, E., Gao, F.B., 2011. MicroRNA-9: functional evolution of a conserved small regulatory RNA. RNA Biol. 8 (4), 557–564.
Zera, A.J., Harshman, L.G., Williams, T.D., 2007. Evolutionary endocrinology: the developing synthesis between endocrinology and evolutionary genetics. Annu. Rev. Ecol. Evol. Syst. 38 (1), 793–817.
Zhang, H., Kolb, F.A., Jaskiewicz, L., et al., 2004. Single processing center models for human Dicer and bacterial RNase III. Cell 118 (1), 57–68.

CHAPTER FOUR

Epigenetic Regulation of Longevity in Insects

Irina A. Kozeretska*, Svitlana V. Serga*, Alexander K. Koliada†, Alexander M. Vaiserman†
*Taras Shevchenko National University of Kyiv, Kyiv, Ukraine
†D.F. Chebotarev Institute of Gerontology, NAMS, Kyiv, Ukraine

Contents

Abstract

When studying aging, an important issue is that it is a complex process influenced by a large number of environmental and genetic factors. The effects of these factors are difficult to investigate because they influence and modify each other. Therefore, it is difficult to examine these factors using complex mammalian models like rodents. Thereby, a large body of biogerontological research is based on simple model organisms such as insects. Such models are particularly useful in an exceedingly complex field such as epigenetics of aging. A high degree of conservation exists between insect and mammalian genomes in terms of both epigenetic mechanisms and signalling pathways associated

Advances in Insect Physiology, Volume 53
ISSN 0065-2806
http://dx.doi.org/10.1016/bs.aiip.2017.03.001

with aging processes. Insect models proved to be very valuable for the study of epigenetic mechanisms mediating the influence of environmental factors on development, aging, neurodegeneration, cancer and infectious disorders. Therefore, the identification of epigenetic processes underlying aging and aging-related pathological conditions in insect models would be an important step towards the further development of treatment strategies to promote human health span and longevity.

1. INTRODUCTION

Aging is considered to be associated with an increase in the number of cells that undergo senescence. This makes an organism progressively less capable of withstanding stress and maintaining homeostasis, and therefore more susceptible to disease (Dillin et al., 2014). It is a complex process that depends on the interaction of multiple genetic and environmental processes. In humans, longevity has been shown to be moderately heritable (~25%) (Christensen et al., 2006). Several studies also indicate that exceptional longevity is an inherited phenotype (Atzmon et al., 2005). A number of human genes likely associated with longevity and thereby potentially used as targets for medical intervention have been identified (de Magalhaes et al., 2012). However, because of the complexity of longevity-associated phenotypic traits, the nature of longevity inheritance is still poorly understood (Christensen et al., 2006).

Insects represent a useful model for studying the genetics of aging. Many insects have two or more distinct phenotypes determined by the same genotype, such as male and female adults, winged and nonwinged aphids or castes of social insects (Simpson et al., 2011; Srinivasan and Brisson, 2012). Moreover, the benefits of using insect models for studying biological mechanisms of aging include its relatively short life span, ease of maintenance, environmental and genetic manipulations that alter life span, availability of stocks containing altered genes, sequence of the full genome in several insect species and clear distinction between developmental and adult stages (Helfand and Rogina, 2003). In addition, almost all cells in adult insects are postmitotic. Therefore, the age-related decline in cellular functions may be examined without interference from newly dividing cells (Grotewiel et al., 2005).

Key postulated cellular and molecular mechanisms of aging include reactive oxygen species-mediated oxidative damage to macromolecules, such as DNA, proteins and lipids (Sanz, 2016), genome instability (Vijg and Suh, 2013), accumulation of advanced glycosylation end products, telomere shortening (Koliada et al., 2015) and cell senescence (Campisi and Robert, 2014). Recently, a crucial role of epigenetic mechanisms regulating gene expression, including DNA methylation, histone modifications (methylation, acetylation, phosphorylation, etc.) and changes of noncoding microRNAs (miRNAs), in aging processes has been highlighted (Pal and Tyler, 2016). These research findings suggest that rather than being genetically predetermined, the rate of aging and longevity are determined epigenetically rather than by primary DNA sequence, and that nutrition and other environmental influences may affect aging by modulating the epigenetic information. Moreover, epigenetic transcriptional reprogramming can lead to the formation of different longevity phenotypes which sometimes are heritable across generations (Vaiserman, 2013).

Presently, a role of epigenetic processes in aging and longevity is being actively studied in a number of experimental models. Highlighting the age-related processes of epigenetic regulation in eukaryotic organisms is very important for the development of strategies to increase the human health span and longevity. The development of insect models in such studies allows to reduce the costs and to promote research of basic aging-related processes, which are conserved across wide taxonomic ranges (Mukherjee et al., 2015). In this chapter, mechanisms of epigenetic gene regulation that control aging and longevity of different insect species are summarized and discussed with an emphasis on the insect models most frequently used in biogerontological research, such as the honey bee, *Apis mellifera* and the fruit fly, *Drosophila melanogaster.*

2. MECHANISMS OF EPIGENETIC REGULATION ACROSS INSECT SPECIES

In most eukaryotes, DNA methylation is the key epigenetic mechanism for regulating gene expression. This mechanism is performed by linking of a methyl chemical group to cytosine (Yamada and Chong, 2017). In different animal species, such linking is often performed in the CpG dinucleotide contexts. Methylation transforms cytosine into methylcytosine, which does not alter the DNA sequence but substantially affects its interaction with proteins. Such chemical alterations are maintained during DNA replication and are thus somatically heritable. DNA methylation is performed by a group of evolutionarily conserved enzymes referred to as DNA methyltransferases (DNMTs). DNMT1 class methyltransferases perform symmetrical methylation de novo of the newly synthesized

DNA strand during replication and use the old strand as a template (Goll and Bestor, 2005). The role of the DNA methyltransferase-2 (DNMT2) enzymes is controversial, as it is unclear whether they function mostly as nuclear DNA or cytoplasmic tRNA methyltransferases. It is also unclear whether DNMT2 impacts development, as mice and fruit flies mutant for *dnmt2* do not exhibit altered phenotypes (Kunert et al., 2003; Okano et al., 1998; Rai et al., 2007). Finally, DNMT3 are methyltransferases that perform de novo DNA methylation (Liu et al., 2003).

DNA methylation is widespread across insect species, although exceptions exist. For example, in Diptera (flies and mosquitoes) such as *D. melanogaster*, only 1% of the genome is methylated (Takayama et al., 2014). This methylation usually occurs in specific 5 bp sequence motifs that are rich in CA and CT duplets but depleted of guanine. Methylation of the gene body appears to associate with a lower expression level, and most genes containing methylated regions within the coding parts have developmental or transcriptional functions. DNMT2 is the only DNMT known from the *Drosophila* genome. However, fly strains deficient for DNMT2 retain the DNA methylation, which suggests the presence of other, possibly novel, methyltransferases (Takayama et al., 2014). In other insect species, including ants, bees, wasps, sawflies, cockroaches and termites, 5-methylcytosine is likely the most common form of the epigenetic DNA modification, although insects with DNA methylation do not always possess the ordinary enzymatic machinery (Bewick et al., 2017). Anyhow, CpG methylation in insects is much lower than in humans. Unlike humans, it tends to be primarily concentrated in exons (Glastad et al., 2011). For instance, out of the 10 million CpG sites in the honey bee genome, only 70,000 (0.7%) are found to be methylated (Lyko et al., 2010), while in humans the figure is close to 70% (Strichman-Almashanu et al., 2002). DNA methylations in CpG contexts has been shown to vary during insect development, as shown, e.g. in *A. mellifera* and the red flour beetle, *Tribolium castaneum*, with the highest levels observed in the embryos (Drewell et al., 2014; Feliciello et al., 2013).

The interaction of proteins with DNA is also regulated by changes in chromatin compaction through the histone tail modification. Acetylated histone tails are more typical of relatively loose DNA packaging in the nucleosomes, which increases the nucleic acid accessibility to various transcription factors and thus promotes gene expression. Deacetylation, on the other hand, is accompanied by opposite processes. These histone modifications are controlled by two enzyme families: histone deacetylases (HDACs) and histone acetyltransferases (HATs). Experiments with HAT and HDAC inhibitors both in mammals and insects have revealed a significant role of the balance between HAT and HDAC activity in gene regulation during both normal development and disease progression (Mukherjee et al., 2012).

miRNAs are another important group of epigenetic regulators. These are short RNA molecules (18–22 nucleotides in length) that regulate gene expression posttranscriptionally. The miRNA 'seed sequence' (nucleotides 2–8 at the 5′ end) usually binds to a complementary site in the 3′ untranslated region of mRNAs and inhibits the translational of the target mRNA or lead to its degradation. In insects, miRNA biogenesis goes through several stages and starts with transcription of the miRNA loci (Ylla et al., 2016). The RNA-induced silencing complexes (RISCs) are constructed into large multiprotein effectors, binding to target transcripts and trigger their destruction. In *Drosophila*, miRNAs have been found to play various roles in longevity-related processes, such as development, apoptosis and maintenance of the longevity-influencing intracellular bacterial endosymbiont, *Wolbachia* (Lucas and Raikhel, 2013).

3. DEVELOPMENTAL EPIGENETIC PROGRAMMING OF CASTE-SPECIFIC DIFFERENCES IN LONGEVITY IN SOCIAL INSECTS

Among all insect models which are used to research the aging processes, social insects are likely the most attractive model systems for deeper insight into this topic since they demonstrate a large plasticity of aging processes across different castes (Maleszka, 2008; Yan et al., 2015). For example, in *A. mellifera* unfertilized eggs develop into males (drones), while the fertilized eggs develop into female queens or workers (Remolina and Hughes, 2008). Thus, queens develop from eggs that are genetically not different from those developing into workers. Queens are, however, much larger in size, have specialized anatomy, develop significantly faster and live much longer than worker bees. The caste switching is determined by hormonal signals triggered by the quantity and quality of nutrition during the third larval instar stage (Winston, 1987). Moreover, workers are sterile, whereas queens are reproductive. Queens develop from larvae fed with a nutritional mixture consisting of essential amino acids, lipids, proteins, vitamins and other compounds (royal jelly) until they enter metamorphosis (Drapeau et al., 2006), while workers develop from larvae fed with a royal jelly until the late instar stage (stage of mature larva) and then fed a worker jelly (a mixture of glandular secretions, pollen and honey). The life span of worker bees is about 15–38 days, while queens have a life span of about 1–2 years (Remolina and Hughes, 2008). In another social insect species including ants, wasps and termites, queens and workers also sometimes demonstrate up to a 100-fold difference in longevity, with reproductive queens having longer life span than nonreproductive workers (Keller and Jemielity, 2006; Remolina et al., 2007). Through these features, social insects provide a useful model to identify candidate pathways involved in control of aging and longevity. In particular, the genome-wide study of gene expression in castes of social insects markedly differing in life span is likely a promising approach for the screening of genes involved in life span determination (Corona et al., 2005).

From studies performed on *A. mellifera*, it became evident that the queen phenotype is driven by epigenetic reprogramming in developing larvae by particular nutritional components contained in royal jelly (Foret et al., 2012; Welch and Lister, 2014). Epigenetic modifications induced by specific environmental cues during the early development have been demonstrated to be persistent throughout the entire life cycle, thereby causing a process known as 'developmental programming' (Lillycrop et al., 2014; Vaiserman, 2014). Epigenetic modifications can be induced in a context-dependent manner in response to both external and internal stimuli and can lead to context-dependent programming effects through the persistent effects on gene regulatory cascades (Dickman et al., 2013). Feeding *A. mellifera* female larvae with royal jelly leads, in general, to reduced levels of global DNA methylation and correlated alteration in gene expression along with elevated juvenile hormone (JH) levels. Such changes collectively cause formation of highly reproductive and long-lived queen phenotype. Larvae fed with less-nutritious worker jelly develop into short-lived and functionally sterile worker bees. In the subsections later, we summarize the evidence for the role of epigenetic mechanisms in developmental determination of the caste fate in various social insect species.

3.1 DNA Methylation

DNA methylation is a key mechanism involved in shifting the caste developmental trajectories in *A. mellifera* and other social insects (Elango et al., 2009; Maleszka, 2016; Standage et al., 2016). The methylation of CpG islands in promoter gene regions is associated with transcriptional silencing (Klose and Bird, 2006). This mechanism of epigenetic regulation is, however, not universal across insect species. For example, low to zero DNA methylation levels were detected in *T. castaneum* and *D. melanogaster* (Lyko and Maleszka, 2011). Recent research, however, highlighted the role of DNA methylation in different social insect species including the honey bee (Maleszka, 2016). All three main subfamilies of DNMT enzymes, namely DNMT1, DNMT2 and DNMT3, are presented in *A. mellifera* (Maleszka, 2008). In most insects studied to date, DNA methylation has been found to be predominantly restricted to coding exons and absent in the promoter regions (Maleszka, 2016; Patalano et al., 2012).

In A. mellifera, the downregulation of DNMT3, which is a key driver in epigenetic remodelling, was found to cause a profound shift in the caste fate of developing larvae. Silencing the *Dnmt3* expression in newly hatched larvae by small interfering RNAs (siRNAs) caused changes in the larval developmental trajectories similar to those induced by royal jelly (Kucharski et al., 2008). The decreased levels of DNA methylation were observed in several genes in the heads of queen-destined larvae regardless of whether they were hive-reared or generated via siRNA-caused DNMT3 silencing (Kucharski et al., 2008; Maleszka, 2008). Differential DNA methylation patterns have been observed between worker and queen brains/heads (Foret et al., 2012; Lyko et al., 2010). Genes involved in insulin and JH pathways were shown to be significantly overrepresented among the differentially methylated genes. Several CpG sites in the *hexamerin 110* gene encoding a storage protein have been shown to be differentially methylated between the worker and queen larvae (Ikeda et al., 2011). It has also been found that the larval dietary conditions may influence various methylation sites inside the *dynactin p62*, a conserved gene responsive to nutritional cues (Shi et al., 2011). In a subsequent study by Shi et al. (2013), significant differences in age dynamics of the global DNA methylation levels were obtained among castes. Most of these differentially methylated genes are involved in the biologically important processes linked to development, reproduction and metabolic regulation.

Evidence for the role of DNA methylation in the process of caste differentiation has also been provided in several ant species. Ants were demonstrated to have a full set of DNMTs and their genomes contain methylcytosines (Bonasio et al., 2012). The caste- and allele-specific differences in methylation correlating with the allele-specific expression was obtained in the Florida carpenter ant, *Camponotus floridanus*, and the Jerdon's jumping ant, *Harpegnathos saltator* (Bonasio et al., 2012).

3.2 Alternative Splicing

Alternative splicing is a process by which different combinations of exons from the same gene can be joined together to form various mature mRNA isoforms and protein products. It has been proposed to be a potential mechanism for linking DNA methylation and caste-specific gene regulation across social insect species (Foret et al., 2012). Several caste-related patterns of CpG methylation were shown to be enriched in exon regions, particularly in splicing sites (Lyko et al., 2010). The association between DNA methylation and splicing sites was particularly evident for the genes belonging to the histone gene family. Also in two ant species interaction between DNA methylation and alternative splicing in determining caste-specific developmental pathways was confirmed (Bonasio et al., 2012). However, some authors suggest that morphological and behavioural differences between castes might rather be attributed to the production of caste-specific protein isoforms than to transcriptional alterations per se (Glastad et al., 2011).

3.3 Histone Modifications

Histone modifications such as acetylation, methylation, phosphorylation, ubiquitination and sumoylation are another crucial mechanism of epigenetic regulation in social insects. This mechanism of epigenetic control operates in a coordinated manner with DNA methylation to regulate gene expression. Generally, methylation of DNA in promoter regions represses transcription, while histone acetylation is linked with an activation of transcription (Musselman et al., 2012). By profiling the genome-wide localization of histone H3 modifications, caste-specific differences in histone methylation patterns were observed between the female worker and male castes in the ant *C. floridanus* (Simola et al., 2013). The level of acetylation of lysine 27 of histone H3 (H3K27ac) has been shown to be a strong predictor for caste identity. The majority of genes identified to have different H3K27ac levels between castes are associated with muscle development, sensory responses and neuronal regulation. Moreover, HDAC inhibitors present in royal jelly appear necessary for the development of bees towards queens (Spannhoff et al., 2011).

3.4 Regulation by miRNAs

Epigenetic regulation by miRNAs is suggested to be another important mechanism for social insect caste differentiation. Significant age- and caste-associated differences in the transcriptional patterns of miRNAs have been observed (Weaver et al., 2007). Most genes in proximity to miRNAs were demonstrated to be linked to gene ontology terms, such as 'physiological process', 'nucleus' and 'response to stress'. Interestingly, worker jelly was found to be significantly enriched in complexity and abundance of miRNAs relative to royal jelly (Guo et al., 2013). Most of these miRNAs are known to be involved in regulation of mRNAs functionally related to the development of the insects' central nervous system. Worker and queen larvae contain differentially expressed miRNAs. These miRNAs coincide with both composition and relative expression levels of miRNAs present in worker jelly. They have been found to be expressed two- to fourfold higher in worker larvae as compared to queen larvae. The supplementation of royal jelly with particular miRNAs resulted in significant changes in the levels of mRNA expression in queen-destined larvae and also in morphological traits of the emerging insects in a way characteristic to the worker phenotype (Guo et al., 2013).

3.5 Caste-Specific Difference in Gene Expression Patterns

Collectively, changes in epigenetic regulation result in modulation of gene expression, thereby causing caste-biased phenotypes, such as long-lived queens and short-lived workers (Chen et al., 2012; Evans and Wheeler, 1999, 2000). Most genes exhibiting differential patterns of expression between the queen- and worker-destined larvae were demonstrated to be involved in metabolic pathways (Barchuk et al., 2007; Corona et al., 1999; Cristino et al., 2006; Evans and Wheeler, 1999, 2000). Most metabolic enzymes were exhibited to be upregulated in queen-destined larvae seemingly reflecting their increased growth rate during the late larval stage (Begna et al., 2011). Some genes coding for proteins responsive for RNA processing and translation were also demonstrated to be upregulated in young queen larvae (Corona et al., 1999; Evans and Wheeler, 1999, 2000). In the genome-wide expressional profiling of the DNMT3-silenced honey bee larvae, most genes differentially expressed among worker and queen larvae were linked to hormonal regulation, energy transfer, protein turnover, posttranslational modification, lipid transport, ribosomal biogenesis and other physiometabolic processes (Kucharski et al., 2008). In a transcriptome comparison by Chen et al. (2012), over 70% among the 4500 genes shown to be differentially expressed between the worker- and queen-destined larvae were found to be more highly expressed in queen than in worker larvae assuming that general levels of transcriptional activity during differentiation are higher in queen larvae compared to worker ones. The genome-wide transcriptional analysis conducted on the brains from the same-aged virgin honey bee queens and both sterile and reproductive workers has revealed significant differences in the expression levels for ~2000 genes between the queen and both worker bee groups, and much smaller differences between the sterile and reproductive worker bees (Grozinger et al., 2007). Importantly, several groups of genes shown to be specifically involved in longevity-associated pathways in other species were found to be differentially expressed between the worker and queen larvae. Among them, there were genes responsible for the regulation of the hypoxia pathway, such as *HIFα/Sima*, *HIFβ/Tango* and *PHD/Fatig* (Azevedo et al., 2011), as well as genes potentially involved in the prevention and repair of oxidative damage (Aamodt, 2009).

Caste-specific patterns of gene expression were also identified in social insect species other than honey bee. These patterns appeared very similar in *A. mellifera* and the bumblebee, *Bombus terrestris* (Pereboom et al., 2005). For example, substantial age- and caste-related expression changes of candidate genes associated with taxonomically widespread aging-related pathways, such as *Dnmt3*, foraging, vitellogenin and coenzyme Q biosynthesis protein 7, were demonstrated in *B. terrestris* (Lockett et al., 2016). Also in several ant species such as *Temnothorax longispinosus* (Feldmeyer et al., 2014) and *Atta vollenweideri* (Koch et al., 2013) caste-specific gene expression patterns were identified. The enhanced expression levels of genes responsible for somatic repair were observed in long-lived queens compared to workers in ant *Lasius niger* (Lucas et al., 2016). In another study with *L. niger*, 16 genes have been found to be differentially expressed between adult queen and worker insects (Gräff et al., 2007). Among them, three genes upregulated in queens are known to be linked to maintenance and repair of the soma, generally considered to be crucial mechanisms of longevity determination. Moreover, genes encoding sirtuin deacetylases and telomerase were shown to be upregulated in longer-lived *H. saltator* reproductives (Bonasio et al., 2010).

Up to now, only one study was solely focused on genetic pathways potentially involved in determining the social insect longevity (Corona et al., 2005). In this research, the predictions of oxidative stress theory of aging were tested. This theory postulates that production of highly reactive free radicals and other reactive oxygen species leads to oxidative damage of cellular components; thereby, the accumulation of oxidative damage is considered to be a proximate cause of aging (Sanz, 2016). To examine these assumptions, Corona et al. (2005) determined the levels of expression of eight genes encoding antioxidant defence system enzymes and also five mitochondrial proteins in *A. mellifera*. The expression of antioxidant genes decreased with age in queens, but not in workers. Therefore, the extraordinary longevity of queen bees may be unlikely explained by increased antioxidant capacity per se. Similar data indicating that increased level of expression of antioxidant genes is most likely not necessary for the evolution of the extended life expectancy in social insects, have been also obtained in the ant *L. niger* (Parker et al., 2004). Specifically, adult queens had equal or even lower levels of expression of the CuZnSod gene in thorax, head and abdomen compared to those in short-lived female workers and male ants. A schematic representation of hypothetical regulatory pathways responsible for determining caste fate in *A. mellifera* is presented in Fig. 1.

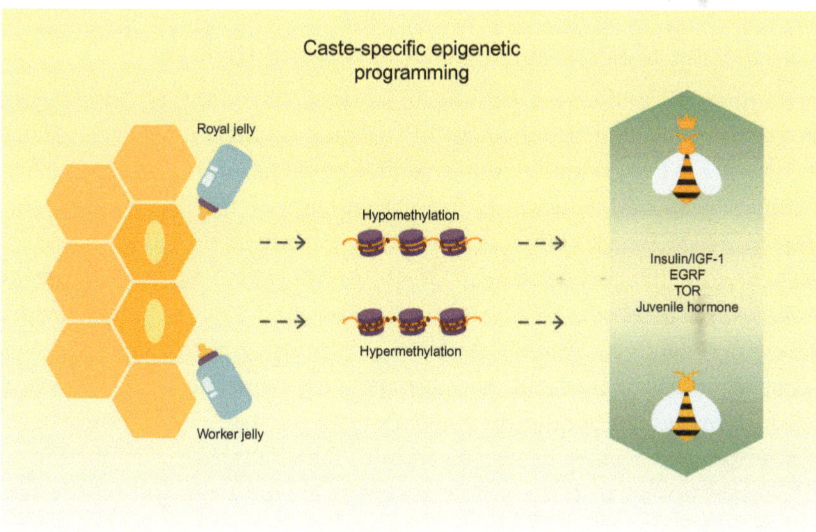

Fig. 1 Schematic representation of hypothetical mechanisms involved in developmental programming of caste-specific differences in longevity in *Apis mellifera*. Under this hypothetical model, feeding of larvae with royal jelly results in reduced levels of global DNA methylation and correlated alterations of gene expression and splicing. These changes, in turn, trigger the endocrine response manifested in the activation of juvenile hormone, target of rapamycin (TOR), and epidermal growth factor receptor (EGFR) signalling along with complicated and ambiguous changes in the insulin/insulin growth factor 1 (IGF-1) pathway in the queen larvae. Collectively, these processes seem to be contributing to the evolution of long-lived queen phenotype.

The extraordinary large differences in longevity between the genetically identical worker and queen female castes substantially challenges our understanding of the processes mediating aging (Kramer et al., 2016) and provides a unique perspective in investigating factors underlying variations in life span both within and among species.

4. MODIFICATION OF AGING AND LONGEVITY IN *DROSOPHILA* BY MODULATING EPIGENETIC PATHWAYS

Since DNA methylation is practically absent in adult fruit flies, it is generally believed that histone modifications are the primary mechanism of epigenetic regulation in this model organism. However, while age-related changes in histone acetylation and methylation in mammals have been described repeatedly, studies of this phenomenon in *D. melanogaster* are rather scarce.

4.1 Histone Methyltransferases and Demethylases

A number of studies on various organisms, including *D. melanogaster*, showed age-dependent changes of histone methylation levels. These changes are mainly related to the general decrease of amount of heterochromatin. Significant decrease in the enrichment of the heterochromatin-repressive H3K9me3, H3K9me2 and heterochromatin protein 1 marks have been found with age (Larson et al., 2012). Another study showed an increase in the overall level of H3K9me3, but decrease at pericentric heterochromatin regions. Such extensive alterations in repressive chromatin state were associated with age-related changes in gene expression (Wood et al., 2010). However, it should be noted that the increase in H3K9me3 levels was observed in the heads of female flies, while the reduction in H3K9me2 was seen in the whole bodies of mixed male and female flies. It is therefore possible that the result depends on the cell type or sex-specific alterations in histone methylation. Additionally, substantial reduction of histone modifications linked to active transcription, such as H3K4me3 and H3K36me3, at transcriptional start sites and over genes has been observed in older flies. Such significant changes in methylation of histones can likely lead to change in expression or activity of histone methyltransferases and demethylases. Moreover, it was shown that heterozygous mutations in two core subunits of Polycomb Repressive Complex 2, the histone H3 lysine 27 (H3K27)-specific methyltransferase *E(z)* and its partner, the

H3-binding protein ESC, result in increased longevity and reduced levels of trimethylated H3K27 (H3K27me3) in adult flies (Siebold et al., 2010).

4.2 Histone Acetyltransferases

In several studies, evidence has been obtained that aging and longevity in *D. melanogaster* may be affected by modulation of HAT activity. Histone acetylation sites H4K12, H3K9, H3K9K14 and H3K23 tend to become hyperacetylated and sites H4K8 and H3K18 become hypoacetylated throughout the aging process in *D. melanogaster* (Peleg et al., 2016). Decreasing the activity of the acetyl-CoA-synthesizing enzyme ATP citrate lyase or the level of the H4K12-specific HAT, *Chameau*, resulted in alleviation of the aging-associated changes and promoted the flies' longevity (Peleg et al., 2016). The overexpression of HAT Tip60 rescued the axonal transport insufficiency and improved memory in a *D. melanogaster* model of Alzheimer's disease (Johnson et al., 2013; Lorbeck et al., 2011; Xu et al., 2014).

4.3 Histone Deacetylases

Some studies also suggest that the aging process in *D. melanogaster* can be influenced by HDAC activity. For instance, the NAD^+-dependent HDAC, *dSir2*, was repeatedly shown to be largely involved in calorie restriction-dependent extension of life span in fruit flies (Rogina and Helfand, 2004). Knockdown of *dSir2* in the flies' adult fat body influences the fat mobilization and survival in conditions of starvation. Moreover, the knockdown of *dSir2* in the adult fat body leads to increase of expression level of insulin-like peptide, *dilp5*, thereby mediating the systemic effects of insulin signalling (Banerjee et al., 2012a). Knockdown of *dSir2* in the adult fat body was also shown to regulate the flies' longevity in a diet-dependent manner (Banerjee et al., 2012b).

4.4 HDAC Inhibitors

In the last few years, a novel class of drugs has been proposed targeting epigenetic pathways (epigenetic drugs). The potential reversibility of epigenetic aberrations makes them attractive targets for therapeutic drug development. Currently, candidate epigenetic drugs are inhibitors of DNA methylation, HATs, HDACs, histone methyltransferases, histone demethylases and bromodomains (Cacabelos and Torrellas, 2015; Heerboth et al., 2014). Among them, HDAC inhibitors (HDACIs) are likely the most promising in the field of biogerontology. HDACIs include four chemical classes that substantially vary in biological activity, structure and specificity: cyclic peptides, hydroxamic acids, short chain fatty acids and synthetic benzamides (Lakshmaiah et al., 2014). Since the transcription levels of many genes decrease with age (Seroude et al., 2002), the restoration of the transcriptional activity by means of HDACIs may likely delay the age-related functional declines. Furthermore, inhibition of HDAC activity can lead to upregulation of genes implicated in response to stress and inflammation, i.e. pathways commonly associated with the regulation of life span (Kourtis and Tavernarakis, 2011). In *D. melanogaster*, each HDAC was shown to regulate transcription of a unique set of genes (Cho et al., 2005), and differential sensitivity of HDACs to HDACIs has been observed. Longevity-modulating effects of HDACIs have been studied mainly in insect experimental models such as *D. melanogaster* (Vaiserman and Pasyukova, 2015). Most studies have been primarily devoted on the life-extending potential of synthetic HDACIs, although HDACIs contained in natural compounds may be promising as well. Examples hereof are sulforaphane, curcumin and diallyl disulfide extracted from broccoli, turmeric and garlic, respectively. Experimental data supporting the health span-promoting and life span-extending properties of different HDACIs are reviewed in the subsections later.

4.4.1 Phenylbutyrate

Sodium 4-phenylbutyrate (PBA) was found to inhibit class I and II HDACs, thereby leading to elevated gene expression, reduced cellular proliferation, induction of apoptosis and the enhanced cell differentiation in neoplastic cell populations (Iannitti and Palmieri, 2011). In *D. melanogaster,* the life-extending potential of the sodium salt of PBA was demonstrated in the study by Kang et al. (2002). Feeding with PBA substantially extended both mean and maximal life span by up to 30%–50%, without diminution of locomotor activity and resistance to stress. The treatment for a limited period, either early or late in adult life, has been shown to extend the flies' longevity as well. This result was not due to caloric restriction, known to extend life span in different model organisms, or due to the decreased reproductive activity. The effect of PBA has been also accompanied by marked changes in the acetylation level of histones H3 and H4 and either down- or upregulation of several hundreds of genes, as it was evident from the DNA microarray-based global transcriptional analysis. The general trend was upregulation of genes involved in detoxification and chaperone activity, including several genes that have previously been found to be involved in life span determination in *D. melanogaster*, and downregulation of genes involved in different metabolic pathways. These data support the hypothesis that life span extension may be caused by overall generalized changes in epigenetic regulation (Vaiserman, 2011).

4.4.2 Sodium Butyrate

Sodium butyrate (SB) is a short chain fatty acid and has an HDAC inhibition activity. SB was shown to influence the processes of cell growth, differentiation and apoptosis in both normal and transformed cells (Buommino et al., 2000; Khan and Jena, 2014). It thus also has a longevity promoting potential.

One-off treatment with SB resulted in significant increase of both mean and maximum life span (by 25.8% and 11.5%, respectively) of fruit flies (Zhao et al., 2005a). Subsequently, life-extending ability of SB treatment in *D. melanogaster* was demonstrated by other authors (McDonald et al., 2013; St Laurent et al., 2013; Vaiserman et al., 2012, 2013). In the McDonald et al. (2013) study, SB-induced life span improvement was accompanied by an increase of the flies' locomotor activity. The obtained effects were dose dependent: treatment with SB in concentrations varying from 10 to 40 mM demonstrated a potential to increase life span, whereas treatments in doses equal or higher than 100 mM decreased longevity (Vaiserman et al., 2012; Zhao et al., 2005a). In some cases, the effect observed depended on whether the line used was short- or long-lived (McDonald et al., 2013; Zhao et al., 2005a). Remarkably, the life-extending effect obtained was unlikely due to the decreased reproductive performance, because no reduction in reproductive activity was revealed in SB-treated females (Vaiserman et al., 2012). The treatment with SB caused elevated levels of acetylation of histone H3 (Zhao et al., 2005a,b, 2006, 2007), whereas the level of acetylation of histone H4 remained unchanged (Zhao et al., 2007). Histone H3 with elevated acetylation levels was found at the promoter regions of the *hsp22*, *hsp70* and *hsp26* genes (Zhao et al., 2005b, 2006, 2007). SB also affected the chromatin structure at the site of cytogenetic location of the *hsp70* gene on the polythene chromosome (Chen et al., 2002). The enhanced levels of expression of *hsp22*, *hsp26* and *hsp70* genes were accordingly found in the SB-treated flies (Chen et al., 2002; Zhao et al., 2005a,b, 2006, 2007). Collectively, these findings suggest that alteration in histone acetylation and, thereafter, in the expression levels of chaperone genes may contribute to the life-extending effects of SB and other HDACIs in *D. melanogaster*. Other mechanisms, however, may also contribute to these effects. For example, the treatment with

SB-supplemented food rescued the early mortality of the flies with the pesticide rotenone-induced Parkinson's disease (St Laurent et al., 2013). The SB-mediated rescue of rotenone-induced Parkinson's disease was associated with elevated dopamine levels in the flies' brains. HDACIs targeting HDAC3 and HDAC1 have been shown to be able to ameliorate polyglutamine-elicited phenotypes in a *Drosophila* model of another age-related neurodegenerative disorder, Huntington's disease, as well (Jia et al., 2012). Upregulation of inducible expression of *sir2* gene known to be substantially involved in the life span extension in *D. melanogaster* (Frankel et al., 2011) was also observed after SB treatment (Vaiserman et al., 2012).

Effects of SB on life span were shown to be depended on the stage and/or age when treatment was applied. Increased life span was observed after treatment at the larval stage only (Vaiserman et al., 2012, 2013; Zhao et al., 2005a), whereas treatment at both the larval and the adult stage or exclusively throughout the adult stage either decreased life span (McDonald et al., 2013), or increased it (Vaiserman et al., 2012, 2013) or had no effect on longevity (Zhao et al., 2005a). Furthermore, life span-modulating effects of SB were shown to be sex specific (Vaiserman et al., 2012, 2013). To explain these inconsistencies in the effects of SB and other epigenetic modulators on the flies' life span, a hypothesis was proposed suggesting phase separation in the adult life of fruit fly and other gradually aging organisms into a health span, a transition phase, and a senescent span (Arking et al., 2002). In analysis conducted by Arking (2009) in *D. melanogaster* and other model organisms, it has been shown that these life stages are characterized by different gene expression patterns. The health span is characterized by a tightly regulated gene expression pattern which leads to a maximized tissue function and to a minimized inflammatory and other damage response; the transition phase is characterized by a gradual decline of the cellular regulatory capacity, and the senescent span is characterized by a gradual deregulation of the gene expression pattern (Arking, 2009). Consistently with this hypothesis, the flies' life span was increased when flies were fed with SB during transition or senescent spans, but it was decreased when SB was administered throughout the entire adult life span or the health span only (McDonald et al., 2013). Similar results were demonstrated in flies fed with another compound known to inhibit HDAC activity, curcumin (Arking, 2015). To summarize, despite the complexity and partial inconsistency of results, SB demonstrated a high potential as a life-extending agent in *Drosophila*.

4.4.3 Trichostatin A

Trichostatin A (TSA) is another widely used HDACI demonstrating a broad spectrum of epigenetic activities, including inhibition of the cell cycle since the beginning of the growth stage and promotion of the expression of apoptosis-associated genes (Vanhaecke et al., 2004). TSA is recognized as a promising anticancer drug candidate. Possible mechanisms of action of this compound are induction of terminal differentiation, cell cycle arrest and apoptosis in different cancer cell lines and thereby inhibition of tumorigenesis.

The epigenetic and phenotypic effects of the TSA treatment are very similar to those shown for the SB treatment in *D. melanogaster*. An increase of both mean and maximum life span was observed due to both one-off and continuous treatment with TSA (Tao et al., 2004; Zhao et al., 2005a). TSA treatment was effective both at the larval (Zhao et al., 2005a) and adult (Tao et al., 2004) stages and influenced the longevity of both short- and long-lived *D. melanogaster* lines, but to a different extent. Life span improvement affected both males and females and was in some cases accompanied by increase in locomotor activity. These life-extending effects induced by the TSA treatment were accompanied by the hyperacetylation of core histone H3 in the promoter and coding regions of some chaperone genes, such as *hsp22*, *hsp26* and *hsp70*, along with upregulation, in most cases, of both basal and inducible expression of these genes (Chen et al., 2002; Tao et al., 2004; Zhao et al., 2005a,b, 2006, 2007). Moreover, the modified chromatin morphology at the locus of *hsp22* was revealed (Tao et al., 2004). The authors suggested that the expression of chaperones can stimulate the repair mechanisms, reduce the level of accumulation of damage and improve the cell stress resistance to create cellular and physiological environments that are favourable for longevity. To summarize, TSA similarly to SB demonstrated a high potential as a life-extending agent in *D. melanogaster*.

4.4.4 Suberoylanilide Hydroxamic Acid

One more HDACI that was shown to be able to extend life in fruit flies, is suberoylanilide hydroxamic acid (SAHA). In in vitro studies, SAHA was found to have similar effects as does SB although at much lower effective doses (Zhou et al., 2011). This compound was demonstrated to induce growth arrest in transformed cells (Yin et al., 2007), and it was shown to be effective in preventing Huntington disease in various animal models (Steffan et al., 2001).

Effects of administration with SAHA throughout *D. melanogaster* health span, transition phase and senescent span have been studied (McDonald et al., 2013). Treatment with SAHA during the transition or senescent spans resulted in decreased mortality rate and extended longevity compared to the control, while supplementation during the entire adult life span or during the health span only led to decreased longevity in the normal-lived strain. The analysis of mortality curves indicated that there were no significant effects of the SAHA administration until the age of ~50 days. When the long-lived strain was administered with SAHA by the same scheme, mostly deleterious effects were detected. Remarkably, the SAHA-treated normal-lived *D. melanogaster* strain showed the late-life-extending effects similar to those seen in the same study for SB. The fact that two different HDACIs, SB and SAHA, had similar effects on mortality rate during the senescent span indicates the similarity of mechanisms that underlie beneficial effects for this class of HDACIs. The authors suggested that HDACIs may significantly influence the mortality rate throughout the senescent phase by reducing the vulnerability of treated individuals, in a manner similar to that of dietary restriction. Indeed, as it was mentioned earlier, genetic alterations in genes encoding HDACs and nutrition regiments partially interact in the course of longevity control (Frankel et al., 2015; Rogina and Helfand, 2004).

In conclusion, HDACIs may likely affect several pathways involved in regulating gene expression patterns associated with healthy aging. The induction of these patterns of gene expression throughout senescence when they are not normally present may likely underlie the life-extending effects of HDACIs.

5. TRANSGENERATIONAL EPIGENETIC INHERITANCE OF LIFE SPAN AND LONGEVITY-ASSOCIATED TRAITS IN *DROSOPHILA*

Compelling evidence for the role of epigenetic mechanisms in life span determination has been obtained in recent studies demonstrating the possibility of nongenomic transmission of longevity phenotypes across generations. Traditionally, epigenetic marks, such as DNA methylation, are not thought to be inherited into subsequent generations, since they are assumed to be completely erased and then reestablished in each generation. Accumulating evidence, however, indicates that many effects may be transmitted across generations via nongenetic factors. Such effects are commonly referred to as 'intergenerational' or 'transgenerational' effects (Fig. 2).

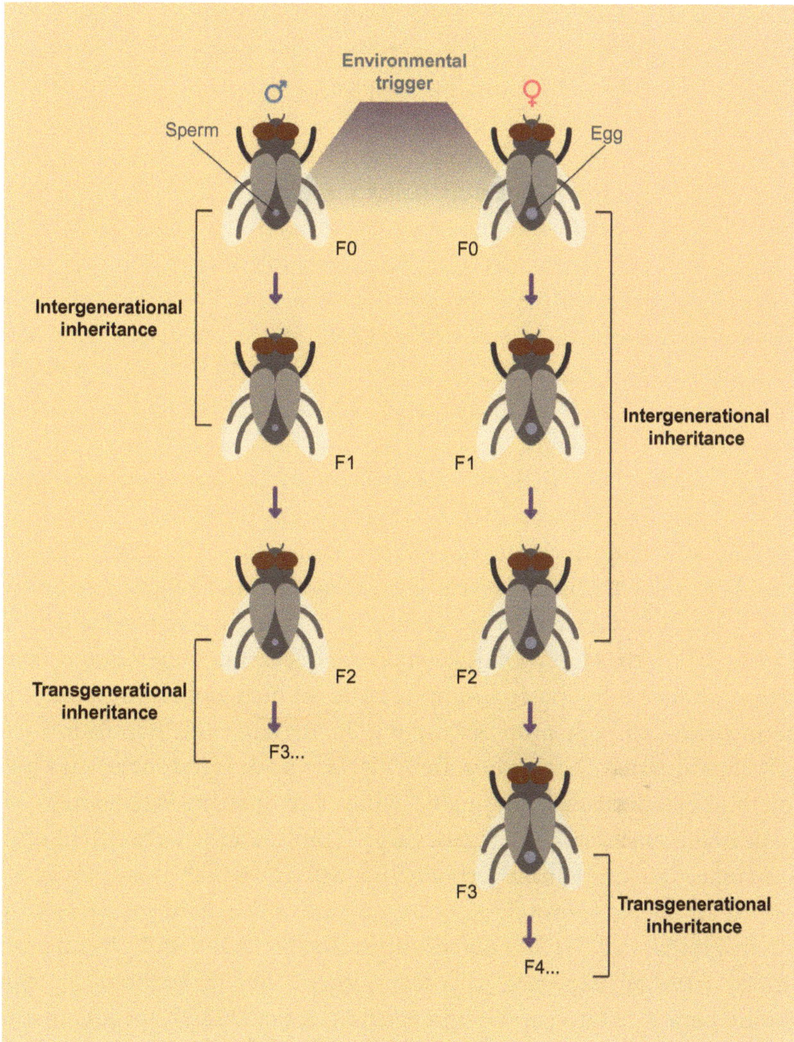

Fig. 2 Schematic representation of intergenerational and transgenerational effects in *D. melanogaster*. In the case of intergenerational inheritance, parental exposures directly influence the offspring's germ cells. By contrast, transgenerational effects occur in the F3 and subsequent generations arising from the F2 generation germline that has not been directly exposed to the triggering environmental stimuli.

The clear distinction between the genetic (hard) and epigenetic (soft) inheritance is that epigenetic modifications, unlike the genetic ones, are reversible and may persist for few (usually 3–4) generations only in the absence of triggering stimulus (Lim and Brunet, 2013). Many recent investigations highlighted the central role of epigenetic mechanisms in mediating such effects. These processes are believed to include altered patterns of DNA methylation and histone posttranslational modifications, and also changed expression of noncoding RNAs causing modified local accessibility to the genetic material and modulated gene expression (Pal and Tyler, 2016). In some recent studies, the importance of nongenomic transgenerational effects in inheritance of age-related characteristics has been highlighted (for a review, see Li and Casanueva, 2016). However, transgenerational effects on life span have hitherto only been reported rarely. Most papers reviewing and discussing these effects are solely focused on findings from the nematode *Caenorhabditis elegans* (Benayoun and Brunet, 2012; Berger, 2012; Pang and Curran, 2012), although similar data have been obtained from insect species as well. High-sugar maternal diet changed the offspring larval body composition for at least two generations in *D. melanogaster*, thereby causing development of obese-like phenotype in adult offspring maintained under high-calorie dietary conditions (Buescher et al., 2013). The levels of expression of metabolic genes (known to be substantially related to longevity) have been largely modified in the offspring of affected flies. Transgenerational effects on longevity and reproduction induced by posteclosion nutritional manipulations with low-, intermediate- or high-protein contents in the ancestral population have been demonstrated (Xia and de Belle, 2016). Both low-protein and high-protein diets decreased life span, while the intermediate-protein diet extended life span until the F3 generation. Similar effects were also observed with respect to the female reproductive activity which was found to be reduced in the low-protein groups and enhanced in the intermediate-protein groups in F0–F2 generations. In a subsequent study by the same authors, ancestral exposure to a low-protein diet throughout the posteclosion period resulted in the upregulation of the H3K27-specific methyltransferase, *E(z)*, thereby causing increased H3K27 trimethylation (H3K27me3) levels (Xia et al., 2016). These changes were accompanied by a longevity shortening in F0 generation as well as in the F2 offspring generation. Both RNAi-mediated knockdown of *E(z)* and pharmacological inhibition of its enzymatic function by a histone methyltransferase inhibitor Tazemetostat (EPZ-6438) resulted in lower levels of H3K27me3 across several generations. Moreover, the exposure to EPZ-6438 entirely mitigated the longevity-shortening effect of the parental exposure to a low-protein diet.

Cross-generational effects have also been shown in fruit flies with regard to factors other than nutrition. Cross-life stage and cross-generational adaptive plasticity induced by gamma irradiation at the egg stage was observed (Vaiserman et al., 2004). This cross-generational plasticity exhibited in enhanced resistance to starvation and heat shock stresses, as well as in extended life span in the F0 and F1 generations of *D. melanogaster*. Cross-generational impact of radiation exposure in ancestral F0 generation on the life span in F1 and F2 generations was also demonstrated (Shameer et al., 2015). Paternal exposure to small-to-moderate radiation doses (1–10 Gy) lead to an increased life span in both male and female offspring, while exposure to higher doses (40 and 50 Gy) resulted in shortening the descendant longevity. These transgenerational effects persisted up to F2 generation and completely disappeared in the F3 generation.

Collectively, these findings suggest that particular nongenetic factors, apparently epigenetic in origin, can influence longevity in *D. melanogaster* not only in affected generation but also in several subsequent generations.

6. CONCLUSIONS

Aging is a complex process influenced by a large number of environmental and genetic factors and their interactions that are far from understood. The effects of these factors are difficult to separate because they influence and modify each other. Therefore, it is difficult to investigate each of these factors separately, especially by using complex mammalian models like rodents, and a large part of modern biogerontological research is based on simpler model organisms, mainly yeast, nematodes and insects (Parrella and Longo, 2010). The use of simple model organisms including insects seems particularly useful in an exceedingly complex field such as epigenetics of aging. These models are inexpensive to maintain, have short generation times and no associated ethical issues. Moreover, a high degree of conservation exists between insect and mammalian genomes in terms of both epigenetic mechanisms and signalling pathways associated with development, immunity and disease.

The importance of epigenetic mechanisms including DNA methylation, histone modifications and the expression of miRNAs in aging and longevity of many insect species has been highlighted repeatedly. In particular, social insects provide many amazing examples of phenotypic plasticity controlled by epigenetic mechanisms, producing different longevity phenotypes from the same genotype by transcriptional reprogramming during development.

Thus, insects can provide very valuable models to investigate the epigenetic factors mediating the influence of environmental factors on development, aging, neurodegeneration, cancer and infectious disorders (Mukherjee et al., 2015). Therefore, the identification of epigenetic processes underlying aging and aging-related pathological conditions, such as inflammation, neurodegeneration and cancer, in insect models seems to be an important step towards the further development of treatment strategies to promote human health span and longevity.

REFERENCES

Aamodt, R.M., 2009. Age-and caste-dependent decrease in expression of genes maintaining DNA and RNA quality and mitochondrial integrity in the honeybee wing muscle. Exp. Gerontol. 44 (9), 586–593.

Arking, R., 2009. Overview of the genomic architecture of longevity. In: Sell, C., Lorenzini, A., Brown-Borg, H.M. (Eds.), Life Span Extension: Single Cell Organisms to Man. Humana Press, Springer, Dordrecht, pp. 59–73.

Arking, R., 2015. Independent chemical regulation of health and senescent spans in *Drosophila.* Invertebr. Reprod. Dev. 59 (Suppl. 1), 28–32.

Arking, R., Novoseltseva, J., Hwangbo, D.S., Novoseltsev, V., Lane, M., 2002. Different age-specific demographic profiles are generated in the same normal-lived *Drosophila* strain by different longevity stimuli. J. Gerontol. A Biol. Sci. Med. Sci. 57 (11), B390–B398.

Atzmon, G., Rincon, M., Rabizadeh, P., Barzilai, N., 2005. Biological evidence for inheritance of exceptional longevity. Mech. Ageing Dev. 126, 341–345.

Azevedo, S.V., Caranton, O.A., de Oliveira, T.L., Hartfelder, K., 2011. Differential expression of hypoxia pathway genes in honey bee (*Apis mellifera* L.) caste development. J. Insect Physiol. 57 (1), 38–45.

Banerjee, K.K., Ayyub, C., Sengupta, S., Kolthur-Seetharam, U., 2012a. *dSir2* deficiency in the fat body, but not muscles, affects systemic insulin signaling, fat mobilization and starvation survival in flies. Aging (Albany NY) 4 (3), 206–223.

Banerjee, K.K., Ayyub, C., Ali, S.Z., Mandot, V., Prasad, N.G., Kolthur-Seetharam, U., 2012b. *dSir2* in the adult fat body, but not in muscles, regulates life span in a diet-dependent manner. Cell Rep. 2 (6), 1485–1491.

Barchuk, A.R., dos Santos Cristino, A., Kucharski, R., da Fontoura Costa, L., Simıes, Z.L.P., Maleszka, R., 2007. Molecular determinants of caste differentiation in the highly eusocial honeybee *Apis mellifera.* BMC Dev. Biol. 7, 70.

Begna, D., Fang, Y., Feng, M., Li, J., 2011. Mitochondrial proteins differential expression during honeybee (*Apis mellifera* L.) queen and worker larvae caste determination. J. Proteome Res. 10 (9), 4263–4280.

Benayoun, B.A., Brunet, A., 2012. Epigenetic memory of longevity in *Caenorhabditis elegans.* WormBook. 1 (1), 77–81.

Berger, S.L., 2012. Transgenerational inheritance of longevity: epigenetic mysteries abound. Cell Metab. 15 (1), 6–7.

Bewick, A.J., Vogel, K.J., Moore, A.J., Schmitz, R.J., 2017. Evolution of DNA methylation across insects. Mol. Biol. Evol. 34, 654–665.

Bonasio, R., Zhang, G., Ye, C., Mutti, N.S., Fang, X., Qin, N., et al., 2010. Genomic comparison of the ants *Camponotus floridanus* and *Harpegnathos saltator.* Science 329 (5995), 1068–1071.

Bonasio, R., Li, Q., Lian, J., Mutti, N.S., Jin, L., Zhao, H., et al., 2012. Genome-wide and caste-specific DNA methylomes of the ants *Camponotus floridanus* and *Harpegnathos saltator*. Curr. Biol. 22 (19), 1755–1764.

Buescher, J.L., Musselman, L.P., Wilson, C.A., Lang, T., Keleher, M., Baranski, T.J., Duncan, J.G., 2013. Evidence for transgenerational metabolic programming in *Drosophila*. Dis. Model. Mech. 6, 1123–1132.

Buommino, E., Pasquali, D., Sinisi, A.A., Bellastella, A., Morelli, F., Metafora, S., 2000. Sodium butyrate/retinoic acid costimulation induces apoptosis-independent growth arrest and cell differentiation in normal and ras-transformed seminal vesicle epithelial cells unresponsive to retinoic acid. J. Mol. Endocrinol. 24 (1), 83–94.

Cacabelos, R., Torrellas, C., 2015. Epigenetics of aging and Alzheimer's disease: implications for pharmacogenomics and drug response. Int. J. Mol. Sci. 16 (12), 30483–30543.

Campisi, J., Robert, L., 2014. Cell senescence: role in aging and age-related diseases. Interdiscip. Top. Gerontol. 39, 45–61.

Chen, T., Sun, H., Lu, J., Zhao, Y., Tao, D., Li, X., Huang, B., 2002. Histone acetylation is involved in hsp70 gene transcription regulation in *Drosophila melanogaster*. Arch. Biochem. Biophys. 408 (2), 171–176.

Chen, X., Hu, Y., Zheng, H., Cao, L., Niu, D., Yu, D., et al., 2012. Transcriptome comparison between honey bee queen- and worker-destined larvae. Insect Biochem. Mol. Biol. 42 (9), 665–673.

Cho, Y., Griswold, A., Campbell, C., Min, K.T., 2005. Individual histone deacetylases in Drosophila modulate transcription of distinct genes. Genomics 86 (5), 606–617.

Christensen, K., Johnson, T.E., Vaupel, J.W., 2006. The quest for genetic determinants of human longevity: challenges and insights. Nat. Rev. Genet. 7, 436–448.

Corona, M., Estrada, E., Zurita, M., 1999. Differential expression of mitochondrial genes between queens and workers during caste determination in the honeybee *Apis mellifera*. J. Exp. Biol. 202 (Pt 8), 929–938.

Corona, M., Hughes, K.A., Weaver, D.B., Robinson, G.E., 2005. Gene expression patterns associated with queen honey bee longevity. Mech. Ageing Dev. 126, 1230–1238.

Cristino, A.S., Nunes, F.M.F., Lobo, C.H., Bitondi, M.M.G., Simoes, Z.L.P., Costa, L.D., et al., 2006. Caste development and reproduction: a genome-wide analysis of hallmarks of insect eusociality. Insect Mol. Biol. 15 (5), 703–714.

de Magalhaes, J.P., Wuttke, D., Wood, S.H., Plank, M., Vora, C., 2012. Genome–environment interactions that modulate aging: powerful targets for drug discovery. Pharmacol. Rev. 64, 88–101.

Dickman, M.J., Kucharski, R., Maleszka, R., Hurd, P.J., 2013. Extensive histone post-translational modification in honey bees. Insect Biochem. Mol. Biol. 43, 125–137.

Dillin, A., Gottschling, D.E., Nystrom, T., 2014. The good and the bad of being connected: the integrons of aging. Curr. Opin. Cell Biol. 26, 107–112.

Drapeau, M.D., Albert, S., Kucharski, R., Prusko, C., Maleszka, R., 2006. Evolution of the Yellow/Major Royal Jelly Protein family and the emergence of social behavior in honeybees. Genome Res. 16, 1385–1394.

Drewell, R.A., Bush, E.C., Remnant, E.J., Wong, G.T., Beeler, S.M., Stringham, J.L., Lim, J., Oldroyd, B.P., 2014. The dynamic DNA methylation cycle from egg to sperm in the honey bee *Apis mellifera*. Development 141, 2702–2711.

Elango, N., Hunt, B.G., Goodisman, M.A., Yi, S.V., 2009. DNA methylation is widespread and associated with differential gene expression in castes of the honeybee, *Apis mellifera*. Proc. Natl. Acad. Sci. U.S.A. 106 (27), 11206–11211.

Evans, J.D., Wheeler, D.E., 1999. Differential gene expression between developing queens and workers in the honey bee, *Apis mellifera*. Proc. Natl. Acad. Sci. U.S.A. 96 (10), 5575–5580.

Evans, J.E., Wheeler, D.E., 2000. Expression profiles during honeybee caste determination. Genome Biol. 2, 1.

Feldmeyer, B., Elsner, D., Foitzik, S., 2014. Gene expression patterns associated with caste and reproductive status in ants: worker-specific genes are more derived than queen-specific ones. Mol. Ecol. 23 (1), 151–161.

Feliciello, I., Parazajder, J., Akrap, I., Ugarkovic, D., 2013. First evidence of DNA methylation in insect *Tribolium castaneum*: environmental regulation of DNA methylation within heterochromatin. Epigenetics 8, 534–541.

Foret, S., Kucharski, R., Pellegrini, M., Feng, S., Jacobsen, S.E., Robinson, G.E., et al., 2012. DNA methylation dynamics, metabolic fluxes, gene splicing, and alternative phenotypes in honey bees. Proc. Natl. Acad. Sci. U.S.A. 109 (13), 4968–4973.

Frankel, S., Ziafazeli, T., Rogina, B., 2011. *dSir2* and longevity in *Drosophila*. Exp. Gerontol. 46 (5), 391–396.

Frankel, S., Woods, J., Ziafazeli, T., Rogina, B., 2015. RPD3 histone deacetylase and nutrition have distinct but interacting effects on *Drosophila* longevity. Aging (Albany NY) 7 (12), 1112–1129.

Glastad, K.M., Hunt, B.G., Yi, S.V., Goodisman, M.A., 2011. DNA methylation in insects: on the brink of the epigenomic era. Insect Mol. Biol. 20, 553–565.

Goll, M.G., Bestor, T.H., 2005. Eukaryotic cytosine methyltransferases. Annu. Rev. Biochem. 74, 481–514.

Gräff, J., Jemielity, S., Parker, J.D., Parker, K.M., Keller, L., 2007. Differential gene expression between adult queens and workers in the ant *Lasius niger*. Mol. Ecol. 16 (3), 675–683.

Grotewiel, M.S., Martin, I., Bhandari, P., Cook-Wiens, E., 2005. Functional senescence in *Drosophila melanogaster*. Ageing Res. Rev. 4, 372–397.

Grozinger, C.M., Fan, Y., Hoover, S.E., Winston, M.L., 2007. Genome-wide analysis reveals differences in brain gene expression patterns associated with caste and reproductive status in honey bees (*Apis mellifera*). Mol. Ecol. 16 (22), 4837–4848.

Guo, X., Su, S., Skogerboe, G., Dai, S., Li, W., Li, Z., et al., 2013. Recipe for a busy bee: microRNAs in honey bee caste determination. PLoS One 8 (12), e81661.

Heerboth, S., Lapinska, K., Snyder, N., Leary, M., Rollinson, S., Sarkar, S., 2014. Use of epigenetic drugs in disease: an overview. Genet. Epigenet. 6, 9–19.

Helfand, S.L., Rogina, B., 2003. Molecular genetics of aging in the fly: is this the end of the beginning? Bioessays 25, 134–141.

Iannitti, T., Palmieri, B., 2011. Clinical and experimental applications of sodium phenylbutyrate. Drugs R&D 11 (3), 227–249.

Ikeda, T., Furukawa, S., Nakamura, J., Sasaki, M., Sasaki, T., 2011. CpG methylation in the hexamerin 110 gene in the European honeybee, *Apis mellifera*. J. Insect Sci. 11, 74.

Jia, H., Pallos, J., Jacques, V., Lau, A., Tang, B., Cooper, A., et al., 2012. Histone deacetylase (HDAC) inhibitors targeting HDAC3 and HDAC1 ameliorate polyglutamine-elicited phenotypes in model systems of Huntington's disease. Neurobiol. Dis. 46 (2), 351–361.

Johnson, A.A., Sarthi, J., Pirooznia, S.K., Reube, W., Elefant, F., 2013. Increasing Tip60 HAT levels rescues axonal transport defects and associated behavioral phenotypes in a *Drosophila* Alzheimer's disease model. J. Neurosci. 33 (17), 7535–7547.

Kang, H.L., Benzer, S., Min, K.T., 2002. Life extension in *Drosophila* by feeding a drug. Proc. Natl. Acad. Sci. U.S.A 99 (2), 838–843.

Keller, L., Jemielity, S., 2006. Social insects as a model to study the molecular basis of ageing. Exp. Gerontol. 41, 553–556.

Khan, S., Jena, G.B., 2014. Protective role of sodium butyrate, a HDAC inhibitor on beta-cell proliferation, function and glucose homeostasis through modulation of p38/ERK MAPK and apoptotic pathways: study in juvenile diabetic rat. Chem. Biol. Interact. 213, 1–12.

Klose, R.J., Bird, A.P., 2006. Genomic DNA methylation: the mark and its mediators. Trends Biochem. Sci. 31, 89–97.

Koch, S.I., Groh, K., Vogel, H., Hannson, B.S., Kleineidam, C.J., Grosse-Wilde, E., 2013. Caste-specific expression patterns of immune response and chemosensory related genes in the leaf-cutting ant, *Atta vollenweideri*. PLoS One 8 (11), e81518.

Koliada, A.K., Krasnenkov, D.S., Vaiserman, A.M., 2015. Telomeric aging: mitotic clock or stress indicator? Front. Genet. 6, 82.

Kourtis, N., Tavernarakis, N., 2011. Cellular stress response pathways and ageing: intricate molecular relationships. EMBO J. 30 (13), 2520–2531.

Kramer, B.H., van Doorn, G.S., Weissing, F.J., Pen, I., 2016. Lifespan divergence between social insect castes: challenges and opportunities for evolutionary theories of aging. Curr. Opin. Insect Sci. 16, 76–80.

Kucharski, R., Maleszka, J., Foret, S., Maleszka, R., 2008. Nutritional control of reproductive status in honeybees via DNA methylation. Science 319 (5871), 1827–1830.

Kunert, N., Marhold, J., Stanke, J., Stach, D., Lyko, F., 2003. A Dnmt2-like protein mediates DNA methylation in *Drosophila*. Development 130 (21), 5083–5090.

Lakshmaiah, K.C., Jacob, L.A., Aparna, S., Lokanatha, D., Saldanha, S.C., 2014. Epigenetic therapy of cancer with histone deacetylase inhibitors. J. Cancer Res. Ther. 10 (3), 469–478.

Larson, K., Yan, S.-J., Tsurumi, A., Liu, J., Zhou, L., Gaur, K., Guo, D., Eickbush, T.H., Li, W.X., 2012. Heterochromatin formation promotes longevity and represses ribosomal RNA synthesis. PLoS Genet. 8 (1), e1002473.

Li, C., Casanueva, O., 2016. Epigenetic inheritance of proteostasis and ageing. Essays Biochem. 60 (2), 191–202.

Lillycrop, K.A., Hoile, S.P., Grenfell, L., Burdge, G.C., 2014. DNA methylation, ageing and the influence of early life nutrition. Proc. Nutr. Soc. 73 (3), 413–421.

Lim, J.P., Brunet, A., 2013. Bridging the transgenerational gap with epigenetic memory. Trends Genet. 29 (3), 176–186.

Liu, K., Wang, Y.F., Cantemir, C., Muller, M.T., 2003. Endogenous assays of DNA methyltransferases: evidence for differential activities of DNMT1, DNMT2, and DNMT3 in mammalian cells in vivo. Mol. Cell. Biol. 23, 2709–2719.

Lockett, G.A., Almond, E.J., Huggins, T.J., Parker, J.D., Bourke, A.F., 2016. Gene expression differences in relation to age and social environment in queen and worker bumble bees. Exp. Gerontol. 77, 52–61.

Lorbeck, M., Pirooznia, K., Sarthi, J., Zhu, X., Elefant, F., 2011. Microarray analysis uncovers a role for Tip60 in nervous system function and general metabolism. PLoS One 6 (4), e18412.

Lucas, K., Raikhel, A., 2013. Insect microRNAs: biogenesis, expression profiling and biological functions. Insect Biochem. Mol. Biol. 43, 24–38.

Lucas, E.R., Privman, E., Keller, L., 2016. Higher expression of somatic repair genes in long-lived ant queens than workers. Aging (Albany NY) 8 (9), 1940–1951.

Lyko, F., Maleszka, R., 2011. Insects as innovative models for functional studies of DNA methylation. Trends Genet. 27, 127–131.

Lyko, F., Foret, S., Kucharski, R., Wolf, S., Falckenhayn, C., Maleszka, R., 2010. The honey bee epigenomes: differential methylation of brain DNA in queens and workers. PLoS Biol. 8 (11), e1000506.

Maleszka, R., 2008. Epigenetic integration of environmental and genomic signals in honey bees: the critical interplay of nutritional, brain and reproductive networks. Epigenetics 3 (4), 188–192.

Maleszka, R., 2016. Epigenetic code and insect behavioural plasticity. Curr. Opin. Insect Sci. 15, 45–52.

McDonald, P., Maizi, B.M., Arking, R., 2013. Chemical regulation of mid- and late-life longevities in *Drosophila*. Exp. Gerontol. 48 (2), 240–249.

Mukherjee, K., Fischer, R., Vilcinskas, A., 2012. Histone acetylation mediates epigenetic regulation of transcriptional reprogramming in insects during metamorphosis, wounding and infection. Front. Zool. 9, 25.

Mukherjee, R., Twyman, R.M., Vilcinskas, A., 2015. Insects as model to study the epigenetic basis of disease. Prog. Biophys. Mol. Biol. 118, 69–78.

Musselman, C.A., Lalonde, M.E., Côté, J., Kutateladze, T.G., 2012. Perceiving the epigenetic landscape through histone readers. Nat. Struct. Mol. Biol. 19 (12), 1218–1227.

Okano, M., Xie, S., Li, E., 1998. Dnmt2 is not required for de novo and maintenance methylation of viral DNA in embryonic stem cells. Nucleic Acids Res. 26 (11), 2536–2540.

Pal, S., Tyler, J.K., 2016. Epigenetics and aging. Sci. Adv. 2 (7), e1600584.

Pang, S., Curran, S.P., 2012. Longevity and the long arm of epigenetics: acquired parental marks influence lifespan across several generations. Bioessays 34 (8), 652–654.

Parker, J.D., Parker, K.M., Sohal, B.H., Sohal, R.S., Keller, L., 2004. Decreased expression of Cu-Zn superoxide dismutase 1 in ants with extreme life span. Proc. Natl. Acad. Sci. U.S.A. 101, 3486–3489.

Parrella, E., Longo, V.D., 2010. Insulin/IGF-I and related signaling pathways regulate aging in nondividing cells: from yeast to the mammalian brain. ScientificWorldJournal 10, 161–177.

Patalano, S., Hore, T.A., Reik, W., Sumner, S., 2012. Shifting behaviour: epigenetic reprogramming in eusocial insects. Curr. Opin. Cell Biol. 24 (3), 367–373.

Peleg, S., Feller, C., Forne, I., Schiller, E., Sévin, D.C., Schauer, T., Regnard, C., Straub, T., Prestel, M., Klima, C., Schmitt Nogueira, M., Becker, L., Klopstock, T., Sauer, U., Becker, P.B., Imhof, A., Ladurner, A.G., 2016. Life span extension by targeting a link between metabolism and histone acetylation in *Drosophila*. EMBO Rep. 17 (3), 455–469.

Pereboom, J.J., Jordan, W.C., Sumner, S., Hammond, R.L., Bourke, A.F., 2005. Differential gene expression in queen-worker caste determination in bumble-bees. Proc. Biol. Sci. 272 (1568), 1145–1152.

Rai, K., Chidester, S., Chad, V., Zavala, C.V., Manos, E.J., James, S.R., Karpf, A.R., Jones, D.A., Cairns, B.R., 2007. Dnmt2 functions in the cytoplasm to promote liver, brain, and retina development in zebrafish. Genes Dev. 21, 261–266.

Remolina, S.C., Hughes, K.A., 2008. Evolution and mechanisms of long life and high fertility in queen honey bees. Age (Dordr.) 30 (2–3), 177–185.

Remolina, S.C., Hafez, D.M., Robinson, G.E., Hughes, K.A., 2007. Senescence in the worker honey bee *Apis mellifera*. J. Insect Physiol. 53 (10), 1027–1033.

Rogina, B., Helfand, S.L., 2004. Sir2 mediates longevity in the fly through a pathway related to calorie restriction. Proc. Natl. Acad. Sci. U.S.A. 101 (45), 15998–16003.

Sanz, A., 2016. Mitochondrial reactive oxygen species: do they extend or shorten animal lifespan? Biochim. Biophys. Acta 1857 (8), 1116–1126.

Seroude, L., Brummel, T., Kapahi, P., Benzer, S., 2002. Spatio-temporal analysis of gene expression during aging in *Drosophila melanogaster*. Aging Cell 1 (1), 47–56.

Shameer, P.M., Sowmithra, K., Harini, B.P., Chaubey, R.C., Jha, S.K., Shetty, N.J., 2015. Does exposure of male *Drosophila melanogaster* to acute gamma radiation influence egg to adult development time and longevity of F1–F3 offspring? Entomol. Sci. 18, 368–376.

Shi, Y.Y., Huang, Z.Y., Zeng, Z.J., Wang, Z.L., Wu, X.B., Yan, W.Y., 2011. Diet and cell size both affect queen-worker differentiation through DNA methylation in honey bees (*Apis mellifera*, Apidae). PLoS One 6, e18808.

Shi, Y.Y., Yan, W.Y., Huang, Z.Y., Wang, Z.L., Wu, X.B., Zeng, Z.J., 2013. Genomewide analysis indicates that queen larvae have lower methylation levels in the honey bee (*Apis mellifera*). Naturwissenschaften 100 (2), 193–197.

Siebold, A.P., Banerjee, R., Tie, F., Kiss, D.L., Moskowitz, J., Harte, P.J., 2010. Polycomb repressive complex 2 and trithorax modulate *Drosophila* longevity and stress resistance. Proc. Natl. Acad. Sci. U.S.A. 107 (1), 169–174.

Simola, D.F., Ye, C., Mutti, N.S., Dolezal, K., Bonasio, R., Liebig, J., et al., 2013. A chromatin link to caste identity in the carpenter ant *Camponotus floridanus*. Genome Res. 23 (3), 486–496.

Simpson, S.J., Sword, G.A., Lo, N., 2011. Polyphenism in insects. Curr. Biol. 21, R738–R749.

Spannhoff, A., Kim, Y.K., Raynal, N.J., Gharibyan, V., Su, M.B., Zhou, Y.Y., et al., 2011. Histone deacetylase inhibitor activity in royal jelly might facilitate caste switching in bees. EMBO Rep. 12, 238–243.

Srinivasan, D.G., Brisson, J.A., 2012. Aphids: a model for polyphenism and epigenetics. Genet. Res. Int. 2012, 431531.

St Laurent, R., O'Brien, L.M., Ahmad, S.T., 2013. Sodium butyrate improves locomotor impairment and early mortality in a rotenone-induced *Drosophila* model of Parkinson's disease. Neuroscience 246, 382–390.

Standage, D.S., Berens, A.J., Glastad, K.M., Severin, A.J., Brendel, V.P., Toth, A.L., 2016. Genome, transcriptome and methylome sequencing of a primitively eusocial wasp reveal a greatly reduced DNA methylation system in a social insect. Mol. Ecol. 25 (8), 1769–1784.

Steffan, J.S., Bodai, L., Pallos, J., Poelman, M., McCampbell, A., Apostol, B.L., Kazantsev, A., Schmidt, E., Zhu, Y.Z., Greenwald, M., Kurokawa, R., Housman, D.E., Jackson, G.R., Marsh, J.L., Thompson, L.M., 2001. Histone deacetylase inhibitors arrest polyglutamine-dependent neurodegeneration in *Drosophila*. Nature 413 (6857), 739–743.

Strichman-Almashanu, L.Z., Lee, R.S., Onyango, P.O., Perlman, E., Flam, F., Frieman, M.B., Feinberg, A.P., 2002. A genome-wide screen for normally methylated human CpG islands that can identify novel imprinted genes. Genome Res. 12, 543–554.

Takayama, S., Dhahbi, J., Roberts, A., Mao, G., Heo, S.J., Pachter, L., Martin, D.I., Boffelli, D., 2014. Genome methylation in *D. melanogaster* is found at specific short motifs and is independent of DNMT2 activity. Genome Res. 24 (5), 821–830.

Tao, D., Lu, J., Sun, H., Zhao, Y.M., Yuan, Z.G., Li, X.X., Huang, B.Q., 2004. Trichostatin A extends the lifespan of *Drosophila melanogaster* by elevating hsp22 expression. Acta Biochim. Biophys. Sin. (Shanghai) 36 (9), 618–622.

Vaiserman, A.M., 2011. Hormesis and epigenetics: is there a link? Ageing Res. Rev. 10 (4), 413–421.

Vaiserman, A.M., 2013. Persistence of developmentally programmed effects across generations: impact on longevity. Non-Genetic Inheritance 1, 33–37.

Vaiserman, A.M., 2014. Early-life nutritional programming of longevity. J. Dev. Orig. Health Dis. 5 (5), 325–338.

Vaiserman, A.M., Pasyukova, E.G., 2015. Life extension in *Drosophila* by histone deacetylase inhibitors. In: Vaiserman, A.M., Moskalev, A.A., Pasyukova, E.G. (Eds.), Life Extension: Lessons from *Drosophila*. Springer, Dordrecht, pp. 245–264.

Vaiserman, A.M., Koshel, N.M., Mechova, L.V., Voitenko, V.P., 2004. Cross-life stage and cross-generational effects of gamma irradiations at the egg stage on *Drosophila melanogaster* life histories. Biogerontology 5 (5), 327–337.

Vaiserman, A.M., Koliada, A.K., Koshel', N.M., Simonenko, A.V., Pasiukova, E.G., 2012. Effect of the histone deacetylase inhibitor sodium butyrate on the viability and life span in *Drosophila melanogaster*. Adv. Gerontol. 25 (1), 126–131.

Vaiserman, A.M., Koshel, N.M., Zabuga, O.G., Kolyada, A.K., Roshina, N.V., Pasyukova, E.G., 2013. Determination of geroprotective potential of sodium butyrate in *Drosophila melanogaster*: long-term effects. Adv. Gerontol. 26, 111–116.

Vanhaecke, T., Papeleu, P., Elaut, G., Rogiers, V., 2004. Trichostatin A-like hydroxamate histone deacetylase inhibitors as therapeutic agents: toxicological point of view. Curr. Med. Chem. 11 (12), 1629–1643.

Vijg, J., Suh, Y., 2013. Genome instability and aging. Annu. Rev. Physiol. 75, 645–668.

Weaver, D.B., Anzola, J.M., Evans, J.D., Reid, J.G., Reese, J.T., Childs, K.L., et al., 2007. Computational and transcriptional evidence for microRNAs in the honey bee genome. Genome Biol. 8 (6), R97.

Welch, M., Lister, R., 2014. Epigenomics and the control of fate, form and function in social insects. Curr. Opin. Insect Sci. 1, 31–38.

Winston, M.L., 1987. The Biology of the Honey Bee. Harvard University Press, Cambridge, MA.

Wood, J.G., Hillenmeyer, S., Lawrence, C., Chang, C., Hosier, S., Lightfoot, W., Mukherjee, E., Jiang, N., Schorl, C., Brodsky, A.S., Neretti, N., Helfand, S.L., 2010. Chromatin remodeling in the aging genome of *Drosophila*. Aging Cell 9 (6), 971–978.

Xia, B., de Belle, J.S., 2016. Transgenerational programming of longevity and reproduction by post-eclosion dietary manipulation in *Drosophila*. Aging (Albany NY) 8, 1115–1134.

Xia, B., Gerstin, E., Schones, D.E., Huang, W., de Belle, J.S., 2016. Transgenerational programming of longevity through E(z)-mediated histone H3K27 trimethylation in *Drosophila*. Aging (Albany NY) 8 (11), 2988–3008.

Xu, S., Wilf, R., Menon, T., Panikker, P., Sarthi, J., Elefant, F., 2014. Epigenetic control of learning and memory in *Drosophila* by Tip60 HAT action. Genetics 198 (4), 1571–1586.

Yamada, L., Chong, S., 2017. Epigenetic studies in developmental origins of health and disease: pitfalls and key considerations for study design and interpretation. J. Dev. Orig. Health Dis. 8 (1), 30–43.

Yan, H., Bonasio, R., Simola, D.F., Liebig, J., Berger, S.L., Reinberg, D., 2015. DNA methylation in social insects: how epigenetics can control behavior and longevity. Annu. Rev. Entomol. 60, 435–452.

Yin, D., Ong, J.M., Hu, J., Desmond, J.C., Kawamata, N., Konda, B.M., Black, K.L., Koeffler, H.P., 2007. Suberoylanilide hydroxamic acid, a histone deacetylase inhibitor: effects on gene expression and growth of glioma cells in vitro and in vivo. Clin. Cancer Res. 13 (3), 1045–1052.

Ylla, G., Fromm, B., Piulachs, M.-D., Belles, X., 2016. The microRNA toolkit of insects. Sci. Rep. 6, 37736.

Zhao, Y., Sun, H., Lu, J., Li, X., Chen, X., Tao, D., Huang, W., Huang, B., 2005a. Lifespan extension and elevated hsp gene expression in *Drosophila* caused by histone deacetylase inhibitors. J. Exp. Biol. 208 (Pt. 4), 697–705.

Zhao, Y., Lu, J., Sun, H., Chen, X., Huang, W., Tao, D., Huang, B., 2005b. Histone acetylation regulates both transcription initiation and elongation of hsp22 gene in *Drosophila*. Biochem. Biophys. Res. Commun. 326 (4), 811–816.

Zhao, Y.M., Chen, X., Sun, H., Yuan, Z.G., Ren, G.L., Li, X.X., Lu, J., Huang, B.Q., 2006. Effects of histone deacetylase inhibitors on transcriptional regulation of the hsp70 gene in *Drosophila*. Cell Res. 16 (6), 566–576.

Zhao, Y., Lu, J., Sun, H., Chen, X., Huang, B., 2007. Roles of histone acetylation modification in basal and inducible expression of hsp26 gene in *D. melanogaster*. Mol. Cell. Biochem. 306 (1–2), 1–8.

Zhou, Q., Dalgard, C.L., Wynder, C., Doughty, M.L., 2011. Histone deacetylase inhibitors SAHA and sodium butyrate block G1-to-S cell cycle progression in neurosphere formation by adult subventricular cells. BMC Neurosci. 12, 50.

CHAPTER FIVE

Epigenetic Influences on Diapause

Julie A. Reynolds
The Ohio State University, Columbus, OH, United States

Contents

Abstract

Diapause is an endogenously regulated dormant state that allows insects and other animals to survive seasons of inimical conditions. Diapause is characterized by altered gene expression profiles, which mediate developmental arrest, metabolic depression, increased stress resistance, and other physiological and biochemical changes associated with this dormant state. In this chapter, I explore epigenetic processes that may regulate gene expression and contribute to the diapause phenotype. Only a few studies have explored the influence of DNA methylation, histone modification, and small RNA silencing on diapause. However, there is evidence that histone methylation and histone acetylation influence diapause in a wide range of species. In addition, microRNAs, siRNAs, and piRNAs appear to regulate gene expression before, during, and after diapause. Together, the accumulating evidence supports the inclusion of multiple epigenetic pathways in the "toolbox" of processes that regulate dormant states in insects and other animals.

Advances in Insect Physiology, Volume 53
ISSN 0065-2806
http://dx.doi.org/10.1016/bs.aiip.2017.03.003

1. OVERVIEW OF DIAPAUSE

Diapause is an endogenously regulated dormant state that provides a means for insects to survive seasons of adverse environmental conditions and allows populations to synchronize periods of active development and reproduction with seasons of optimal resource availability (Košťál, 2006; Lees, 1955; Tauber et al., 1986). Diapause can occur at any stage of development; and examples of embryonic, larval/nymphal, pupal, and adult diapauses have thoroughly been studied and described. However, the capacity for diapause is usually limited to a single, genetically determined developmental stage for each species. For some species, diapause is obligatory, and every individual becomes dormant at the same, genetically determined, stage in the life cycle. For many other species, diapause is a plastic response (i.e. facultative) and constitutes an alternative developmental pathway that is programmed by token signals (e.g. changes in day length, temperature, or food quality) that indicate a season of harsh conditions is coming (Tauber et al., 1986).

Unlike quiescence, a type of dormancy that is an immediate response to a change in the environment, insects that enter diapause anticipate the decline in environmental conditions and prepare by seeking an appropriate shelter and/or sequestering additional metabolic energy reserves. In addition, while quiescence ends as soon as the environment returns to "normal", diapause persists until the diapause programme is terminated by endogenously controlled processes that are not completely understood (Košťál, 2006; Tauber et al., 1986).

Some aspects of the diapause phenotype are variable and species specific, but in general, diapause is characterized by developmental arrest and metabolic restructuring that includes downregulation of processes that consume energy (e.g. cell growth and development, transcription, and translation), suppression of oxidative processes that produce energy (e.g. oxidative phosphorylation and glycolysis), and upregulation of processes (e.g. production of cryoprotectants molecules an heat-shock proteins) that enhance stress resistance (Denlinger, 2002; Hahn and Denlinger, 2011; Hand et al., 2011; Lees, 1955; Tauber and Tauber, 1976). Diapause-related changes in biochemistry and physiology have been described for a number of insect species including crickets (Reynolds and Hand, 2009b), mosquitoes (Reynolds et al., 2012; Zhou and Miesfeld, 2009), and butterflies (Lehmann et al., 2016) and have been the subject of a number of recent reviews (Hahn and Denlinger, 2011; Hand et al., 2011). Together, the physiological and biochemical changes that occur during diapause conserve energy stores and maximize survivorship during an extended period of "suspended animation".

Extensive changes in gene expression underlie the biochemical and physiological changes that define the diapause phenotype for each species (Denlinger, 2002). A growing body of literature documents diapause-related changes in gene expression for a variety of species including crickets (Reynolds and Hand, 2009b), locusts (Tu et al., 2015), moths (Wadsworth and Dopman, 2015), spider mites (Bryon et al., 2013), bees (Amsalem et al., 2015; Yocum et al., 2015), mosquitoes (Poelchau et al., 2011, 2013a,b), and several species of higher Diptera (Flannagan et al., 1998; Kubrak et al., 2014; Ragland et al., 2010, 2011). However, the regulatory networks that regulate diapause-relevant changes in gene expression remain unknown.

Epigenetic processes, including DNA methylation, histone modifications, and small RNA interference, can alter the phenotype independent of changes in the genome and may have a role in regulating changes in gene expression necessary for diapause initiation, maintenance, and termination. My goal for this chapter is to summarize what is currently known about epigenetic processes that may regulate diapause or, at least, contribute to the diapause phenotype. The majority of the discussion focuses on insects, but I have also included some details about epigenetic regulation of other animals that enter diapause, or have a diapause-like dormancy during some part of their life cycle.

Diapause is not a static state, rather it is defined by distinct prediapause, diapause, and postdiapause phases that are also divided into segments defined by Koštál (2006) and summarized in Table 1. When possible, I have described the function of each epigenetic process as it relates to a particular phase of diapause. Briefly, prediapause is the phase when insects are receptive to diapause-inducing environmental cues and the developmental switch between "direct development" and "diapause" trajectories occurs. Following diapause induction, there may be a preparatory segment during which individuals programmed for diapause entry may search for suitable locations to overwinter or otherwise prepare for diapause. During the diapause phase, development is arrested and metabolic suppression occurs. In some cases, "Diapause" is divided into induction, maintenance, and termination phases, but most studies refer to this phase as simply "diapause" or "diapause maintenance". Finally discussions about epigenetic regulation of diapause

Table 1 Summary of the Phases of Diapause (Koštál, 2006)

Prediapause	Induction	Genetically determined ontogenetic stage when environmental cues are perceived and the developmental pathway is switched from the direct development to diapause destined
	Preparation	A period of direct development between diapause induction and initiation when physiological and behavioural preparations for diapause take place
Diapause	Initiation	Direct development is arrested and metabolism is suppressed even though environmental conditions may be favourable to support both
	Maintenance	Developmental arrest persists and metabolic rate remains relatively low and constant
	Termination	Specific changes in the environment lessen the intensity of diapause and the potential for direct development is restored, but environmental conditions generally are not conducive for continued development
Postdiapause	Postdiapause quiescence	Development and metabolism remain inhibited because of unfavourable environmental conditions

termination generally consider events right before and soon after diapause is artificially terminated with applications of hexane or another chemical.

2. DNA METHYLATION AND DIAPAUSE INDUCTION

DNA methylation is the process of adding methyl groups to cytosine nucleotides on DNA, thereby changing gene activity. DNA methylation is the process most often considered in discussions about epigenetics; yet this pathway has received little attention by diapause researchers, most likely because many insects used as models for studying diapause (e.g. *Sarcophaga bullata*, *Culex pipiens*) lack robust systems for DNA methylation (Bewick et al., 2017). However, the role of DNA methylation in diapause induction has been studied in two insects: the silkworm, *Bombyx mori* (Egi et al., 2016), and the jewel wasp, *Nasonia vitripennis* (Pegoraro et al., 2016). *B. mori* has a maternally regulated diapause; females reared under short days during the larval stage produce offspring that enter diapause as embryos (Yamashita, 1996). A functional DNA methylation system has been found in *B. mori* (Xiang et al., 2010); and a methyltransferase gene is upregulated in diapausing embryos of *B. mori* (Sasibhushan et al., 2013). However, DNA methylation may not regulate diapause induction in this species because feeding the DNA methylation inhibitor, 5-azacytidine, to larval *B. mori* did not change the diapause incidence (i.e. the proportion of embryos entering diapause) in the offspring of treated animals regardless of whether they were reared in short-day or long-day conditions (Egi et al., 2016). The authors of this study concluded that feeding to 5-azacytidine to photosensitive larvae does not influence diapause induction. However, additional experiments are needed to determine whether DNA methylation influences other phases of diapause in *B. mori* or any other species.

The only insect for which DNA methylation of diapause regulation has been confirmed is *N. vitripennis*. This parasitoid wasp lives in temperate zones; like *B. mori*, *N. vitripennis* has a maternally regulated diapause, and the progeny of females exposed to short-day lengths enter diapause as a larva (Saunders, 1965). Sequencing the *Nasonia* genome revealed a functioning DNA methylation tool kit and substantial methylation in the genome (Werren et al., 2010). DNA methylation in *N. vitripennis* is influenced by photoperiod, and females reared in short-day, diapause-inducing conditions have different genome-wide methylation profiles than females reared in long-day, diapause-averting conditions (Pegoraro et al., 2016). These photoperiod-related changes in DNA methylation influence diapause induction and subsequent diapause programming because inhibiting DNA methylation changes the diapause response. RNA interference of *DNA methyltransferase 1a* or chemical inhibition with 5-azacytidine caused females to produce diapausing offspring regardless of day length, and knocking down *Dnmt3* increased diapause incidence in wasps reared in long-day, diapause-averting, conditions (Pegoraro et al., 2016). Together, these results provide the first example of DNA methylation influencing photoperiodism (i.e. responses to cycles of light and dark) and diapause incidence for any insect. It will be interesting to discover whether seasonal timing in other insects is also regulated by DNA methylation.

3. HISTONE MODIFICATIONS THAT MAY REGULATE INSECT DIAPAUSE

Within cells, compact chromatin structures are formed by wrapping DNA around nucleosomes composed of histone proteins H2A, H2B, H3, and H4. Histones H3 and H4 each have a tail that can be covalently modified to increase or decrease the interaction between the DNA strand

and the nucleosome and alter the transcription potential of a specific region of DNA. Increasing DNA–histone interaction generally creates a compact chromatin structure that is transcriptionally silent; decreasing the interaction between histones and DNA creates a region of relaxed chromatin that is transcriptionally active. Histone acetylation, methylation, and phosphorylation have all been reported for insects, and currently these are the best-understood types of histone modification. Histone methylation and acetylation have been studied in the context of insect diapause and will be discussed.

3.1 Changes in Histone Methylation Are Associated With Diapause

Transcriptional evidence implicating histone methylation as a regulator of diapause induction comes from studies with two higher Diptera, *Chymomyza costata* (Poupardin et al., 2015) and *S. bullata* (Reynolds and Denlinger, unpublished data). *C. costata* is a drosophilid that enters diapause as mature third-instar larvae in response to short-day lengths. Diapause induction in this species occurs during the first 3 days of the third larval instar, and larvae begin to exhibit the diapause phenotype between 10 and 40 days after the moult to this instar (Košťál et al., 2000a,b). Transcriptome profiling of *C. costata* early in the third larval instar (i.e. photosensitive stage) identified genes and processes that were differentially regulated in short day-reared flies, compared to long day-reared animals. *Dpy-30* was one gene that was downregulated in response to short-day conditions and may mediate the switch from the direct development to diapause (Poupardin et al., 2015). *Dpy-30* encodes a subunit of the Set1C/COMPASS complex that methylates lysine 4 of histone H3 (H3K4) in species ranging from yeast to mammals (Ardehali et al., 2011; Mohan et al., 2011). Methylation of H3K4 is typically associated with transcriptionally active genes (Kouzarides, 2007), and depletion of DPY-30 leads to downregulation of genes involved in cell cycle progression, cell growth and proliferation, and DNA replication and repair (Simboeck et al., 2013). Many of the biological processes regulated by DPY-30-regulated genes are arrested during diapause in *C. costata*, and but further research is required to see if changes in H3K4 methylation result from the decrease in *dpy-30* expression.

S. bullata is a sarcophagid flesh fly that enters diapause during the pupal stage in response to short-day lengths experienced as in utero embryos or first-instar larvae (Denlinger, 1971, 1972). Two histone demethylases, LSD1 and Su(var)3–9, which are upregulated 1.5- and 2.5-fold,

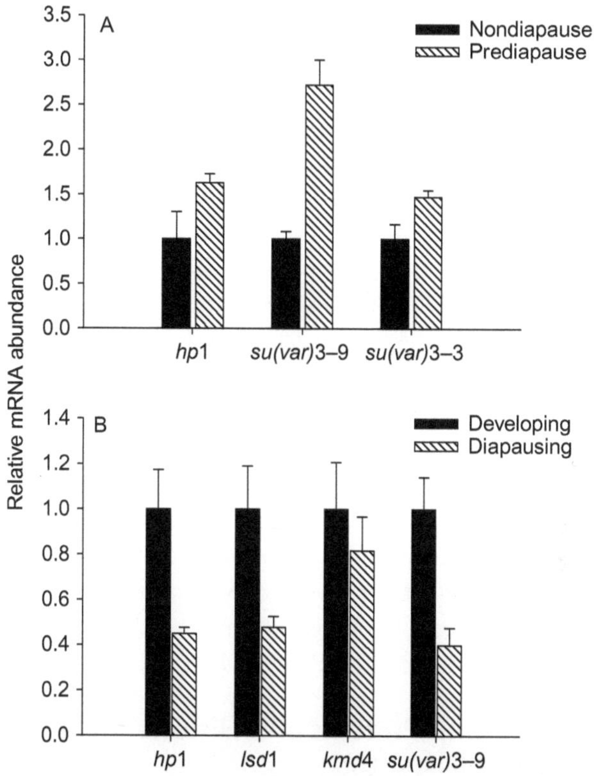

Fig. 1 mRNA profiles for genes encoding histone methyltransferase genes (A) prediapause first-instar larvae and (B) diapausing and developing pupae. Transcription was measured with qRT-PCR as described in Reynolds et al. (2013).

respectively, in prediapause larvae (Fig. 1), may have roles in programming diapause in *S. bullata*. LSD1 is a lysine-specific demethylase that targets H3K4 (Di Stefano et al., 2007), and Su(var)3–9 removes methyl groups from H3K9 (Czermin et al., 2001). H3K4 and H3K9 methylation are often associated with gene activation, so increasing the activity of LSD1 and Su(var) 3–9 could remove methyl groups and turn off genes when diapause is induced. However, this possibility has not been experimentally tested, and for now the function of LSD1 and Su(var)3–9 in programming diapause remains unknown.

It is possible that Su(var)3–9 has a role in regulating lipid metabolism during the preparation phase of diapause. Methylation of H3K9 regulates lipid metabolism during the dauer stage, a diapause-like dormant state, in the nematode *Caenorhabditis elegans* (Meister et al., 2011). Su(var)3–9 has also been recently implicated in the regulation of transposable elements (TEs) in *Drosophila melanogaster*, and overexpression of *su(var)3–9* reduces TE activity and extends the lifespan (Wood et al., 2016). It is possible that Su(var)3–9 and demethylation of H3K9 may have a similar role in diapausing *S. bullata* as well as in other insects.

LSD1 (a.k.a. Su(var)3–3) is a highly conserved lysine-specific demethylase that removes methyl groups from H3K4 and negatively regulates gene transcription. Inactivation of LSD1 results in genome-wide changes to the levels of H3K4me and H3K4me2 and leads to upregulation of a subset of genes (Di Stefano et al., 2007). A study in *C. elegans* has suggested that one role of LSD1 is to act as an epigenetic eraser (Katz et al., 2009). Dimethylated H3K4 is associated with transcriptionally active genes and may function as a molecular bookmark, or epigenetic place holder that maintains developmentally important transcriptional patterns rather than establishing them de novo (Katz et al., 2009; B. Li et al., 2007). First-instar larvae of *S. bullata* that have been programmed for diapause have 1.5-fold more *lsd1* mRNA than larvae not programmed for diapause (Fig. 1, Reynolds and Denlinger, unpublished data), and additional research is required to discover what role this demethylase plays in diapause induction.

3.2 Polycomb Group Proteins and Histone Methylation Influences on Diapause

Polycomb group (PcG) proteins work cooperatively to epigenetically silence gene expression by remodelling chromatin structure through post-translational modification of histones. PcG proteins are well conserved evolutionarily, and two PcG complexes have been identified and studied in insects: Polycomb Repressive Complex 1 (PRC1), which ubiquitinates histone 2A on lysine 119 (H2AK119Ubi), and PRC2, which trimethylates histone H3 at lysine 27 (H3K27me3). H3K27me3 is generally associated with inactive gene promoters and is indicative of gene silencing (Lu et al., 2013). Many genes that are silenced by PRC2 through H3K27 methylation in *Drosophila* are involved in developmental processes (Schwartz et al., 2006; Tolhuis et al., 2006), making PRC2 a candidate regulator of insect diapause. To date, PRC2 has been studied in this context for two species, *Megachile rotundata* (Yocum et al., 2015) and *Helicoverpa armigera* (Lu et al., 2013).

For *M. rotundata*, the alfalfa leaf cutter bee, changes in mRNA abundance provide transcriptional evidence that PRC2 has a role in diapause termination and/or postdiapause development (Yocum et al., 2015). Diapause in this species occurs during the prepupal stage and is terminated by exposure to cold. Five genes that encode members of the PRC2 complex B lymphoma Mo-MLV insertion region 1 (BMI-1), enhancer of polycomb [E(Pc)], posterior sex comb (PSC), additional sex comb (ASX), and sex comb on midleg (SCM) (Yocum et al., 2015) are upregulated after cold exposure terminates diapause. Diapause termination is commonly associated with rapid downregulation of genes that were upregulated during diapause (Denlinger, 2002), and it is likely that the upregulation of these PRC2 complex genes facilitates the repression of genes that are upregulated to elicit the diapause phenotype. Future research will hopefully identify genes repressed by PRC2 in postdiapause bees and provide information about the role they play in postdiapause development.

PRC2 also has a role in regulating diapause in *H. armigera*, also known as the cotton bollworm, which enters diapause in the pupal stage in response to short-day lengths and low temperatures experienced during the larval stage (Lu et al., 2013). In *H. armigera*, PRC2 regulates diapause entry by mediating the production of prothoracicotropic hormone (PTTH), a neuropeptide that regulates the production of ecdysone, a key regulator of developmental timing and diapause entry in insects (Denlinger et al., 2005; Gilbert et al., 2002). In *H. armigera*, the expression of Extra sex combs (Esc), a component of PRC2, is regulated by day length; and the brains of individuals exposed to long-day lengths (i.e. diapause-averting conditions) have higher levels of *esc* mRNA than those reared in short-day, diapause-inducing conditions. Elevated *esc* in nondiapausing pupal brains results in increased H3K27me3 along with the promoters of genes encoding PTTH. Surprisingly, the increase in H3K27me3 activates *ptth* expression rather than suppressing it as expected and promotes continued development. Reducing *esc* mRNA abundance with double-stranded RNA or inhibiting PRC2 chemically with 3-deazaneplanocin A reduced H3K27me3 methylation and PTTH production lead to developmental arrest in individuals that had been programed for continued development. Thus, in *H. armigera* H3K27me3 stimulates rather than represses PTTH production and prevents diapause entry (Lu et al., 2013). Additional studies with other insects will help to determine whether this role for PRC2 is unique to *H. armigera* or is conserved in other insect species.

3.3 Histone Acetylation in Prediapause, Diapause, and Postdiapause Insects

Transcriptional evidence that histone acetylation influences diapause comes from studies with *Allonemobius socius* (Reynolds and Hand, 2009b), *C. pipiens*

(Hickner et al., 2015), and *S. bullata* (Reynolds et al., 2016). *A. socius* is a ground cricket with a maternally regulated diapause; the progeny of female adults exposed to short-day lengths enter diapause as early embryos, approximately 4–6 days postoviposition. Prior to diapause entry, mRNA expression of *reptin*, which encodes part of the Tip60 histone acetylation complex (Qi et al., 2006), is upregulated twofold in embryos programmed to enter diapause compared to ontogenetically matched embryos not programmed for diapause (Reynolds and Hand, 2009a). The Tip60 complex can generate silent chromatin through acetylation of histone H4 (Qi et al., 2006), which could result in developmental arrest. Once the embryos of *A. socius* enter diapause, *reptin* is downregulated by more than 70%, suggesting that its role is not necessary after diapause is established (Reynolds and Hand, 2009b).

C. pipiens is a mosquito that has an adult reproductive diapause initiated in females following exposure to short-day lengths during the late fourth larval instar and pupal stages (Spielman and Wong, 1973). Transcriptional changes associated with diapause induction in *C. pipiens* were discovered using a whole-transcriptome microarray that identified differentially regulated genes in photosensitive pupae reared in either diapause-inducing (i.e. short day) or diapause-averting (i.e. long day) environments. The study found significant overexpression of genes involved in chromatin-remodelling processes in pupae reared in diapause-inducing conditions compared to those in diapause-averting, long-day conditions (Hickner et al., 2015). Genes overexpressed in short-day pupae included components of the NuRD and Sin3 chromatin-remodelling complexes (Hickner et al., 2015). NuRD is an ATP-dependent multiprotein complex that promotes chromatin condensation and silences gene transcription (reviewed by Clapier and Cairns, 2009). The Sin3 complex is also a multiprotein complex that includes histone deacetylase 1 (HDAC 1) and HDAC 2 and has a role in repressing transcription. The Sin3 complex has been implicated in diapause-relevant biological processes including metabolism (Liu and Pile, 2017), cell cycle progression (Swaminathan and Pile, 2010), stress resistance, and lifespan (Barnes et al., 2014).

Histone deacetylation may contribute to lifespan extension in diapausing *C. pipiens*, at least in part. The basis for this hypothesis is a fourfold increase in the mRNA expression of Pax-interacting protein (Pax) in diapausing mosquitoes (Sim et al., 2015). Pax is a downstream target of the transcription factor FOXO, a prominent diapause regulator in this species (Sim and Denlinger, 2008). Pax is involved in the localization of Sirtuin 1 (Sir2), an NAD-dependent HDAC, and other silencing proteins that regulate lifespan extension in *D. melanogaster* and *C. elegans* (Latham and Dent, 2007; Lechner et al., 2000; Rogina and Helfand, 2004; Tissenbaum and Guarente, 2001). Future studies should provide information about the role of Pax and Sir2 in mediating lifespan extension in diapausing *C. pipiens* and possibly other insect species.

In *S. bullata*, diapause is associated with a 75% reduction in the total amount of acetylated histone H3 (Reynolds et al., 2016). Histone deacetylation is commonly associated with gene silencing, and the reduction in histone H3 acetylation may contribute to the large-scale downregulation of gene transcription that characterizes diapause in this, and other, species (Denlinger, 2002). Surprisingly, the decrease in total histone H3 acetylation in diapausing pupae was accompanied by a significant decrease in the enzyme activity of HDACs and changes in the mRNA expression of several genes encoding HDACs and histone acetyltransferases (HATs). *Hdac3*, *hdac6*, *sirt1*, and *sirt2* were underexpressed in diapausing pupae as was *gnc5*, a gene that encodes a HAT. Genes encoding Tip60 and Reptin were both increased. The measured changes in the transcription of HDAC- and HAT-encoding genes in diapausing pupae suggest that histone deacetylation does not occur during diapause, but likely takes place prior to diapause entry. The timing for diapause-relevant histone deacetylation is currently unknown, but it may begin during, or shortly after, the photosensitive stage during diapause induction. Genes that encode histone deacetylases Hdac1/Rpd3, Hdac3, Hdac6, and Sirt2 are 1.5- to 2.5-fold more abundant in larvae programmed to enter diapause than in larvae programmed for direct development (Reynolds et al., 2016). Further experiments will allow the identification of genes regulated by changes in histone acetylation during diapause and how this histone modification regulates the developmental trajectory in *S. bullata* and other species.

3.4 Heterochromatin Protein 1 in Diapausing *S. bullata*

The heterochromatin protein 1 (HP1) protein family is a conserved group of epigenetic factors that contributes to the formation of heterochromatin domains (e.g. telomeres and centromeres) through interaction with trimethylated lysine 9 of histone H3 (H3K9me3). HP1 can also bind to regions of euchromatin and, in cooperation with other proteins and/or RNA components, either activate or repress transcription in a context-dependent manner (Meister et al., 2011). For example, in *Drosophila*,

HP1 acts as a positive regulator of transcription by interacting with lysine demethylase 4A (dKDM4A; Meister et al., 2011) and facilitating demethylation of H3K36.

HP1 regulates some features of dauer formation, a diapause-like state in the nematode *C. elegans* (Meister et al., 2011). In *C. elegans*, HPL-2, a homologue of HP1, interacts with TGF-β and insulin/insulin-like growth factor-1-like receptor signalling pathways (Meister et al., 2011). Loss-of-function mutants lacking HPL-2 are more likely to enter the dauer stage than their wild-type counterparts under identical conditions, due to an increased sensitivity to dauer hormone, which initiates the dauer programme. HPL-2 mutants also remain dormant longer, and the addition of HPL-2 facilitates dauer exit. Thus, HPL-2 regulates both entry into and exit from the dauer stage. HPL-2 also regulates genes involved in other aspects of the dauer stage including the synthesis and degradation of the extracellular matrix, the Hedgehog signalling pathway, cuticle physiology, and lipid metabolism and transport (Meister et al., 2011).

Hp1a is downregulated in diapausing pupae of *S. bullata* by more than ~50% compared to nondiapausing pupae (Reynolds and Denlinger, unpublished data), but it is unclear whether HP1 regulates diapause in this, or any other, insect. There have been no published studies describing a function for HP1 in the initiation, maintenance, or termination of diapause, and experiments to elucidate the role of HP1 in diapausing *S. bullata* have largely been inconclusive (Reynolds and Denlinger, unpublished data). Attempts to knock down *hp1* in flies not programmed for diapause using RNA interference resulted in ~25% reduction in *hp1* mRNA. However, there was no change in the time to adult emergence or lifespan, and there was no change in the transcription of genes that are predicted targets of HP1 (Piacentini et al., 2009). Thus, it remains unclear whether *HP1* downregulation in diapausing *S. bullata* contributes to the diapause phenotype or is simply one gene that is not necessary during diapause development. Future research will help to discover whether HP1a functions in diapausing insects in a manner similar to HPL-2 in *C. elegans*.

4. SMALL RNA REGULATION OF INSECT DIAPAUSE

Small noncoding RNAs are key for posttranscriptional gene regulation and are increasingly recognized as important regulators of numerous biological processes including cell cycle regulation and cell proliferation

(Sun et al., 2008), developmental timing (Bethke et al., 2009; Hammell et al., 2009; Resnick et al., 2010), metabolism (Xu et al., 2003), stress resistance (Flynt et al., 2009; Teleman et al., 2006; Zhang et al., 2011), fecundity (Klattenhoff et al., 2007), and immune function (Kemp et al., 2013). Although there is a debate about whether small RNAs should be considered to be epigenetic pathways (Deans and Maggert, 2015), they are included in this discussion about epigenetic regulation of diapause because they can influence an organism's phenotype independent of changes in DNA sequence and act as a link between environmental downstream physiological and biochemical processes that need to be regulated to maintain cellular homeostasis. In addition, many of the biological processes regulated by small RNAs are processes altered as the diapause phenotype is established, and there is increasing evidence that small RNAs are associated with insect diapause.

Small RNAs are divided into three classes, including microRNAs (miRNAs), small-interfering RNAs (siRNAs), and piwi-associated RNAs (piRNAs). These classes are defined by their size, origin, structure, localization, and mode of action. Briefly, miRNAs and siRNAs are 20–25 nucleotides long, and they originate from double-stranded RNA precursors that are processed by Dicer1 and Dicer2, respectively. The resulting short RNA fragments are loaded into an RNA-induced silencing complexes (RISCs) and aid in the degradation or translational repression of their targets. In *Drosophila*, miRNAs typically associate with RISCs that contain Argonaute-1 (Ago-1) at their catalytic centre, and siRNAs generally bind to RISCs containing Ago-2 (Tomari et al., 2007). piRNAs are 22–30 nucleotides long and originate from single-stranded RNAs including TEs (see reviews by Juliano et al., 2011; Thompson and Lin, 2009). Biogenesis of piRNAs requires Spindle-E, an RNA helicase (Handler et al., 2013; Kennerdell et al., 2002; Olivier et al., 2010); and mature piRNAs are loaded into an RISC-containing Piwi, Aubergine (Aub), or Ago-3. Piwi–piRNA complexes regulate gene expression in the nucleus, while Aub–piRNA and Ago3–piRNA complexes remain in the cytoplasm where they regulate expression of TEs (Handler et al., 2013).

Transcriptional evidence suggests that all three classes of small RNAs are associated with insect diapause and have the potential to regulate diapause induction, maintenance, and termination. Differential regulation of *ago-2* in postdiapausing *M. rotundata*, as well as changes in *drosha* and *pasha*, which encode components of the siRNA biogenesis pathway, suggests that siRNAs regulate diapause in this species (Yocum et al., 2015). However, it is not clear whether these genes are more or less abundant in postdiapause bees, so it is difficult to interpret these changes in gene expression and infer a role for siRNAs in diapausing *M. rotundata*.

Genes encoding some components of the siRNA and piRNA pathways are also differentially regulated in prediapause stages of *C. costata* and *S. bullata*. In prediapause larvae of *C. costata*, *ago-2* is downregulated in larvae exposed to a diapause-inducing short-day photoperiod compared to larvae reared in diapause-averting long-day conditions (Poupardin et al., 2015). Reduced transcription of *ago-2* suggests that reducing the activity of the siRNA pathway may be important for initiating the switch from the direct development to diapause in this species. The functional significance for *ago-2* downregulation in *C. costata* remains untested. However, one possibility is that reducing Ago-2, which negatively regulates the transcription of target genes by binding to specific chromatin sites near gene promoters (Taliaferro et al., 2013), allows for upregulation of genes necessary for diapause entry.

There is also transcriptional evidence that siRNAs play a role in diapause induction in *S. bullata*, but, in contrast with *C. costata*, *ago-2* transcription is increased rather than decreased. Prediapause first-instar larvae (i.e. those reared under short-day conditions during the photosensitive stage) had more *ago-2* mRNA than those programmed for direct development (Reynolds et al., 2013). The gene encoding R2D2, which is another part of the siRNA-associated RISC (Liu et al., 2009), was also upregulated in prediapause embryos of *S. bullata*, reinforcing the idea that siRNA-mediated gene silencing is important for diapause induction in this species.

Another class of small-regulator RNAs, the piRNA pathway, also appears to be important for diapause induction in *S. bullata*. Genes encoding Piwi and Spindle-E were both upregulated more than twofold in prediapause first-instar larvae (Reynolds et al., 2013). Spin-E is an RNA helicase that is a required component of the piRNA biogenesis pathway (Handler et al., 2013; Kennerdell et al., 2002), and Piwi is one of three possible catalytic proteins found within the pi-RISC, that mature piRNAs are loaded into, for transport into the nucleus (Handler et al., 2013). Within the nucleus, Piwi–piRNA complexes can regulate gene expression through their interactions with HP1, which is also encoded by a gene that is upregulated in prediapause *S. bullata* (Reynolds and Denlinger, unpublished data; Fig. 1), and other proteins that promote formation of heterochromatin, which is transcriptionally silent (McCue and Slotkin, 2012; Yin and Lin, 2007). Whether there is an increased abundance of mature piRNAs, and their functional significance, in prediapause larvae of *S. bullata* remains to be tested, but it may be that upregulation of *piwi* and *spin-E* leads to an increased proportion of heterochromatin in diapause-destined *S. bullata* that initiates long-lasting changes necessary for the preparation for diapause and for diapause entry once these flies reach the pupal stage.

Ago-1, which encodes the Ago protein associated with miRNA-mediated silencing, is upregulated more than twofold in diapausing pupae of *S. bullata* (Fig. 2; Reynolds et al., 2013). This differential regulation of

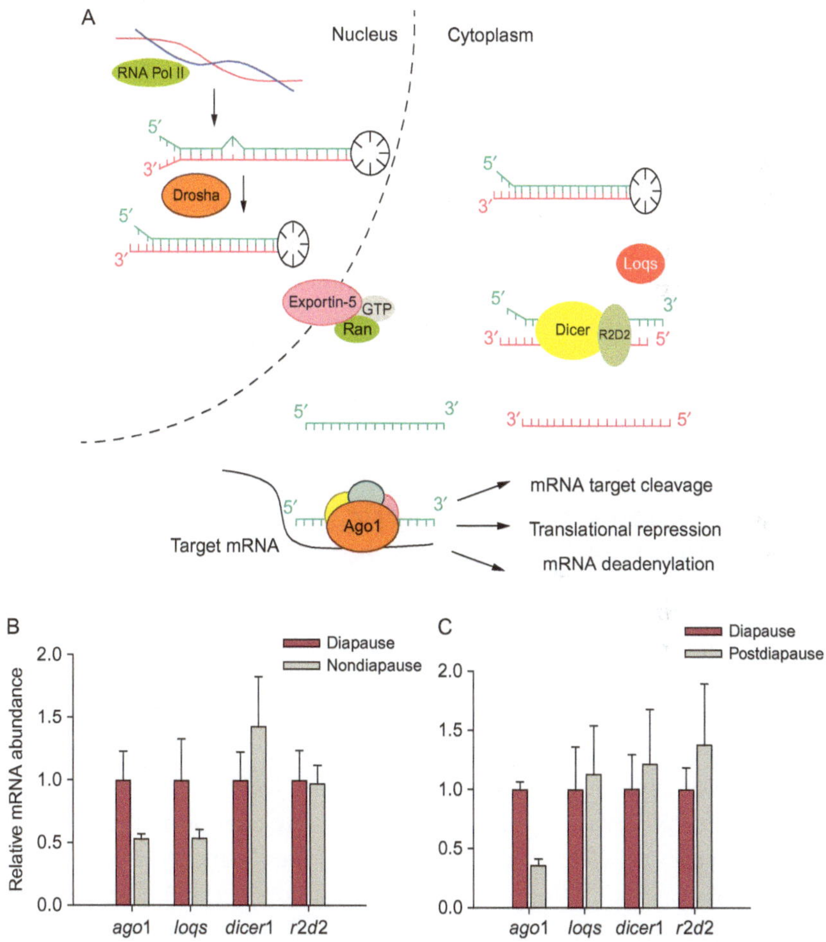

Fig. 2 (A) Generalized microRNA biosynthesis pathway and expression profiles genes encoding pathway components (B) diapause vs nondiapause pupae and (C) diapause vs postdiapause pharate adults (*bottom*). *Adapted from Reynolds, J.A., Clark, J., Diakoff, S.J., Denlinger, D.L., 2013. Transcriptional evidence for small RNA regulation of pupal diapause in the flesh fly,* Sarcophaga bullata. *Insect Biochem. Mol. Biol. 43, 982–989.*

miRNA machinery is accompanied by significant changes in the abundance of 10 miRNAs. Two miRNAs, miR-1-3p and miR-289-5p, were overexpressed in diapausing pupae, while miR-9c-5p, miR-13b-3p, miR-31a-5p, miR-92b-3p, miR-275-3p, miR-276a-5p, miR-277-3p, and miR-305-5p were less abundant in diapausing pupae compared to their nondiapausing counterparts (Reynolds et al., 2017). Putative targets of these differentially regulated miRNAs include numerous diapause-relevant genes and pathways identified in transcriptome and proteome studies with diapausing *Sarcophaga crassipalpis*, a species closely related to *S. bullata* (A.Q. Li et al., 2007; Pavlides et al., 2011; Ragland et al., 2010). KEGG pathways that may be regulated by miRNAs during diapause include DNA replication, glycerolipid metabolism, and pyruvate metabolism (Fig. 3). Gene Ontology defined processes that may be regulated by differentially regulated miRNAs include cellular carbohydrate metabolic process (GO:0044262), defence response (GO:0006952), and nuclear prereplicative process (GO:0005656). Independently, or together, the miRNAs that are differentially regulated in diapausing *S. bullata* have the potential to regulate numerous diapause-relevant pathways, and it will be interesting to see which miRNA:mRNA interactions have confirmed roles in maintaining diapause.

5. EVIDENCE FOR EPIGENETIC REGULATION OF A MATERNAL EFFECT IN *S. bullata*

S. bullata is an established model for studying diapause, and many aspects of the diapause phenotype have been characterized for this species and for *S. crassipalpis*, a close relative (see Denlinger, 2002; Ragland et al., 2010). One aspect that makes *S. bullata* an especially interesting model for studying epigenetic regulation of diapause is a maternal effect that mediates diapause induction in this species (Henrich and Denlinger, 1982; Rockey et al., 1989). Maternal regulation of diapause is common among insects and has been well described for *A. socius* (Huestis and Marshall, 2006), *B. mori* (Yamashita, 1996), *N. vitripennis* (Saunders, 1965), the mosquito *Aedes albopictus* (Poelchau et al., 2011), and others. For most insects, the mother is sensitive to diapause-inducing cues that lead to changes in the maternal provisioning of the eggs. For example, female *A. albopictus* mosquitoes exposed to short-day lengths during the pupal stage produce larger eggs that contain more lipids than females reared in long-day conditions (Reynolds et al., 2012).

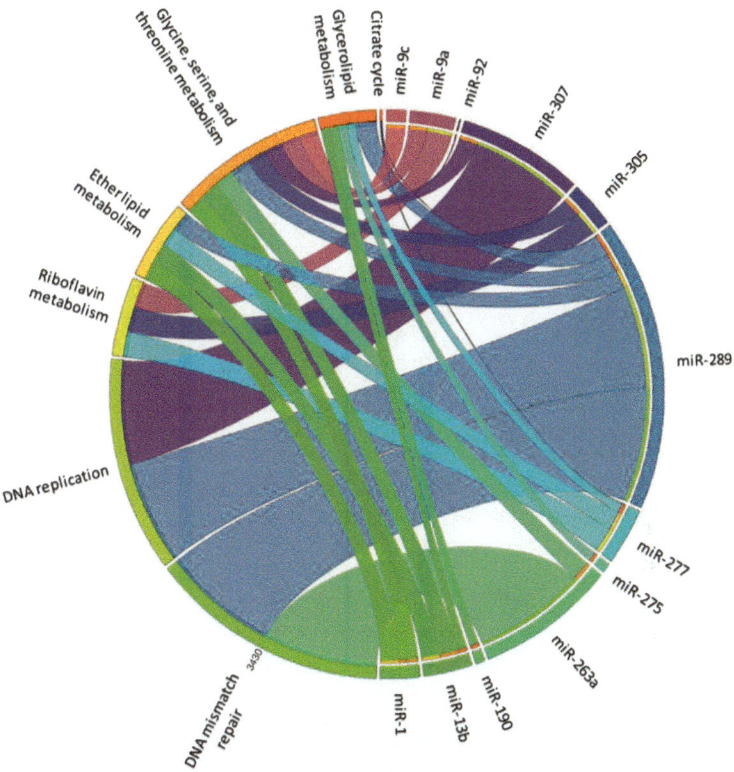

Fig. 3 Circos plot illustrating interactions between miRNAs that are differentially regulated during pupal diapause in *S. bullata* and KEGG pathways enriched in diapausing vs nondiapausing *S. crassipalpis* pupae (Reynolds et al., 2017). Pathways represented include DNA replication (dme03030), DNA mismatch repair (dme03430), glycerolipid metabolism (dme00561), ether lipid metabolism (dme00565), citrate cycle (dme00020), and riboflavin metabolism (dme00740). Ribbons connect pathways and predicted miRNA regulators. Pathway/miRNA interactions were predicted using DIANA miRPath v.3; enriched pathways were identified by Ragland et al. (2010).

The maternal effect in *S. bullata* is unique because in this species the mother's diapause history determines the capacity for diapause in her offspring, and diapause induction is blocked in the progeny of any female who experienced diapause herself (Fig. 4; Henrich and Denlinger, 1982; Rockey et al., 1989). The mechanism that regulates this maternal block of diapause is unknown, but expression of the maternal effect appears to be mediated by factors produced in the brain and/or ovaries of females that experienced diapause sometime during the first 3 days following adult eclosion (Rockey et al., 1989). Some clues about the regulation of this maternal

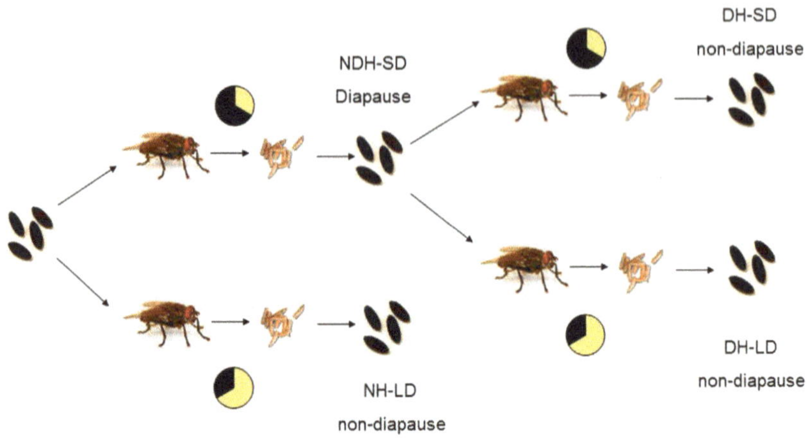

Fig. 4 Developmental timeline for diapause destined and nondiapause *S. bullata*. A maternal effect in this species blocks diapause entry in the progeny of any female that experienced diapause.

effects come from comparing the transcription of candidate genes in first-instar larvae from mothers that did not enter diapause (i.e. have no diapause history, NDH) to first-instar larvae from mothers that experienced diapause (i.e. have a diapause history, DH). Several genes that were upregulated in first-instar larvae from NDH mothers were much less abundant in larvae from DH mothers even though both groups were reared in diapause-inducing conditions. Thus, it appears that some mechanism, possibly a change in chromatin structure, silences genes that need to be activated in order to switch the developmental programme from the direct development to diapause destined. Genes that are upregulated in short day-reared NDH larvae but not in DH larvae encode components of epigenetic pathways and include *hdac1/rpd3*, *hdac3*, *hdac6*, *piwi*, *spin-E*, *ago2*, *r2d2*, *kmd4a*, and *lsd1* (Reynolds et al., 2013, 2016; unpublished data). Additional research is needed to discover the significance of these genes in programming diapause and regulating the maternal effect.

miRNAs may also regulate the maternal block of diapause in *S. bullata* (Reynolds et al., 2017). Three miRNAs (let-7-5p, miR-125-5p, and miR-190-5p) were more abundant in developing red-eye pharate adults that experienced pupal diapause (i.e. have a diapause history) than same stage flies that did not enter diapause. The genes targeted by these miRNAs are likely important for establishing the block of diapause in the next generation. There are also six miRNAs, miR-9c-5p, miR-100-5p, miR-125-5p, let-7, miR-236-5p, and miR-92b-3p, differentially regulated in developing pupae that could not enter diapause because of the maternal effect (i.e. they were progeny of DH mothers) compared to developing pupae that did not enter diapause because of their rearing environment (i.e. were reared in diapause-averting conditions). The functional significance of these miRNAs, as it relates to diapause, has not been tested for this or any species. However, the abundance of let-7, miR-100, and miR-125 is tied to PTTH and ecdysone production, and these miRNAs may regulate diapause entry through these hormones (Chen et al., 2014).

6. EPIGENETIC INFLUENCES ON DORMANCY IN OTHER ANIMALS

The ability to arrest development and enter a state of "suspended animation" at some point in the life cycle is widespread evolutionarily, and some form of dormancy has been observed in all major animal phyla except for Echinodermata (Hand, 1991). The focus of this review has been on diapause in insects, but it is useful to consider epigenetic regulation of dormancy in other animals as well. Specifically, histone modification and small RNA interference have been studied in dormant nematode larvae and in diapausing brine shrimp and fish embryos as reviewed below.

6.1 Epigenetic Regulators of Dauer Formation in *C. elegans*

Many species of nematodes, including *C. elegans* can enter a diapause-like state known as the dauer stage. Like diapause, the dauer stage is an alternative developmental pathway that enhances survival in harsh environments and is characterized by adaptive changes in morphology, behaviour, metabolism, and stress resistance. Unlike diapause, which is a preparatory measure, entering the dauer stage is an immediate response to environmental decline that is triggered by crowding, starvation, and elevated temperatures (Golden and Riddle, 1982, 1984). However, there are some conserved signalling pathways (e.g. insulin signalling) and specific changes in gene expression that mediate both dauer formation and diapause entry (Kimura et al., 1997; Ragland et al., 2010) and indicate types of dormancy utilize a common genetic tool box. Together, these shared features make this well-studied model species useful for developing hypotheses about epigenetic regulation of insect diapause.

Both histone modification and small RNA interference have been implicated in the regulation of dauer formation. Transcriptional evidence for the role of histone modification in dauer arrest comes from a serial analysis of gene expression (SAGE) experiment that identified a significant increase in the transcription of histone H1 in dauer larvae compared to their developing L3 counterparts. Increased histone H1 is thought to increase chromatin packaging and lead to reduced gene expression and/or increased chromatin stability during the dauer stage (Jones et al., 2001). However, there have been no studies that tested this hypothesis. More compelling evidence for histone regulation of *C. elegans* dauer formation comes from a study by Delaney et al. (2017) that illustrates the connection between environmental cues, histone modification, and developmental plasticity. Specifically, these authors showed that histone methyltransferase SET-4, which specifically methylates lysine 20 of histone H4 (H4K20), represses expression of *ins-9*, a gene that encodes an insulin-like peptide found specifically in the ASI sensory neurons that play a central role in detecting the environmental cues that trigger dauer entry. This disruption of the insulin signalling pathway promotes DAF-16/FoxO activity, which is necessary for expression of the dauer phenotype.

Dauer-related changes in histone modifications can persist past the period of dormancy and may leave a "memory" of the altered life history. Specifically, genome-wide levels of methylated and acetylated H3K4 (i.e. indicators of transcriptionally active euchromatin) were reduced in post-dauer adults and correlated with changes in the expression of genes involved in cell cycle regulation, transcription, metabolism, and reproduction. Thus, dauer-related changes of histone modifications in *C. elegans* likely not only mediate short-term survival during environmental insult but also benefit individuals through extended adult lifespan and increased fecundity (Hall et al., 2010, 2013).

Small interfering RNAs, especially miRNAs, have a regulatory role in dauer formation in *C. elegans* and other nematodes. The best-studied miRNA:mRNA interaction includes the let-7 miRNA family, the DAF-12 receptor, and the transcription factor Hunchback-like-1 (*hbl-1*), which regulates developmental timing. DAF-12 receptor and its steroid ligand stimulate the production of let-7 family miRNAs, which repress *hbl-1* expression and allow development to progress from L2 to L3. In the absence of the DAF-12 ligand, DAF-12 receptor prevents let-7 production, allowing expression of *hbl-1* and inducing dauer production. Together, DAF-12, let-7, and *hbl-1* form a hormone-coupled switch that regulates dormancy in *C. elegans* and possibly insects entering diapause as well (Hammell et al., 2009).

6.2 Epigenetic Regulators of Embryonic Diapause in Brine Shrimp

Brine shrimp in the genus *Artemia* live in hypersaline bodies of water (e.g. the Great Salt Lake, Utah, USA). During seasons of favourable environmental conditions, *Artemia* spp. are ovoviviparous and release free-swimming nauplius larvae, but during seasons of unfavourable conditions, *Artemia* spp. produce encysted embryos that enter diapause (Drinkwater and Clegg, 1990; Hand et al., 2016). Diapause in *Artemia* is characterized by severe metabolic depression (Patil et al., 2013; Reynolds and Hand, 2004) and cell cycle arrest (Clegg and Conte, 1980; Drinkwater and Clegg, 1990). Accumulating evidence suggests that cell cycle arrest in diapausing *Artemia* is controlled through histone modifications (Dai et al., 2017; Zhou et al., 2013) and miRNAs (Zhao et al., 2015). Several histone marks including monoacetylation of histone H3 lysine 56 (H3K56ac), trimethylation of histone H3 lysine 9 (H3K9me1), mono- and trimethylation of H3 lysine 79 (H3K79me1 and H3K79me3), and trimethylation of histone H4 lysine 20 (H4K20me3) are all more abundant in diapausing embryos than pre- or postdiapause embryos (Dai et al., 2017; Zhou et al., 2013). The functional significance of these modifications is not clear, but H3K56ac and H4K20me3 may regulate cell cycle and developmental arrest. H3K56ac is known to regulate cell cycle progression in yeast and mammals (Zhou et al., 2013), and in *Artemia* chemical inhibition of Sirtuin1, the HDAC that removes acetyl groups from H3K56, leads to cell cycle and developmental arrest in postdiapause, developing embryos (Zhou et al., 2013). H4K20me3 also regulates cellular quiescence (Jorgensen et al., 2013). Knockdown of Ar-SETD4, histone methyltransferase, reduces the level of H4K20me3 and prevents diapause entry (Dai et al., 2017). Together, these results suggest that multiple histone modifications regulate the cell cycle during diapause.

A recent study identified 107 miRNAs that were differentially expressed in diapausing *Artemia*, and two of these, miR-100 and miR-34, have been implicated as cell cycle regulators in *Artemia* (Zhao et al., 2015). miR-100 targets polo-like kinase 1, a regulator of the mitogen-activated protein kinase kinase–extracellular signal-regulated kinase–ribosomal S6 kinase 2 (MEK–ERK–RSK2) pathway. miR-100, which is downregulated in diapausing embryos, activates the MEK–ERK–RSK2 pathway resulting in cell cycle arrest. Simultaneous upregulation of miR-34 targets cyclin K and suppresses RNA polymerase II activity.

6.3 Histone Modification in Diapausing *Austrofundulus* Embryos

The annual killifish, *Austrofundulus limnaeus*, is one of the best-studied examples of diapause in vertebrates and is becoming a model for studying epigenetic regulation of dormancy. *A. limnaeus* inhabits temporary ponds in Africa and South America and survives the dry season as diapausing embryos embedded in the sediment (Berois et al., 2012; Podrabsky and Hand, 1999, 2000; Wourms, 1972). *A. fundulus* can have three distinct phases of dormancy (DI, DII, DIII). DI and DIII are directly induced by environmental factors, while DII is endogenously regulated (Podrabsky et al., 2010). DII begins approximately 24–26 days postfertilization and is characterized by arrested development at the 20-somite stage, reduced heart rate, depressed metabolism, and increased stress resistance (Podrabsky and Hand, 1999, 2000, 2015). The molecular processes that regulate DII are unknown, but a recent study provides evidence that histone modifications regulate gene expression and developmental arrest in DII embryos (Toni and Padilla, 2016). Phosphorylation of serine 10 of histone 3 (H3S10p), an indicator of gene activity and/or mitotic nuclei (Prigent and Dimitrov, 2003), is reduced in DII embryos and is consistent with the reduced number of mitotic cells in DII embryos (Podrabsky and Culpepper, 2012).

Levels of methylated lysines 4 and 27 on histone H3 (H3K4 and H3K27) also change during diapause. H3K27me1, a positive regulator of gene expression (Ferrari et al., 2014), is decreased in DII embryos, while H3K27me3, which is generally an indicator of transcriptionally repressed chromatin (but see exception in Lu et al., 2013), is increased soon after diapause entry (Toni and Padilla, 2016). Levels of H3K4me1 and H3K4me2 do not change during diapause, but there is a significant reduction H3K4me3, a histone mark found on the promoters of active genes (Lauberth et al., 2013; Santos-Rosa et al., 2004) upon diapause entry. Taken together, these changes in H3S10p, H3K27me, and H3K4me indicate that histone methylation has a significant role regulating gene expression and cell cycle progression in diapausing *A. limnaeus* embryos.

7. CONCLUSIONS AND FUTURE DIRECTIONS

Diapause is an interesting developmental phenomenon. The normal developmental programme comes to a halt and the animal remains in a state of "suspended animation" for an extended period during which cellular structures and physiological homeostasis are preserved. As such, diapause is a power tool for exploring fundamental questions about developmental timing, metabolism, and stress resistance; and understanding the molecular basis of diapause provides novel insight into questions on ageing, obesity, cancer, and disease transmission. In addition, understanding diapause provides information on seasonal biology that will enhance predictive models and improve strategies for managing populations of both beneficial insects (e.g. *M. rotundata* and other pollinators) and medically and/or economically important insect pests (e.g. *A. albopictus*, which may be a vector of the Zika virus, and *H. armigera*, an important agricultural pest).

Our understanding of the molecular basis of diapause has expanded dramatically during the last several years because tools for analysing insect genomes, transcriptomes, and metabolomes have become more accessible. However, the study of epigenetic mechanisms that regulate insect diapause is still in its infancy; and questions about how DNA methylation, histone modifications, and small RNAs influence diapause induction, maintenance, and/or termination are just beginning to be answered.

Most of what we currently know about epigenetic regulation of insect diapause comes from mRNA expression (i.e. transcriptome) studies that have identified changes in transcript abundance for a variety of genes that encode components of histone modification and small RNA pathways. However, the functional relevance of these changes is largely unknown, and additional experiments will be necessary to discover how these processes contribute to developmental arrest, metabolic depression, increased stress resistance, and other features of the diapause phenotype in insects and other animals.

ACKNOWLEDGEMENTS

I would like to thank Dr David Denlinger for helpful suggestions during the preparation of this manuscript. This work was supported in part by National Science Foundation Grant IOS-1354377.

REFERENCES

Amsalem, E., Galbraith, D.G., Cnaani, J., Teal, P.E.A., Grozinger, C.M., 2015. Conservation and modification of genetic and physiological toolkits underpinning diapause in bumble bee queens. Mol. Ecol. 24, 5596–5615.

Ardehali, M.B., Mei, A., Zobeck, K.L., Caron, M., Lis, J.T., Kusch, T., 2011. *Drosophila* Set1 is the major histone H3 lysine 4 trimethyltransferase with role in transcription. EMBO J. 30, 2817–2828.

Barnes, V.L., Bhat, A., Unnikrishnan, A., Heydari, A.R., Arking, R., Pile, L.A., 2014. SIN3 is critical for stress resistance and modulates adult lifespan. Aging 6, 645–660.

Berois, N., Arezo, M.J., Papa, N.G., Clivio, G.A., 2012. Annual fish: developmental adaptations for an extreme environment. Wiley Interdiscip. Rev. Dev. Biol. 1, 595–602.

Bethke, A., Fielenbach, N., Wang, Z., Mangelsdorf, D.J., Antebi, A., 2009. Nuclear hormone receptor regulation of microRNAs controls developmental progression. Science 324, 95–98.

Bewick, A.J., Vogel, K.J., Moore, A.J., Schmitz, R.J., 2017. Evolution of DNA methylation across Insects. Mol. Biol. Evol. 34, 654–665.

Bryon, A., Wybouw, N., Dermauw, W., Tirry, L., van Leeuwen, T., 2013. Genome wide gene-expression analysis of facultative reproductive diapause in the two-spotted spider mite *Tetranychus urticae*. BMC Genomics 14, 815.

Chen, W., Liu, Z., Li, T., Zhang, R., Xue, Y., Zhong, Y., Bai, W., Zhou, D., Zhao, Z., 2014. Regulation of *Drosophila* circadian rhythms by miRNA let-7 is mediated by a regulatory cycle. Nat. Commun. 5, 5549.

Clapier, C.R., Cairns, B.R., 2009. The Biology of chromatin remodeling complexes. Annu. Rev. Biochem. 78, 273–304.

Clegg, J.S., Conte, F.P., 1980. A review of the cellular and developmental biology of Artemia. In: Persoone, G., Sorgeloos, P., Roels, O., Jaspers, E. (Eds.), The Brine Shrimp Artemia. Universa Press, Wetteren, Belgium, pp. 11–54.

Czermin, B., Schotta, G., Hulsmann, B.B., Brehm, A., Becker, P.B., Reuter, G., Imhof, A., 2001. Physical and functional association of SU(VAR)3-9 and HDAC1 in *Drosophila*. EMBO Rep. 2, 915–919.

Dai, L., Ye, S., Li, H.-W., Chen, D.-F., Want, H.-L., Jia, S.-N., Lin, C., Yang, J.-S., Yang, F., Nagasawa, H., Yang, W.-J., 2017. SETD4 regulates cell quiescence and catalyzes the trimethylation of H4K20 during diapause formation in Artemia. Mol. Cell. Biol. 37, e00453. http://dx.doi.org/10.1128/MCB.00453-16.

Deans, C., Maggert, K.A., 2015. What do you mean, "Epigenetic"? Genetics 199, 887–896. http://dx.doi.org/10.1534/genetics.114.173492.

Delaney, C.E., Chen, A.T., Graniel, J.V., Dumas, K.D., Hu, P.J., 2017. A histone H4 lysine 20 methyltransferase couples environmental cues to sensory neuron control of developmental plasticity. Development 144, 1273–1282. http://dx.doi.org/10.1242/dev.145722.

Denlinger, D.L., 1971. Embryonic determination of pupal diapause in the flesh fly *Sarcophaga crassipalpis* Macquart. J. Insect Physiol. 17, 1815–1822.

Denlinger, D.L., 1972. Induction and termination of pupal diapause in *Sarcophaga* (Diptera: Sarcophagidae). Biol. Bull. 142, 11–24.

Denlinger, D.L., 2002. Regulation of diapause. Annu. Rev. Entomol. 47, 93–122.

Denlinger, D.L., Yocum, G.D., Rinehart, J.P., 2005. Hormonal control of diapause. In: Gilbert, L.I., Iatrou, K., Gill, S. (Eds.), Comprehensive Insect Molecular Science. Elsevier, Amsterdam, pp. 615–650.

Di Stefano, L., Ji, J.Y., Moon, N.S., Herr, A., Dyson, N., 2007. Mutation of *Drosophila* Lsd1 disrupts H3-K4 methylation, resulting in tissue-specific defects during development. Curr. Biol. 17, 808–812.

Drinkwater, L.E., Clegg, J.S., 1990. Environmental biology of cyst diapause. In: Browne, R.A., Sorgeloos, P., Trotman, C.A.N. (Eds.), Artemia Biology. CRC Press, Boca Raton, FL, pp. 93–117.

Egi, Y., Tsubouchi, H., Sakamoto, K., 2016. Does DNA methylation play a role in photoperiodic diapause of moths? J. Entomol. Zool. Stud. 4, 458–460.

Ferrari, K.J., Scelfo, A., Jammula, S., Cuomo, A., Barozzi, I., Stützer, A., Fischle, W., Bonaldi, T., Pasini, D., 2014. Polycomb-dependent H3K27me1 and H3K27me2 regulate active transcription and enhancer fidelity. Mol. Cell 53, 49–62.

Flannagan, R.D., Tammariello, S.P., Joplin, K.H., Cikra-Ireland, R.A., Yolcum, G.D., Denlinger, D.L., 1998. Diapause-specific gene expression in pupae of the flesh fly *Sarcophaga crassipalpis*. Proc. Natl. Acad. Sci. U.S.A. 95, 5616–5620.
Flynt, A.X., Thatcher, E.J., Burkewitz, K., Li, N., Liu, Y., Patton, J.G., 2009. miR-8 microRNAs regulate the response to osmotic stress in zebrafish embryos. J. Cell Biol. 185, 115–127.
Gilbert, L.I., Rybczynski, R., Warren, J.T., 2002. Control and biochemical nature of the ecdysteroidogenic pathway. Annu. Rev. Entomol. 47, 883–916.
Golden, J.W., Riddle, D.L., 1982. A pheromone influences larval development in the nematode *Caenorhabditis elegans*. Science 218, 578–580.
Golden, J.W., Riddle, D.L., 1984. The *Caenorhabditis elegans* dauer larvae: developmental effects of pheromone, food, and temperature. Dev. Biol. 102, 368–378.
Hahn, D.A., Denlinger, D.L., 2011. Energetics of insect diapause. Annu. Rev. Entomol. 56, 103–121.
Hall, S.E., Beverly, M., Russ, C., Nusbaum, C., Sengupta, P., 2010. A cellular memory of developmental history generates phenotypic diversity in *C. elegans*. Curr. Biol. 20, 149–155.
Hall, S.E., Chirn, G.W., Lau, N.C., Sengupta, P., 2013. RNAi pathways contribute to developmental history-dependent phenotypic plasticity in *C. elegans*. RNA 19, 306–319.
Hammell, C.M., Karp, X., Ambros, V., 2009. A feedback circuit involving let-7-family miRNAs and DAF-12 integrates environmental signals and developmental timing in *Caenorhabditis elegans*. Proc. Natl. Acad. Sci. U.S.A. 106, 18668–18673.
Hand, S.C., 1991. Metabolic dormancy in aquatic invertebrates. In: Gilles, R. (Ed.), In: Advances in Comparative and Environmental Physiology, vol. 8. Springer-Verlag, New York, USA, pp. 1–50.
Hand, S.C., Menze, M.A., Borcar, A., Patil, Y., Covi, J.A., Reynolds, J.A., Toner, M., 2011. Metabolic restructuring during energy-limited states: insights from *Artemia franciscana* embryos and other animals. J. Insect Physiol. 57, 584–594.
Hand, S.C., Denlinger, D.L., Podrabsky, J.E., Roy, R., 2016. Mechanisms of animal diapause: recent developments from nematodes, crustaceans, insects, and fish. Am. J. Physiol. Regul. Integr. Comp. Physiol. 310, R1193–R1211.
Handler, D., Meixner, K., Pizka, M., Kauss, K., Schmied, C., Gruber, F.S., Brennecke, J., 2013. The genetic makeup of the *Drosophila* piRNA pathway. Mol. Cell 50, 1–16.
Henrich, V.C., Denlinger, D.L., 1982. A maternal effect that eliminates pupal diapause in progeny of the flesh fly *Sarcophaga bullata*. J. Insect Physiol. 28, 881–884.
Hickner, P.V., Mori, A., Zeng, E., Tan, J.C., Severson, D.W., 2015. Whole transcriptome responses among females of the filariasis and arbovirus vector mosquito *Culex pipiens* implicate TGF-β signaling and chromatin modification as key drivers of diapause induction. Funct. Integr. Genomics 15, 439–447.
Huestis, D.L., Marshall, J.L., 2006. Interaction between maternal effects and temperature affects diapause occurrence in the cricket *Allonemobius socius*. Oecologia 146, 513–520.
Jones, S.J., Riddle, D.L., Pouzyrev, A.T., Velculescu, V.E., Hillier, L., Eddy, S.R., Stricklin, S.L., Baillie, D.L., Waterston, R., Marra, M.A., 2001. Changes in gene expression associated with developmental arrest and longevity in *Caenorhabditis elegans*. Genome Res. 11, 1346–1352.
Jørgensen, S., Schotta, G., Sørensen, C.S., 2013. Histone H4 lysine 20 methylation: key player in epigenetic regulation of genomic integrity. Nucleic Acids Res. 41, 2797–2806.
Juliano, C., Wang, J., Lin, H., 2011. Proteins and the piRNA pathway in diverse organisms. Annu. Rev. Genet. 45, 447–469.

Katz, D.J., Edwards, T.M., Reinke, V., Kelly, W.G., 2009. A *C. elegans* LSD1 demethylase contributes to germline immortality by reprogramming epigenetic memory. Cell 137, 308–320.

Kemp, C., Mueller, S., Goto, A., Barbier, V., Paro, S., Bonnay, F., Dostert, C., Troxler, L., Hetru, C., Meignin, C., Pfeffer, S., Hoffmann, J.A., Imler, J.-L., 2013. Broad RNA interference-mediated antiviral immunity and virus-specific inducible responses in *Drosophila*. J. Immunol. 190, 650–658.

Kennerdell, J.R., Yamaguchi, S., Carthew, R.W., 2002. RNAi is activated during *Drosophila* oocyte maturation in a manner dependent on aubergine and spindle-E. Genes Dev. 16, 1884–1889.

Kimura, K.D., Tissenbaum, H.A., Liu, Y., Ruvkun, G., 1997. daf-2, an insulin receptor-like gene that regulates longevity and diapause in *Caenorhabditis elegans*. Science 277, 942–946.

Klattenhoff, C., Bratu, D.P., McGinnis-Schultz, N., Koppetsch, B.S., Cook, H.A., Theurkauf, W.E., 2007. Drosophila rasiRNA pathway mutations disrupt embryonic axis specification through activation of an ATR/Chk2 DNA damage response. Dev. Cell 12, 45–55.

Koštál, V., 2006. Eco-physiological phases of insect diapause. J. Insect Physiol. 52, 113–127.

Koštál, V., Noguchi, H., Shimada, K., Hayakawa, Y., 2000a. Circadian component influences the photoperiodic induction of diapause in a drosophilid fly, *Chymomyza costata*. J. Insect Physiol. 46, 887–896.

Koštál, V., Shimada, K., Hayakawa, Y., 2000b. Induction and development of winter larval diapause in a drosophilid fly, *Chymomyza costata*. J. Insect Physiol. 46, 417–428.

Kouzarides, T., 2007. Chromatin modifications and their function. Cell 128, 693–705.

Kubrak, O.I., Kucerova, L., Theopold, U., Nassel, D.R., 2014. The sleeping beauty: how reproductive diapause affects hormone signaling, metabolism, immune response and somatic maintenance in *Drosophila melanogaster*. PLoS One 9, e113051.

Latham, J.A., Dent, S.Y., 2007. Cross-regulation of histone modifications. Nat. Struct. Mol. Biol. 14, 1017–1024.

Lauberth, S.M., Nakayama, T., Wu, X., Ferris, A.L., Tang, Z., Hughes, S.H., Roeder, R.G., 2013. H3K4me3 interactions with TAF3 regulate preinitiation complex assembly and selective gene activation. Cell 152, 1021–1036.

Lechner, M.S., Levitan, I., Dressler, G.R., 2000. PTIP, a novel BRCT domain-containing protein interacts with Pax2 and is associated with active chromatin. Nucleic Acids Res. 28, 2741–2751.

Lees, A.D., 1955. Physiology of Diapause in Arthropods. Cambridge University Press, Cambridge, UK.

Lehmann, P., Pruisscher, P., Posledovich, D., Carlsson, M., Käkelä, R., Tang, P., Nylin, S., Wheat, C.W., Wiklund, C., Gotthard, K., 2016. Energy and lipid metabolism during direct and diapause development in a pierid butterfly. J. Exp. Biol. 219, 3049–3060.

Li, A.Q., Popova-Butler, A., Dean, D.H., Denlinger, D.L., 2007. Proteomics of the flesh fly brain reveals an abundance of upregulated heat shock proteins during pupal diapause. J. Insect Physiol 53, 385–391.

Li, B., Carey, M., Workman, J.L., 2007. The role of chromatin during transcription. Cell 128, 707–719.

Liu, M., Pile, L.A., 2017. The transcriptional corepressor SIN3 directly regulates genes involved in methionine catabolism and affects histone methylation, linking epigenetics and metabolism. J. Biol. Chem. 292, 1970–1976.

Liu, Y., Ye, X., Jiang, F., Liang, C., Chen, D., Peng, J., Kinch, L.N., Grishin, N.V., Liu, Q., 2009. C3PO, an endoribonuclease that promotes RNAi by facilitating RISC activation. Science 325, 750–753.

Lu, Y.-X., Denlinger, D.L., Xu, W.-H., 2013. Polycomb complex 2 (PRC2) protein ESC regulates insect developmental timing by mediating H3K27me3 and activating prothoracicotropic hormone gene expression. J. Biol. Chem. 288, 23554–23564.

Meister, P., Schott, S., Bedet, C., Xiao, Y., Rohner, S., Bodennec, S., Hudry, B., Molin, L., Solari, F., Gasser, S.M., Palladino, F., 2011. *Caenorhabditis elegans* heterochromatin protein 1 (HPL-2) links developmental plasticity, longevity and lipid metabolism. Genome Biol. 12, R123.

McCue, A.D., Slotkin, R.K., 2012. Transposable element small RNAs as regulators of gene expression. Trends Genet. 28, 616–623.

Mohan, M., Herz, H.M., Smith, E.R., Zhang, Y., Jackson, J., Washburn, M.P., Florens, L., Eissenberg, J.C., Shilatifard, A., 2011. The COMPASS family of H3K4 methylases in *Drosophila*. Mol. Cell. Biol. 31, 4310–4318.

Olivier, D., Sykora, M.M., Sachidanandam, R., Mechtler, K., Brennecke, J., 2010. An *in vivo* RNAi assay identifies major genetic and cellular requirements for primary piRNA biogenesis in *Drosophila*. EMBO J. 29, 3301–3317.

Patil, Y., Marden, B., Brand, M.D., Hand, S.C., 2013. Hand metabolic downregulation and inhibition of carbohydrate catabolism during diapause in embryos of *Artemia franciscana*. Physiol. Biochem. Zool. 86, 106–118.

Pavlides, S.C., Pavlides, S.A., Tammariello, S.P., 2011. Proteomic and phosphoproteomic profiling during diapause entrance in the flesh fly, *Sarcophaga crassipalpis*. J. Insect Physiol. 57, 635–644.

Piacentini, L., Fanti, L., Negri, R., Del Vescovo, V., Fatica, A., Altieri, F., Pimpinelli, S., 2009. Heterochromatin protein 1 (HP1a) positively regulates euchromatic gene expression through RNA transcript association and interaction with hnRNPs in *Drosophila*. PLoS Genet. 5, e1000670.

Pegoraro, M., Bafna, A., Davies, N.J., Shuker, D.M., Tauber, E., 2016. DNA methylation changes induced by long and short photoperiods in *Nasonia*. Genome Res. 26, 203–210.

Podrabsky, J.E., Culpepper, K.M., 2012. Cell cycle regulation during development and dormancy in embryos of the annual killifish *Austrofundulus limnaeus*. Cell Cycle 11, 1697–1704.

Podrabsky, J.E., Hand, S.C., 1999. The bioenergetics of embryonic diapause in an annual killifish, *Austrofundulus limnaeus*. J. Exp. Biol. 202, 2567–2580.

Podrabsky, J.E., Hand, S.C., 2000. Depression of protein synthesis during diapause in embryos of the annual killifish *Austrofundulus limnaeus*. Physiol. Biochem. Zool. 73, 799–808.

Podrabsky, J.E., Hand, S.C., 2015. Physiological strategies during animal diapause: lessons from brine shrimp and annual killifish. J. Exp. Biol 218, 1897–1906.

Podrabsky, J.E., Garrett, I.D.F., Kohl, Z.F., 2010. Alternative developmental pathways associated with diapause regulated by temperature and maternal influences in embryos of the annual killifish *Austrofundulus limnaeus*. J. Exp. Biol. 213, 3280–3288.

Poelchau, M.F., Reynolds, J.A., Denlinger, D.L., Elsik, C.G., Armbruster, P.A., 2011. A *de novo* transcriptome of the Asian tiger mosquito, *Aedes albopictus*, to identify candidate transcripts for diapause preparation. BMC Genomics 12, 619.

Poelchau, M.F., Reynolds, J.A., Elsik, C.G., Denlinger, D.L., Armbruster, P.A., 2013a. Deep sequencing reveals complex mechanisms of diapause preparation in the invasive mosquito, *Aedes albopictus*. Proc. Biol. Soc. 280, 20130143.

Poelchau, M.F., Reynolds, J.A., Elsik, C.G., Denlinger, D.L., Armbruster, P.A., 2013b. RNA-Seq reveals early distinctions and late convergence of gene expression between diapause and quiescence in the Asian tiger mosquito, *Aedes albopictus*. J. Exp. Biol. 216, 4082–4090.

Poupardin, R., Schöttner, K., Korbelová, J., Provazník, J., Doležel, D., Pavlinic, D., Beneš, V., Koštál, V., 2015. Early transcriptional events linked to induction of diapause revealed by RNAseq in larvae of drosophilid fly, *Chymomyza costata*. BMC Genomics 16, 720.

Prigent, C., Dimitrov, S., 2003. Phosphorylation of serine 10 in histone H3, what for? J. Cell Sci. 116, 3677–3685.

Qi, D., Haining, J., Lilia, T., Mannervick, M., 2006. *Drosophila* Reptin and other TIP60 complex components promote generation of silent chromatin. Genetics 174, 241–251.

Ragland, G.J., Denlinger, D.L., Hahn, D.A., 2010. Mechanisms of suspended animation are revealed by transcript profiling in the flesh fly. Proc. Natl. Acad. Sci. U.S.A. 107, 14909–14914.

Ragland, G.J., Egan, S.P., Feder, J.L., Berlocher, S.H., Hahn, D.A., 2011. Developmental trajectories of gene expression reveal candidates for diapause termination: a key life-history transition in the apple maggot fly *Rhagoletis pomonella*. J. Exp. Biol. 214, 3948–3959.

Resnick, T.D., McCullouch, K.A., Rougvie, A.E., 2010. miRNAs give worms the time of their lives: small RNAs and temporal control in *Caenorhabditis elegans*. Dev. Dyn. 239, 1477–1489.

Reynolds, J.A., Hand, S.C., 2004. Differences in isolated mitochondria are insufficient to account for respiratory depression during diapause in *Artemia franciscana* embryos. Physiol. Biochem. Zool. 77, 366–377.

Reynolds, J.A., Hand, S.C., 2009a. Decoupling development and energy flow during embryonic diapause in the cricket, *Allonemobius socius*. J. Exp. Biol. 212, 2064–2073.

Reynolds, J.A., Hand, S.C., 2009b. Embryonic diapause highlighted by differential expression of mRNAs for ecdysteroidogenesis, transcription and lipid sparing in the cricket *Allonemobius socius*. J. Exp. Biol. 212, 2074–2083.

Reynolds, J.A., Poelchau, M.F., Rahman, Z., Armbruster, P.A., Denlinger, D.L., 2012. Transcript profiling reveals mechanisms for lipid conservation during diapause in the mosquito, *Aedes albopictus*. J. Insect Physiol. 58, 966–973.

Reynolds, J.A., Clark, J., Diakoff, S.J., Denlinger, D.L., 2013. Transcriptional evidence for small RNA regulation of pupal diapause in the flesh fly, *Sarcophaga bullata*. Insect Biochem. Mol. Biol. 43, 982–989.

Reynolds, J.A., Bautista-Jiménez, R., Denlinger, D.L., 2016. Changes in histone acetylation as potential mediators of pupal diapause in the flesh fly, *Sarcophaga bullata*. Insect Biochem. Mol. Biol. 76, 29–37.

Reynolds, J.A., Peyton, J.T., Denlinger, D.L., 2017. Changes in microRNA abundance may regulate diapause in the flesh fly *Sarcophaga bullata*. Insect Biochem. Mol. Biol. 84, 1–14. http://dx.doi.org/10.1016/j.ibmb.2017.03.002.

Rockey, S.J., Miller, B.B., Denlinger, D.L., 1989. A diapause maternal effect in the flesh fly, *Sarcophaga bullata*: transfer of information from mother to progeny. J. Insect Physiol. 35, 553–558.

Rogina, B., Helfand, S.L., 2004. Sir2 mediates longevity in the fly through a pathway related to calorie restriction. Proc. Natl. Acad. Sci. U.S.A. 101, 15998–16003.

Santos-Rosa, H., Bannister, A.J., Dehe, P.M., Geli, V., Kouzarides, T., 2004. Methylation of H3 lysine 4 at euchromatin promotes Sir3p association with heterochromatin. J. Biol. Chem. 279, 47506–47512.

Sasibhushan, S., Rao, C.G.P., Ponnuvel, K.M., 2013. Genome wide microarray based expression profiles during early embryogenesis in diapause induced and non-diapause eggs of polyvoltine silkworm *Bombyx mori*. Genomics 102, 379–387.

Saunders, D.S., 1965. Larval diapause of maternal origin: induction of diapause in *Nasonia vitripennis* (Walk.) (Hymenoptera: Pteromalidae). J. Exp. Biol. 42, 495–508.

Schwartz, Y.B., Kahn, T.G., Nix, D.A., Li, X.Y., Bourgon, R., Biggin, M., Pirrotta, V., 2006. Genome-wide analysis of polycomb targets in *Drosophila melanogaster*. Nat. Genet. 38, 700–705.

Sim, C., Denlinger, D.L., 2008. Insulin signaling and FOXO regulate the overwintering diapause of the mosquito *Culex pipiens*. Proc. Natl. Acad. Sci. U.S.A. 105, 6777–6781.

Sim, C., Kang, D.S., Kim, S., Bai, X., Denlinger, D.L., 2015. Identification of FOXO targets that generate diverse features of the diapause phenotype in the mosquito *Culex pipiens*. Proc. Natl. Acad. Sci. U.S.A. 112, 3811–3816.

Simboeck, E., Gutierrez, A., Cozzuto, L., Beringer, M., Caizzi, L., Keyes, W.M., Di Croce, L., 2013. DPY30 regulates pathways in cellular senescence through ID protein expression. EMBO J. 32, 2217–2230.

Spielman, A., Wong, J., 1973. Environmental control of ovarian diapause in *Culex pipiens*. Ann. Entomol. Soc. Am. 66, 905–907.

Sun, F., Fu, H., Liu, Q., Tie, Y., Zhu, J., Xing, R., Sun, Z., Zheng, X., 2008. Downregulation of CCND1 and CDK6 by mir-34a induces cell cycle arrest. FEBS Lett. 582, 1564–1568.

Swaminathan, A., Pile, L.A., 2010. Regulation of cell proliferation and wing development by *Drosophila* SIN3 and String. Mech. Dev. 127, 96–106.

Taliaferro, J.M., Aspden, J.L., Bradley, T., Marwha, D., Blanchette, M., Rio, D.C., 2013. Two new and distinct roles for Drosophila Argonaute-2 in the nucleus: alternative pre-mRNA splicing and transcriptional repression. Genes Dev. 27, 378–389.

Tauber, M.J., Tauber, C.A., 1976. Insect seasonality: diapause maintenance, termination, and post-diapause development. Annu. Rev. Entomol. 21, 81–107.

Tauber, M.J., Tauber, C.A., Masaki, S., 1986. Seasonal Adaptations of Insects. Oxford University Press, Oxford, UK.

Teleman, A.A., Maitra, S., Cohen, S.M., 2006. *Drosophila* lacking miRNA miR-278 are defective in energy homeostasis. Genes Dev. 20, 417–422.

Thompson, T., Lin, H., 2009. The biogenesis and function of PIWI proteins and piRNAs: progress and prospect. Annu. Rev. Cell Dev. Biol. 25, 355–376.

Tissenbaum, H.A., Guarente, L., 2001. Increased dosage of a sir-2 gene extends lifespan in *Caenorhabditis elegans*. Nature 410, 227–230.

Tolhuis, B., de Wit, E., Muijrers, I., Teunissen, H., Talhout, W., van Steensel, B., van Lohuizen, M., 2006. Genome-wide profiling of PRC1 and PRC2 polycomb chromatin binding in *Drosophila melanogaster*. Nat. Genet. 38, 694–699.

Tomari, Y., Du, T., Zamore, P.D., 2007. Sorting of *Drosophila* small silencing RNAs. Cell 130, 299–308.

Toni, L.S., Padilla, P.A., 2016. Developmentally arrested *Austrofundulus limnaeus* embryos have changes in post-translational modifications of histone H3. Journal of Experimental Biology 219, 544–552.

Tu, X., Wang, J., Hao, K., Whitman, D.W., Fan, Y., Cao, G., Zhang, Z., 2015. Transcriptomic and proteomic analysis of pre-diapause and non-diapause eggs of migratory locust, *Locusta migratoria* L. (Orthoptera: Acridoidea). Sci. Rep. 5, 11402.

Wadsworth, C.B., Dopman, E.B., 2015. Transcriptome profiling reveals mechanisms for the evolution of insect seasonality. J. Exp. Biol. 218, 3611–3622.

Werren, J.H., Richards, S., Desjardins, C.A., Niehuis, O., Gadau, J., Colbourne, J.K., Nasonia Genome Working Group, 2010. Functional and evolutionary insights from the genomes of three parasitoid *Nasonia* species. Science 327, 343–348.

Wood, J.G., Jones, B.C., Jiang, N., Chang, C., Hosier, S., Wickremesinghe, P., Garcia, M., Hartnett, D.A., Burhenn, L., Neretti, N., Helfand, S.L., 2016. Chromatin-modifying genetic interventions suppress age-associated transposable element activation and extend life span in *Drosophila*. Proc. Natl. Acad. Sci. U.S.A. 113, 11277–11282.

Wourms, J.P., 1972. The developmental biology of annual fishes. III. Preembryonic and embryonic diapause of variable duration in the eggs of annual fishes. J. Exp. Zool. 182, 389–414.

Xiang, H., Zhu, J., Chen, Q., Dai, F., Li, X., Li, M., Zhang, H., Zhang, G., Li, D., Dong, Y., Zhao, L., Lin, Y., Cheng, D., Yu, J., Sun, J., Zhou, X., Ma, K., He, Y., Zhao, Y., Guo, S., Ye, M., Guo, G., Li, Y., Li, R., Zhang, X., Ma, L., Kristiansen, K., Guo, Q., Jiang, J., Beck, S., Xia, Q., Wang, W., Wang, J., 2010. Single base-resolution methylome of the silkworm reveals a sparse epigenomic map. Nat. Biotechnol. 28, 516–520.

Xu, P., Vernooy, S.Y., Guo, M., Hay, B.A., 2003. The *Drosophila* microRNA mir-14 suppresses cell death and is required for normal fat metabolism. Curr. Biol. 13, 790–795.

Yamashita, O., 1996. Diapause hormone of the silkworm, *Bombyx mori*: structure, gene expression and function. J. Insect Physiol. 42, 669–679.

Yin, H., Lin, H., 2007. An epigenetic activation role of Piwi and a Piwi-associated piRNA in *Drosophila melanogaster*. Nature 450, 304–308.

Yocum, G.D., Rinehart, J.P., Horvath, D.P., Kemp, W.P., Bosch, J., Alroobi, R., Salem, S., 2015. Key molecular processes of the diapause to post-diapause quiescence transition in the alfalfa leafcutting bee *Megachile rotundata* identified by comparative transcriptome analysis. Physiol. Entomol. 40, 103–112.

Zhang, X., Zabinsky, R., Teng, Y., Cui, M., Han, M., 2011. microRNAs play critical roles in the survival and recovery of *Caenorhabditis elegans* from starvation-induced L1 diapause. Proc. Natl. Acad. Sci. U.S.A. 108, 17997–18002.

Zhao, L.-L., Jin, F., Ye, X., Zhu, L., Yang, J.-S., Yang, W.-J., 2015. Expression profiles of miRNAs and involvement of miR-100 and miR-34 in regulation of cell cycle arrest in *Artemia*. Biochem. J. 470, 223–231.

Zhou, G., Miesfeld, R.L., 2009. Energy metabolism during diapause in *Culex pipiens* mosquitoes. J. Insect Physiol. 55, 40–46.

Zhou, R., Yang, F., Chen, D.F., Sun, Y.X., Yang, J.S., Yang, W.J., 2013. Acetylation of chromatin-associated histone H3 lysine 56 inhibits the development of encysted *Artemia* embryos. PLoS One 8, e68374.

CHAPTER SIX

The Impact of Parasites on Host Insect Epigenetics

Andreas Vilcinskas
Institute for Insect Biotechnology, Justus Liebig University of Giessen, Giessen, Germany
Fraunhofer Institute for Molecular Biology and Applied Ecology, Giessen, Germany

Contents

Abstract

The epigenetic regulation of gene expression has been recognized as an alternative to genetic mutation for the conversion of environmental stressors, such as the presence of parasites, into heritable phenotypic changes. The activation of immune responses in insects involves the orchestrated expression of genes that mediate pathogen recognition and immunity-related signalling, which in turn induces the synthesis of effector molecules such as antimicrobial peptides. This transcriptional reprogramming is regulated by DNA methylation and the acetylation and deacetylation of histones, which operate before transcription begins, and by microRNAs, which control protein synthesis posttranscriptionally. Therefore, research on epigenetic regulation at the molecular, cellular, tissue, organ and organism levels in insects represents an emerging field to understand the regulation of complex parameters such as longevity, fecundity and reproduction. Microbial parasites and parasitic wasps have evolved strategies to interfere with the epigenetic mechanisms of their host insects to favour their own development. The impact of parasites on host insect epigenetics is a growing area of research because it influences phenomena such as host–parasite coevolution, the transmission of diseases by vector insects, and the control of endogenous symbionts. This research field could also pave the way for the development of novel strategies to control pest and vector insects.

Advances in Insect Physiology, Volume 53
ISSN 0065-2806
http://dx.doi.org/10.1016/bs.aiip.2017.05.001

1. INTRODUCTION

DNA sequence mutations and chromosomal alterations that cause heritable phenotypic variation are thought to be the principal molecular mechanisms underlying Charles Darwin's theory of evolution by natural selection, even though most environmental factors do not directly affect the DNA sequence (Laland et al., 2014). More recently, epigenetic regulation has been proposed as an alternative mechanism to convert environmental stimuli into heritable phenotypic changes (Bonduriansky et al., 2012; Jaenisch and Bird, 2003; Skinner, 2015). Epigenetic mechanisms regulate development, response to the environment and in natural variation of gene expression levels. There is growing appreciation that abiotic stressors such as toxins, mild heat shock and starvation, as well as biotic stressors such as infections, cause the transcriptional reprogramming of genes encoding stress or immunity-related proteins in insects (Freitak et al., 2012; Vandegehuchte and Janssen, 2014). It is well established that the chromatin landscape controls the expression of inducible genes, especially immunity-related genes, in vertebrates (Lim et al., 2013). This review addresses two questions: first, whether immunity-related genes in insects are also regulated by epigenetic mechanisms, and second, whether pathogens, parasites and parasitoids can influence the immune system of their insect hosts by interfering with the epigenetic machinery. An increasing body of evidence supports the hypothesis that epigenetic mechanisms such as DNA methylation, histone acetylation/deacetylation and microRNAs (miRNAs) control complex parameters such as development, longevity, reproduction, gender-specific phenotypic variation and immunity in insects (Bingsohn et al., 2016; Lemos et al., 2010; Mukherjee et al., 2015; Peleg et al., 2016; Reynolds et al., 2016). The coordinated actions of these multifaceted epigenetic regulations determine cell development, cell cycle regulation, cell state and cell fate (Chen et al., 2017). Therefore, research on epigenetic regulation at the molecular, cellular, tissue and organ levels in insects represents an emerging field. The epigenetic landscape is also thought to influence the transgenerational transcriptional reprogramming associated with shifts in host resistance during host–parasite coevolution, although it is not clear how such epigenetic information is passed from the parents to the offspring (Cheeseman and Weitzman, 2015; Pigeault et al., 2016; Vilcinskas, 2016).

The effect of parasites on host insect epigenetic regulation has emerged as a promising new field of research with many potential applications. For example, the epigenetics of vector insects such as the mosquitoes *Aedes aegypti* and *Anopheles gambiae* is an attractive topic because DNA methylation, histone acetylation/deacetylation and miRNAs influence the susceptibility of these species to parasites and hence the success of transmission (Hussain et al., 2011; Sharakhov and Sharakhova, 2015; Ye et al., 2013). The epigenetic interaction between insects and their beneficial symbionts, such as those harboured by aphids, is also attracting more attention (Kim et al., 2016). Understanding of the epigenetic regulation of gene expression in host insects could facilitate the development of novel approaches to control pest and vector insects (Will and Vilcinskas, 2013). The translation of basic research in insect epigenetics into applications for plant protection and control of vector insects has become a promising discipline in insect biotechnology.

2. INSECT IMMUNITY

Insects are relatively well protected by their exoskeleton, which is constructed from a mixture of proteins and chitin. Only parasitic fungi such as those of the genera *Beauveria* and *Metarhizium* are equipped with enzymes that can degrade this tough matrix, allowing them to enter insect hosts directly through the integument (Butt et al., 2016; Gillespie et al., 2000). The insect immune system is based on well-characterized layers of innate immunity, and unlike the vertebrate immune system it lacks the adaptive capabilities provided by antibodies and memory cells (Buchmann, 2014). Insects produce a variety of proteins that recognize microbial cell wall components such as peptidoglycans, lipopolysaccharides and β(1,3)-glucans (Steiner, 2004). The binding of microbe-derived molecules to the corresponding host receptors triggers immunity-related signalling cascades such as the Toll, Jak/STAT, IMD and JNK pathways, which control the expression of immunity-related effector molecules (Boutros et al., 2002; Buchon et al., 2014; Hoffmann and Reichhart, 2002; Kaneko and Silverman, 2005).

Cellular immunity in insects involves the phagocytosis of microbial propagules by hemocytes such as plasmatocytes or granular cells circulating in the haemocoel. Multicellular encapsulation and nodule formation occurs if the parasites are too numerous or too large for ingestion by a single haemocyte. The nodules are aggregates of haemocytes that form a multicellular envelope to surround the entrapped parasites, which are often melanized following the activation of the phenoloxidase cascade (Lavine and

Strand, 2002; Parsons and Foley, 2016). Humoral immunity in insects involves the activation of the phenoloxidase cascade to regulate melanization as well as the induced synthesis of proteinase inhibitors and antimicrobial peptides (AMPs). The latter can be classified according to their target range (e.g. antibacterial or antifungal peptides) and according to their structure and properties (Tonk and Vilcinskas, 2017). Many insect AMPs have been identified, some of which are evolutionarily conserved and widely distributed (such as the defensins), whereas other are restricted to particular taxa. For example, coleoptericins are found only in the Coleoptera, attacins, gallerimycins, gloverins and moricins only in the Lepidoptera, and drosomycins and metchnikowins only in the Diptera (Mylonakis et al., 2016). A comparative analysis of insect genomes and immunity-related transcriptomes provides evidence for rapid functional shifts as well as the gain and loss of AMP genes, highlighting the evolutionary plasticity of insect immunity (Vilcinskas, 2013). Insects often synthesize multiple AMPs simultaneously during an immune response resulting in synergistic antimicrobial activity (Rahnamaeian et al., 2015), whereas other AMPs are expressed in a context-specific manner (Jacobs et al., 2016). The complex regulation of individual AMPs depending on environmental influences requires sophisticated regulatory mechanisms. In vertebrates, it is well established that the adaptive immune system is under epigenetic control (Busslinger and Tarakhovsky, 2014). This review considers emerging data, suggesting parasites affect the immune system of infected insect hosts by interfering with multiple epigenetic processes that govern susceptibility and resistance, reflecting the transcriptional reprogramming of immunity-related genes, particularly those encoding AMPs (Dubovskiy et al., 2013, 2016).

3. HISTONE ACETYLATION AND DEACETYLATION

The chromosomes in the nucleus of eukaryotic cells are composed of chromatin, a combination of DNA and proteins organized as repeated elements known as nucleosomes, which are in turn sequestered into higher-order structures. Nucleosomes comprise an octamer of core histone proteins (two each of H2A, H2B, H3 and H4) around which the DNA is wrapped. Nucleosomes are joined like beads on a string by linker DNA paired with linker histones (H1 and H5). The N-terminal tail of each histone protein contains certain amino acids that are sites of posttranslational regulation. The positive charge of the core histones determines the consistency of the chromatin and thereby regulates access by transcription factors (Adcock and Lee, 2006; Bayarsaihan, 2011). The condensed form of chromatin, in which the histones and DNA are tightly packed, is transcriptionally inactive, whereas open chromatin is a looser assembly that allows transcription factors to bind DNA and thus facilities gene expression (Fig. 1). The transition from condensed to open chromatin is regulated by the addition or removal of acetyl groups to the core histones, and this process is controlled by two enzyme complexes with opposing activities. The addition of acetyl groups to histones is catalysed by histone acetyltransferases (HATs), which increase DNA accessibility and promote transcription, whereas the removal of acetyl groups from histones is mediated by histone deacetylases (HDACs), which limit access to the DNA and thus silence gene expression (Marks et al. 2003) (Fig. 1). The acetylation and deacetylation of histones is, therefore, considered as a reversible epigenetic mechanism operating before transcription is initiated (Hamon and Cossart, 2008). Despite the fundamental basis of this epigenetic mechanism, little has been published about its role in insects. Three of the four classes of HDACs found in humans are also found in the fruit fly *Drosophila melanogaster*. A recent study elucidated a pathway connecting changes of intermediate metabolism during aging with the chromatin-mediated regulation of transcription and changes in the activity of associated enzymes that modulate organismal life span. The H4 K12-specific HAT plays a crucial role in this context (Peleg et al., 2016).

The balanced upregulation of HATs and HDACs during development, particularly at the onset of metamorphosis when transcriptional reprogramming is associated with the reorganization of tissues and organs, was first reported in the greater wax moth *Galleria mellonella* (Mukherjee et al., 2012) and has also been demonstrated in the flesh fly *Sarcophaga bullata* (Reynolds et al., 2016). In *G. mellonella*, injecting the larvae with HAT inhibitors (e.g. suberonylanilide hydroxamic acid) caused an early transition to the pupal phase compared with untreated controls, whereas injecting them with HDAC inhibitors (e.g. sodium butyrate) delayed pupation. Sodium butyrate inhibits HDAC classes 1 and 2a and can selectively modify all nucleosomal histones (Dokmanovic et al., 2007). Inhibiting histone deacetylation with sodium butyrate in the honey bee *Apis mellifera* was recently shown to induce genes responsible for detoxification and immunity, thus enhancing resistance against both the insecticide imidacloprid and the microsporidian parasite *Nosema ceranae* (Hu et al., 2017). Recent studies using human epithelial cells suggest that the regulation of AMP expression by inhibiting histone deacetylases is conserved in insects and vertebrates (Fischer et al., 2016).

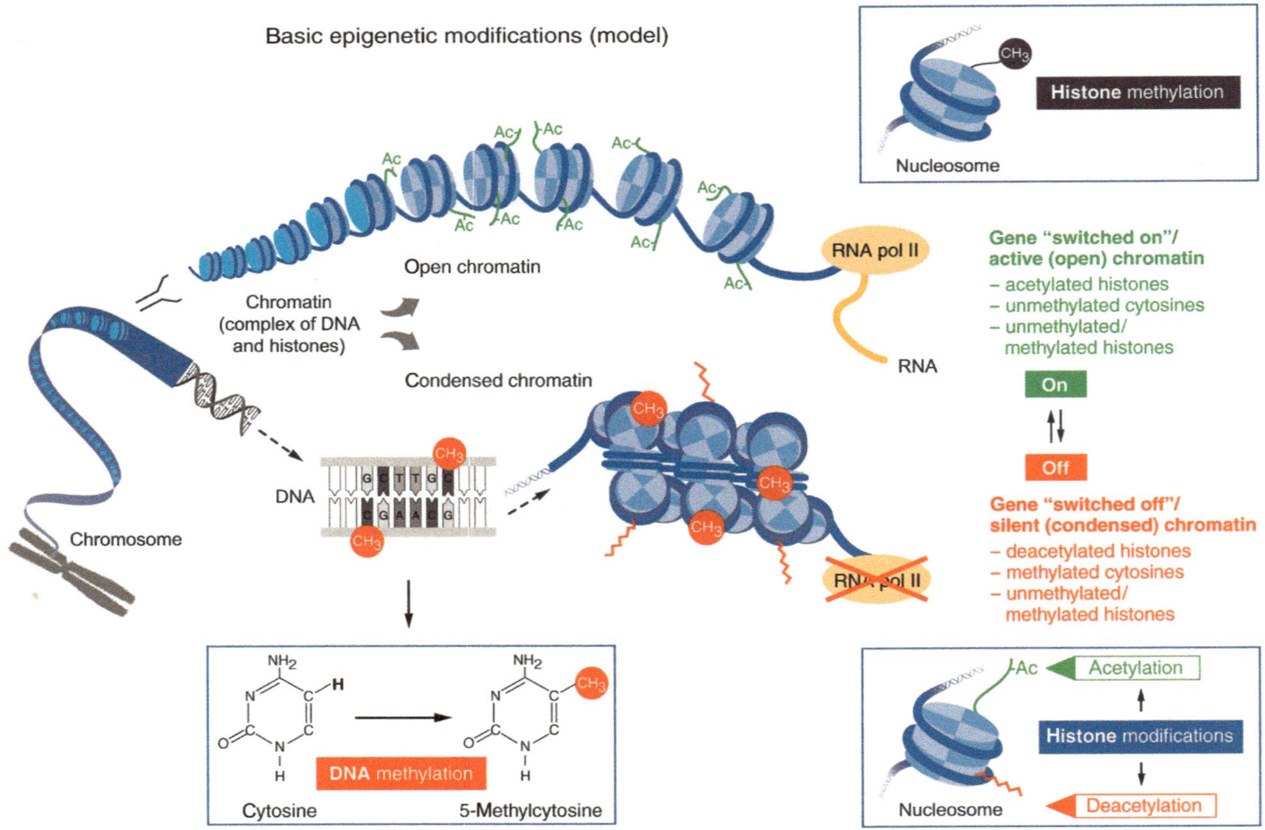

Fig. 1 See legend on opposite page.

4. IMPACT OF PARASITES ON HISTONE ACETYLATION AND DEACETYLATION IN THE HOST

Infections with bacteria such as *Listeria monocytogenes* and *Mycobacterium tuberculosis* cause chromatin remodelling by histone modification in humans (Hamon and Cossart, 2008). Histone modification by *L. monocytogenes* involves two mechanisms. Schmeck et al. (2005) showed that the pathogen activates the p38 and MAPK pathways, promoting the acetylation of lysine 8 on core histone H4, as well as serine 10 phosphorylation and lysine 14 acetylation on core histone H3, whereas Hamon et al. (2007) demonstrated that *L. monocytogenes* dephosphorylates histone H3 and deacetylates histone H4, suppressing a subset of proinflammatory and immunity-related genes (Hamon and Cossart, 2011).

Much less is known about the corresponding effects in insects, but the larvae of the greater wax moth have been established as a powerful model for human pathogens such as *L. monocytogenes* (Mukherjee et al., 2010). This pathogen can cross the blood–brain barrier to induce immune responses in the brains of infected humans and *G. mellonella* larvae (Mukherjee et al., 2013). The latter study also showed that virulent *L. monocytogenes* strains delay development, whereas nonvirulent strains cause precocious metamorphosis. Accordingly, *L. monocytogenes* was shown to shift the balance between the expression of genes encoding HATs and HDACs (Mukherjee et al., 2012). Similar transcriptional reprogramming of these enzyme families was observed when *G. mellonella* larvae were fed on the entomopathogenic bacterium *Serratia entomophila*. Oral infection induced the expression of genes encoding HDAC 8, HDAC 8 isoform 2, HDAC

Fig. 1 Model illustrating basic epigenetic modifications operating at the pretranscriptional level. The chromosomes in the nucleus of eukaryotic cells are composed of chromatin, a complex of DNA and proteins organized as repeated elements known as nucleosomes. The latter comprise a complex of histone proteins around which the DNA is wrapped. Gene expression is switched on in the open chromatin displaying looser assembly of nucleosomes, whereas gene expression is switched off in the condensed form of chromatin with tightly packed histones and DNA. The transition from condensed to open chromatin is regulated by addition or removal of acetyl groups to the core histones. The removal of acetyl groups produces a compact form of chromatin that is inaccessible to RNA polymerase. DNA methylation is mediated by DNA methyltransferases that can transfer methyl groups (CH_3) to cytosine residues in the dinucleotide motif CpG forms 5-methylcytidine. Methylated cytosines silence gene expression whereas unmethylated cytosines promote gene expression.

complex subunit Sap18, HAT Tip60 and the HAT type B catalytic subunit in both the larval midgut and in eggs laid when these larvae became adults (Mukherjee et al., 2015). The same model host was recently used to demonstrate for the first time that the maternal transfer of bacteria from the gut into the developing eggs can mediate transgenerational immune priming in insects (Freitak et al., 2014), and this phenomenon is clearly accompanied by corresponding pathogen-mediated transgenerational epigenetic effects (Mukherjee et al., 2015).

Entomopathogens such as the parasitic fungus *Metarhizium robertsi* (formerly *M. anisopliae*) can also delay the development of infected host insects (Mukherjee et al., 2012). Arrested development promotes the transformation of the infected host into pathogen propagules. This raises the question as to whether pathogens actively manipulate host development to their advantage, or whether developmental delay is an incidental consequence of infection, e.g., due to the absence of essential metabolites or proteins. Interestingly, the injection of heat-inactivated *M. robertsi* propagules into last-instar *G. mellonella* larvae induces precocious metamorphosis whereas the injection of living spores delays the pupal phase, suggesting that successful pathogens can manipulate host development (Mukherjee et al., 2012, 2015). The extent of developmental acceleration or delay was comparable to that induced by the injection of specific HAT or HDAC inhibitors (Mukherjee et al., 2012). These observations suggest that parasites may interfere with histone acetylation and deacetylation in insects to epigenetically regulate host gene expression before transcription.

Evidence that both bacterial and fungal entomopathogens can skew the HAT/HDAC balance associated with developmental timing argues in favour of active manipulation to suppress the expression of immunity-related effector genes, thus favouring the pathogen during infection (Mukherjee et al., 2012, 2017; Vilcinskas and Matha, 1997). It seems that the developmental delay is a collateral effect caused by the pathogen's efforts to suppress the host immune response reflecting the fact that developmental and immunity-related genes are controlled by the same epigenetic mechanisms (Vilcinskas, 2016). This hypothesis is supported by the observation that transient immune responses occur at the onset of metamorphosis in noninfected insects, e.g., in *G. mellonella* larvae where fundamental transcriptional reprogramming is associated with tissue remodelling (Altincicek and Vilcinskas, 2006; Vogel et al., 2011). Both metamorphosis and the invasion of tissues and organs by pathogens are associated with the degradation of the extracellular matrix, and both processes release fragments of collagen IV (the only type of collagen present in insects) which in turn function as endogenous danger signals that induce immunity-related effector genes such as those encoding AMPs (Berisha et al., 2013). Many studies have proposed crosstalk between developmental and immunity-related gene expression, regulated by the balanced expression of HATs and HDACs (Mukherjee et al., 2012). We therefore conclude that pathogens can directly or indirectly manipulate histone acetylation and deacetylation in the infected host. Developmental delay or arrest favours the pathogen by reallocating resources from tissue and organ remodelling in the host to the formation of pathogen propagules.

Interestingly, parasitic fungi such as *M. robertsi* communicate with their host insects during infection. This fungus is able to sense the presence of antifungal peptides, such as metchnikowin which are induced in the host in response to infection, and counteract by the synthesis of proteases which are able to degrade host defence molecules. This parasite can also detect the presence of host serine protease inhibitors operating against fungal serine proteases and respond by the production of chymotrypsin-like proteinases and metalloproteinases which in turn degrade host protease inhibitors. The transcriptional reprogramming in *M. robertsi* upon exposure to host antifungal peptides or serine protease inhibitors is accompanied by imbalanced expression of HATs and HDACs (Mukherjee and Vilcinskas, 2017).

The role of insect histone acetylation and deacetylation in host–pathogen interactions has also been highlighted by the experimental selection of *G. mellonella* for resistance against the economically most important entomopathogen *Bacillus thuringiensis* (Bt), which results in the transcriptional reprogramming of immunity-related genes (Dubovskiy et al., 2016) accompanied by a corresponding shift in the balance between the expression of genes encoding HATs and HDACs (Mukherjee et al., 2017). A *G. mellonella* line selected for resistance against Bt over 30 generations was nearly 11 times more resistant than a nonselected control line. Histone acetylation was enhanced in the midgut and fat bodies of uninfected Bt-resistant larvae, as confirmed by the percentage of acetylated core histones H3 and H4 and the balanced transcriptional induction of HDACs and HATs. The opposite phenomenon was observed when resistant and susceptible larvae were infected with Bt. These results show that the acetylation of core histones H3 and H4 plays a role in the evolution of pathogen resistance in *G. mellonella* (Mukherjee et al., 2017). In another model host, the red flour beetle *Tribolium castaneum*, the oral uptake of specific HAT or HDAC inhibitors results in transgenerational effects affecting development

(Bingsohn et al., 2016). We do not yet know whether particular histone modifications can be passed from parents to offspring, but there are known complex interactions among the mechanisms of histone modification, DNA methylation and posttranscriptional regulation that can yield heritable histone-based epigenetic markers. Taken together, these data indicate that selection for resistance results in transgenerational effects associated with altered histone acetylation, and that pathogens can exploit this epigenetic mechanism to delay or even arrest host development (Vilcinskas, 2016).

Histone acetylation also plays a role in the interactions between host insects and braconid or ichneumonid wasps. These parasitoids are associated with symbiotic polyDNAviruses that are injected into the insect host along with the eggs, resulting in the suppression of host immune responses to benefit the developing wasp larvae (Bae and Kim, 2009; Bitra et al., 2012; Song et al., 2008). The bracovirus associated with the parasitic wasp *Cotesia plutellae* carries transcriptional regulators, including a histone H4 with a longer N-terminal tail than the corresponding host histone, which integrates into the host nucleosome and displaces the incumbent histone, ultimately causing transcriptional reprogramming in the host insect (Gad and Kim, 2008; Hepat et al., 2013). Such sophisticated interventions to manipulate host gene expression may also underlie the observed developmental arrest of the parasitized host.

5. DNA METHYLATION

Like histone modification, DNA methylation is an epigenetic mechanism that operates before the onset of transcription. The addition of a methyl group to a cytosine residue in the dinucleotide sequence CpG forms 5-methylcytosine, which retains the base-pairing capacity of the unmodified nucleoside but modifies its interaction with regulatory proteins (Fig. 1). The transfer of methyl groups is catalysed by enzymes known as DNA methyltransferases (DNMTs). Three families have been recognized, but only DNMT-1 and DNMT-3 are relevant in the context of epigenetic gene regulation before transcription because DNMT-2 preferentially binds tRNA in both vertebrates and insects and therefore has a posttranscriptional role (Durdevic et al., 2013; Goll and Bestor, 2005; Hunt et al., 2013; Jurkowska et al., 2011). DNMT-1 (and DNMT-2) are evolutionarily conserved and widely distributed among insects, whereas DNMT-3 is restricted to some hymenopteran and hemipteran species (Glastad et al., 2011). DNMTs can also be classified as either de novo methyltransferases, which form new methylation marks on unmethylated DNA, or maintenance methyltransferases, which complete symmetrical methylation marks on newly replicated DNA by recognizing the hemimethylated sequences originating from each parent (Bestor, 2000; Klose and Bird, 2006).

The degree of genomic DNA methylation can be quantified at a global level using enzymatic methods, whereas the identification of specific methylated genes requires more sophisticated technologies such as bisulphite sequencing. The application of bisulphite sequencing on a genome-wide scale has revealed that the ratio of methylated to unmethylated CpG is much lower in insects than vertebrates (Lyko et al., 2000). For example, less than 0.5% of cytosine residues in the CpG dinucleotide are methylated in *D. melanogaster* (Marhold et al., 2004) compared to ~5% in vertebrates. The methylated cytosine residues in insects are mostly located within exons (Glastad et al., 2011) and only rarely in transposable elements or repetitive sequences (Regev et al., 1998). Furthermore, genome-wide methylation analysis in the silkworm *Bombyx mori*, the honeybee *A. mellifera* and the parasitic wasp *Nasonia vitripennis*, has shown that 5-methylcytosine is the most common DNA variation (Beeler et al., 2014; Cingolani et al., 2013; Xiang et al., 2013). Studies in *A. mellifera* and *T. castaneum* show that the highest degree of CpG methylation occurs in the embryos then tails off during later developmental stages (Drewell et al., 2014; Feliciello et al., 2013).

Little is currently known about the effect of entomopathogens on host DNA methylation. When *T. castaneum* larvae were fed on diets supplemented with *Escherichia coli*, *Micrococcus luteus* or *Pseudomonas entomophila* to mimic natural exposure to bacterial pathogens, the expression of various immunity-related effector genes was induced in both the fed larvae and eggs laid by the same individuals as adult beetles, thus providing evidence for transgenerational immune priming. The latter prepares the offspring to repel pathogens that the parents have previously encountered by allowing the rapid induction of a specific immune response, associated with the transcriptional reprogramming of immunity-related genes (Milutinovic et al., 2016). Although DNA methylation marks can be transmitted from parents to offspring, the transgenerational immune priming in *T. castaneum* did not involve changes in the overall DNA methylation status of the eggs laid by beetles that were exposed to the bacteria as larvae (Knorr et al., 2015). A comparative analysis of gene expression and DNA methylation marks in *A. mellifera* following a viral infection also yielded inconclusive data (Galbraith et al., 2015). In the latter study, virus infection altered the DNA methylation status of 156 genes including those involved in antiviral responses in humans, but there was no overlap between the differentially methylated genes and the differentially expressed genes.

A recent study addressed the impact of *Wolbachia pipientis* on the genomic methylation status of the mosquito *A. aegypti* (Kim et al., 2016). *W. pipientis* is a maternally inherited intracellular bacterium which has been found in ~40% of investigated arthropod species (Zug and Hammerstein, 2012). *W. pipientis* provides an interesting model because it reduces the fecundity and longevity of its hosts, hence reducing the transmission of viruses and parasites by vector insects. In the case of *A. aegypti*, this could translate into strategies to control vectored human diseases such as yellow fever, dengue fever and Zika fever. Approximately 1000 genes are methylated in the *A. aegypti* genome, typically around the promoter. Infection with a virulent *W. pipientis* strain altered the methylation status in an apparently random manner and did not affect transcription, with the exception of genes involved in membrane-related functions (Ye et al., 2013). These results suggest *W. pipientis* may target certain genes to influence the success of infection because the bacteria become enveloped by host membranes shortly after infection and the manipulation of host gene expression may promote the survival of virulent endosymbionts in a host vacuole. The *A. aegypti* DNMT gene *AedaeDnmt2* is suppressed by *W. pipientis* but induced by the Dengue virus, and the overexpression of *AedaeDnmt2* in a mosquito cell line can suppress *W. pipientis* replication but enhance Dengue virus replication (Hussain et al., 2013). The same authors identified a miRNA (aae-miR-2940) that silences *AedaeDnmt2*, and found that it was expressed solely in mosquitoes infected with *W. pipientis*, providing an interesting example of the sophisticated role of epigenetic mechanisms in insect–microbe interactions (Hussain et al., 2013). Although DNMT2 prefers tRNA substrates it has been shown to regulate antiviral responses in *D. melanogaster* (Durdevic et al., 2013).

DNA methylation in insects can also regulate antifungal immunity. Parasitic fungi such as *M. robertsi* influence DNA methylation in *G. mellonella* in a strain-specific manner (Vilcinskas, 2016). The relative expression of DNMTs fluctuates during infection, supporting the observation that fungal infection elicits immune responses at an early stage but these are subsequently suppressed by the fungus during colonization (Gillespie et al., 2000; Vilcinskas and Matha, 1997).

In the abovementioned experiment involving the selection of *G. mellonella* for Bt resistance over 30 generations, the observed changes in histone modification occurred in concert with changes in DNA methylation. Specifically, the levels of 5-methylcytosine in the midgut and fat body of the resistant larvae were higher than in the corresponding susceptible larvae (Mukherjee and Vilcinskas, 2017). Remarkably, the differences in global DNA methylation were tissue specific. For example, DNA methylation levels in the resistant larvae were higher in the fat bodies than the midguts but there was no corresponding difference between these tissues in the susceptible larvae. In contrast, DNA methylation levels in the resistant larvae were lower when the larvae were exposed to Bt than in unchallenged controls of the same line.

Another recent study has provided the first evidence that parasitoids can also influence host DNA methylation (Kumar and Kim, 2017). The koinobiotic endoparasitic wasp *C. plutellae* lays eggs in small lepidopteran larvae, including caterpillars of the diamondback moth *Plutella xylostella*. Endoparasitoid development is accompanied by the suppression of host transcription and protein expression (Bae and Kim, 2009; Bitra et al., 2012; Song et al., 2008). The global level of DNA methylation was significantly lower in the parasitized larvae compared to the naïve larvae and this was reflected by lower expression levels of both DNMT-1 and DNMT-2. Furthermore, the loss of DNA methylation induced by the parasitoid was shown to suppress haemocyte spreading and activity, suggesting that DNA methylation plays a key role in the regulation of multicellular immune responses (Kumar and Kim, 2017). It remains unclear whether the suppression of DNA methylation in parasitized *P. xylostella* larvae is directly or indirectly imposed by *C. plutellae*.

A very recent study provides evidence that the impact of parasites on host insect epigenetics can be expanded beyond DNA methylation to include methylation on RNA (Castro-Vargas et al., 2017). Using the mealworm *Tenebrio molitor*, it was demonstrated that bacterial and fungal infection does not influence the ratio of methylated cytosine entities in DNA, which confirms our findings from the red flour beetle *T. castaneum* (Knorr et al., 2015), but decreases the percentage of methylated cytosine entities in RNA (5mC) within but not across generations (Castro-Vargas et al., 2017).

6. miRNAs

miRNAs are short noncoding RNAs (~18–24 nucleotides in length) which epigenetically regulate gene expression at the posttranscriptional level. They inhibit the translation of specific mRNAs by base-pairing with the untranslated regions or occasionally in the coding region (Ambros, 2004;

Asgari, 2011; Bartel, 2004). These small untranslated RNA molecules are evolutionarily conserved between different eukaryotic species. There is a growing gap between the rapidly increasing number of miRNAs predicted in silico and those with experimentally validated functions. More than 30% of animal genes may be regulated by miRNAs, and a single miRNA may control hundreds of different target genes (Bartel, 2009; Bushati and Cohen, 2007; Sato et al., 2011). Despite the large number of genes that are differentially expressed during humoral immune responses in insects, the number of miRNAs known to influence immunity-related gene regulation at the posttranscriptional level is relatively low. Fullaondo and Lee (2012) identified seven miRNAs involved in the regulation of immune responses in *D. melanogaster* among which one was found to target mRNAs encoding peptidoglycan receptor proteins. Furthermore, the conserved miRNA known as miR-8 controls immunity-related mRNA translation in the *D. melanogaster* fat body (Choi and Hyun, 2012).

The contribution of miRNAs to vertebrate immune response is recognized although the precise functions of the miRNA network and individual miRNA on cellular processes await further investigations. The hypothesis that miRNAs are involved in both parasite virulence and host immunity is relatively well supported by experimental evidence (Harris et al., 2013; Lu and Liston, 2009). The role of miRNAs in the regulation of the mammalian immune system was confirmed before similar observations in insects (Asgari, 2013; Xiao and Rajewsky, 2009). The contribution of miRNAs to antiviral responses has received particular attention (Hussain and Asgari, 2014). For example, West Nile virus inhibits the expression of Aedal-miR-2940-5p in the vector mosquito *Aedes albopictus* and thus limits viral replication (Slonchak et al., 2014). This miRNA also enhances the expression of metalloprotease m41 gene *FtsH*, which is necessary for efficient West Nile virus replication. Similarly, Dengue virus influences miRNA expression in the vector *A. aegypti* (Campbell et al., 2014).

The potential role of insect miRNAs in antibacterial immune responses has been explored in the model beetle *T. castaneum*. Microarrays containing 455 unique mature miRNA sequences that are thought to be conserved among arthropods were used to measure changes in expression when the larvae were challenged with starvation, mild heat shock or bacterial infection, revealing not only the stressor-specific expression of miRNAs, but also gender-specific expression (Freitak et al., 2012). Following the injection of bacterial peptidoglycans or feeding the larvae on a diet contaminated with bacteria, approximately one-third of all miRNAs were modulated in females, but only one miRNA was modulated in males. This observation supports Bateman's principle, which postulates gender-specific immune responses because females gain fitness through increased longevity, whereas males gain fitness by more frequent mating (Freitak et al., 2012).

G. mellonella has also been used to screen for miRNAs that regulate innate immunity. A microarray spotted with more than 2000 insect miRNA probe sequences was used to identify miRNAs that are modulated by infection with the entomopathogenic bacterium *S. entomophila* and the parasitic fungus *M. robertsi* (Mukherjee and Vilcinskas, 2014). The coexpression of miRNAs with their predicted target mRNAs is required to verify their putative functions. Interestingly, api-miR-263a was more abundant in eggs from female imagoes that were fed as larvae with bacteria, suggesting that miRNAs also mediate transgenerational immune priming and host–parasite coevolution (Mukherjee and Vilcinskas, 2014). Specific miRNAs also contribute to the evolution of Bt resistance in *G. mellonella* (Mukherjee et al., 2017).

When the malaria parasite *Plasmodium falciparum* invades the midgut of its vector *A. gambiae*, the resulting immune response involves the Toll and IMD pathways (Dong et al., 2006). Interestingly, the prevention of miRNA biogenesis in *A. gambiae* results in higher resistance against the parasite, thus providing evidence that miRNAs contribute to the regulation of anti-*Plasmodium* responses (Dennison et al., 2015). In the latter study, higher levels of Anoga-miR-989 and aga-miR-305 were detected in infected midguts compared to controls, suggesting these miRNAs facilitate infection. This was confirmed by the silencing of these miRNAs using specific miRNA-blocking oligonucleotides (antagomirs), which increased resistance towards *P. falciparum* and also inhibited the midgut microbiota. Such studies open new avenues of research for the development of epigenetic approaches to control vector insects. The roles of particular miRNAs are difficult to predict even if putative targets are known. Importantly, miRNAs can also regulate enzymes involved in other epigenetic mechanisms, including DNMTs, HATs and HDACs, providing a complex network of crosstalk among the various epigenetic strategies to control gene expression.

7. CONCLUDING REMARKS

Recent advances in insect physiology provide evidence that epigenetic mechanisms operating before and after transcription contribute to the transgenerational inherited transcriptional reprogramming of immunity-related genes in insects, ultimately providing mechanisms for the evolution of resistance to parasites. In response, microbial parasites have evolved strategies to manipulate host immunity by interfering with these epigenetic mechanisms in order to benefit their own development within the host. The importance of epigenetics in the regulation of insect immune responses and the impact of parasites on host insect epigenetics is far from completely understood. However, the epigenetic interactions between insects and their parasites are gaining attention among the research community because they provide opportunities to develop novel strategies for the control of vector and pest insects.

ACKNOWLEDGEMENTS

The author acknowledges funding provided by the German Research Foundation for the project 'The role of epigenetics in host–parasite coevolution' (VI 219/3-2) which is embedded within the DFG Priority Programme 1399 'Host–parasite coevolution—rapid reciprocal adaptations and its genetic basis'. Further, he thanks the Hessen State Ministry of Higher Education, Research and the Arts (HMWK) via the collaborative research projects granted under the LOEWE Centre for 'Insect Biotechnology and Bioresources'. The author also acknowledges Dr. Richard M. Twyman for professional editing of the manuscript.

REFERENCES

Adcock, I.M., Lee, K.Y., 2006. Abnormal histone acetylase and deacetylase expression and function in lung inflammation. Inflamm. Res. 55, 311–321.

Altincicek, B., Vilcinskas, A., 2006. Metamorphosis and collagen-IV-fragments stimulate innate immune response in the greater wax moth, *Galleria mellonella*. Dev. Comp. Immunol. 30, 1108–1118.

Ambros, V., 2004. The functions of animal microRNAs. Nature 431, 350–355.

Asgari, S., 2011. Role of microRNAs in insect host-microorganism interactions. Front. Physiol. 2, 48.

Asgari, S., 2013. MicroRNA functions in insects. Insect Biochem. Mol. Biol. 43, 388–397.

Bae, S., Kim, Y., 2009. IkB genes encoded in *Cotesia plutellae* bracovirus suppress an antiviral response and enhance baculovirus pathogenicity against the diamondback moth *Plutella xylostella*. J. Invertebr. Pathol. 102 (1), 79–87.

Bartel, D.P., 2004. MicroRNAs: genomics, biogenesis, mechanism, and function. Cell 116, 281–297.

Bartel, D.P., 2009. MicroRNAs: target recognition and regulatory functions. Cell 136, 215–233.

Bayarsaihan, D., 2011. Epigenetic mechanisms in inflammation. J. Dent. Res. 90, 9–17.

Beeler, S.M., Wong, G.T., Zheng, J.M., Bush, E.C., Remnant, E.J., Oldroyd, B.P., Drewell, R.A., 2014. Whole-genome DNA methylation profile of the jewel wasp (*Nasonia vitripennis*). G3 (Bethesda) 4, 383–388.

Berisha, A., Mukherjee, K., Vilcinskas, A., Spengler, B., Römpp, A., 2013. High-resolution mass spectrometry driven discovery of peptidic danger signals in insect immunity. PLoS One 8 (11), e80406.

Bestor, T.H., 2000. The DNA methyltransferases of mammals. Hum. Mol. Genet. 9, 2395–2402.

Bingsohn, L., Knorr, E., Vilcinskas, A., 2016. The model beetle *Tribolium castaneum* can be used as an early warning system for transgenerational epigenetic side effects caused by pharmaceuticals. Comp. Biochem. Physiol. C Toxicol. Pharmacol. 185/186, 57–64.

Bitra, K., Suderman, R.J., Strand, M.R., 2012. Polydnavirus Ank proteins bind NF-κ B homodimers and inhibit processing of Relish. PLoS Pathog. 8, e1002722.

Bonduriansky, R., Crean, A.J., Day, T., 2012. The implications of nongenetic inheritance for evolution in changing environments. Evol. Appl. 5, 192–201.

Boutros, M., Agaisse, H., Perrimon, N., 2002. Sequential activation of signaling pathways during innate immune responses in *Drosophila*. Dev. Cell 3, 711–722.

Buchmann, K., 2014. Evolution of innate immunity: clues from invertebrates via fish to mammals. Front. Immunol. 5, 459.

Buchon, N., Silverman, N., Cherry, S., 2014. Immunity in *Drosophila melanogaster*—from microbial recognition to whole-organism physiology. Nat. Rev. Immunol. 14 (12), 796–810.

Bushati, N., Cohen, S.M., 2007. MicroRNA functions. Annu. Rev. Cell Dev. Biol. 23, 175–205.

Busslinger, M., Tarakhovsky, A., 2014. Epigenetic control of immunity. Cold Spring Harb. Perspect. Biol. 6, a019307.

Butt, T.M., Coates, C.J., Dubovskiy, I.M., Ratcliffe, N.A., 2016. Entomopathogenic fungi: new insights into host-pathogen interactions. Adv. Genet. 94, 307–364.

Campbell, C.L., Harrison, T., Hess, A.M., Ebel, G.D., 2014. MicroRNA levels are modulated in *Aedes aegypti* after exposure to Dengue-2. Insect Mol. Biol. 23, 132–139.

Castro-Vargas, C., Linares-Lopez, C., Lopez-Torres, A., Wrobel, K., Torres-Guzman, J.C., Hernandez, G.A., Wrobel, K., Lanz-Mendoza, H., Contreras-Garduno, J., 2017. Methylation on RNA: a potential mechanism related to immune priming within but not across generations. Front. Microbiol. 8, 473.

Cheeseman, K., Weitzman, J.B., 2015. Host-parasite interactions: an intimate epigenetic relationship. Cell. Microbiol. 17, 1121–1132.

Chen, Z., Li, S., Subramaniam, S., Shyy, J.Y., Chien, S., 2017. Epigenetic regulation: a new frontier for biomedical engineers. Annu. Rev. Biomed. Eng. In press.

Choi, I.K., Hyun, S., 2012. Conserved microRNA miR-8 in fat body regulates innate immune homeostasis in *Drosophila*. Dev. Comp. Immunol. 37, 50–54.

Cingolani, P., Cao, X., Khetani, R.S., Chen, C.C., Coon, M., Sammak, A., Bollig-Fischer, A., Land, S., Huang, Y., Hudson, M.E., Garfinkel, M.D., Zhong, S., Robinson, G.E., Ruden, D.M., 2013. Intronic non-CG DNA hydroxymethylation and alternative mRNA splicing in honey bees. BMC Genomics 14, 666.

Dennison, N.J., BenMarzouk-Hidalgo, O.J., Dimopoulos, G., 2015. MicroRNA-regulation of *Anopheles gambiae* immunity to *Plasmodium falciparum* infection and midgut microbiota. Dev. Comp. Immunol. 49 (1), 170–178.

Dokmanovic, M., Clarke, C., Marks, P.A., 2007. Histone deacetylase inhibitors: overview and perspectives. Mol. Cancer Res. 5, 981–989.

Dong, Y.M., Aguilar, R., Xi, Z.Y., Warr, E., Mongin, E., Dimopoulos, G., 2006. *Anopheles gambiae* immune responses to human and rodent Plasmodium parasite species. PLoS Pathog. 2, 513–525.

Drewell, R.A., Bush, E.C., Remnant, E.J., Wong, G.T., Beeler, S.M., Stringham, J.L., Lim, J., Oldroyd, B.P., 2014. The dynamic DNA methylation cycle from egg to sperm in the honey bee *Apis mellifera*. Development 141, 2702–2711.

Dubovskiy, I.M., Whitten, M.M., Yaroslavtseva, O.N., Greig, C., Kryukov, V.Y., Grizanova, E.V., Mukherjee, K., Vilcinskas, A., Glupov, V.V., Butt, T.M., 2013. Can insects develop resistance to insect pathogenic fungi? PLoS One 8, e60248.

Dubovskiy, I.M., Grizanova, E.V., Whitten, M.M., Mukherjee, K., Greig, C., Alikina, T., Kabilov, M., Vilcinskas, A., Glupov, V.V., Butt, T.M., 2016. Immuno-physiological adaptations confer wax moth *Galleria mellonella* resistance to *Bacillus thuringiensis*. Virulence 7 (8), 860–870.

Durdevic, Z., Hanna, K., Gold, B., Pollex, T., Cherry, S., Lyko, F., 2013. Efficient RNA virus control in *Drosophila* requires the RNA-methyltransferase Dnmt2. EMBO Rep. 14, 269–275.

Feliciello, I., Parazajder, J., Akrap, I., Ugarkovic, D., 2013. First evidence of DNA methylation in insect *Tribolium castaneum*: environmental regulation of DNA methylation within heterochromatin. Epigenetics 8, 534–541.

Fischer, N., Sechet, E., Friedman, R., Amiot, A., Sobhani, I., Nigro, G., Sansonetti, P.J., Speradino, B., 2016. Histone deacetylase inhibition enhances antimicrobial peptide but not inflammatory cytokine expression upon bacterial challenge. Proc. Natl. Acad. Sci. U.S.A. 113 (21), e2993–e3001.

Freitak, D., Knorr, E., Vogel, H., Vilcinskas, A., 2012. Gender- and stressor-specific microRNA expression in *Tribolium castaneum*. Biol. Lett. 8, 860–863.

Freitak, D., Schmidtberg, H., Dickel, F., Lochnit, G., Vogel, H., Vilcinskas, A., 2014. The maternal transfer of bacteria can mediate trans-generational immune priming in insects. Virulence 5, 547–554.

Fullaondo, A., Lee, S.Y., 2012. Identification of putative miRNA involved in *Drosophila melanogaster* immune response. Dev. Comp. Immunol. 36, 267–273.

Gad, W., Kim, Y., 2008. A viral histone H4 encoded by *Cotesia plutellae* bracovirus inhibits haemocyte-spreading behaviour of the diamondback moth, *Plutella xylostella*. J. Gen. Virol. 89, 931–938.

Galbraith, D.A., Yang, X., Niño, E.L., Yi, S., Grozinger, C., 2015. Parallel epigenomic and transcriptomic responses to viral Infection in honey bees (*Apis mellifera*). PLoS Pathog. 11 (3), e1004713.

Gillespie, J., Bailey, A., Cobb, B., Vilcinskas, A., 2000. Fungal elicitors of insect immune responses. Arch. Insect Biochem. Physiol. 44, 49–68.

Glastad, K.M., Hunt, B.G., Yi, S.V., Goodisman, M.A., 2011. DNA methylation in insects: on the brink of the epigenomic era. Insect Mol. Biol. 20, 553–565.

Goll, M.G., Bestor, T.H., 2005. Eukaryotic cytosine methyltransferases. Annu. Rev. Biochem. 74, 481–514.

Hamon, M.A., Cossart, P., 2008. Histone modifications and chromatin remodeling during bacterial infections. Cell Host Microbe 4, 100–109.

Hamon, M.A., Cossart, P., 2011. K+ efflux is required for histone H3 dephosphorylation by *Listeria monocytogenes* listeriolysin O and other pore-forming toxins. Infect. Immun. 79, 2839–2846.

Hamon, M.A., Batsché, E., Régnault, B., Tham, T.N., Seveau, S., Muchardt, C., Cossart, P., 2007. Histone modifications induced by a family of bacterial toxins. Proc. Natl. Acad. Sci. U.S.A. 104 (33), 13467–13472.

Harris, J.F., Micheva-Viteva, S., Li, N., Hong-Gellner, E., 2013. Small RNA-mediated regulation of host–pathogen interactions. Virulence 4, 785–795.

Hepat, R., Song, J.J., Lee, D., Kim, Y., 2013. A viral histone h4 joins to eukaryotic nucleosomes and alters host gene expression. J. Virol. 87, 11223–11230.

Hoffmann, J.A., Reichhart, J.M., 2002. *Drosophila* innate immunity: an evolutionary perspective. Nat. Immunol. 3, 121–126.

Hu, Y.-T., Wu, T.-C., Yang, E.-C., Wu, P.-C., Lin, P.-T., Wu, Y.-L., 2017. Regulation of genes related to immune signaling and detoxification in *Apis mellifera* by an inhibitor of histone deacetylation. Sci. Rep. 7, 41255.

Hunt, B.G., Glastad, K.M., Yi, S.V., Goodisman, M.A., 2013. The function of intragenic DNA methylation: insights from insect epigenomes. Integr. Comp. Biol. 53 (2), 319–328.

Hussain, M., Asgari, S., 2014. MicroRNAs as mediators of insect–host–pathogen interactions and immunity. J. Insect Physiol. 70, 151–158.
Hussain, M., Frentiu, F.D., Moreira, L.A., O'Neill, S.L., Asgari, S., 2011. Wolbachia uses host microRNAs to manipulate host gene expression and facilitate colonization of the dengue vector *Aedes aegypti*. Proc. Natl. Acad. Sci. U.S.A. 108, 9250–9255.
Hussain, M., Walker, T., O'Neill, S.L., Asgari, S., 2013. Blood meal induced microRNA regulates development and immune associated genes in the Dengue mosquito vector, *Aedes aegypti*. Insect Biochem. Mol. Biol. 43 (2), 146–152.
Jacobs, C.G.C., Steiger, S., Heckel, D.G., Wielsch, N., Vilcinskas, A., Vogel, H., 2016. Sex, offspring and carcass determine antimicrobial peptide expression in the burying beetle. Sci. Rep. 6, 25409.
Jaenisch, R., Bird, A., 2003. Epigenetic regulation of gene expression: how the genome integrates intrinsic and environmental signals. Nat. Genet. 33, 245–254.
Jurkowska, R.Z., Jurkowski, T.P., Jeltsch, A., 2011. Structure and function of mammalian DNA methyltransferases. Chembiochem 12, 206–222.
Kaneko, T., Silverman, N., 2005. Bacterial recognition and signalling by the *Drosophila* IMD pathway. Cell. Microbiol. 7, 461–469.
Kim, D., Thairu, M.W., Hansen, A.K., 2016. Novel insights into insect-microbe interactions—role of epigenomics and small RNAs. Front. Plant Sci. 7, 1164.
Klose, R.J., Bird, A.P., 2006. Genomic DNA methylation: the mark and its mediators. Trends Biochem. Sci. 31, 89–97.
Knorr, E., Schmidtberg, H., Arslan, D., Bingsohn, L., Vilcinskas, A., 2015. Translocation of bacteria from the gut to the eggs triggers maternal trans-generational immune priming in *Tribolium castaneum*. Biol. Lett. 11, 20150885.
Kumar, S., Kim, Y., 2017. An endoparasitoid wasp influences host DNA methylation. Sci. Rep. 7, 43287.
Laland, K., Uller, T., Feldman, M., Sterelny, K., Muller, G.B., Moczek, A., Jablonka, E., Odling-Smee, J., Wray, G.A., Hoekstra, H.E., et al., 2014. Does evolutionary theory need a rethink? Nature 514 (7521), 161–164.
Lavine, M.D., Strand, M.R., 2002. Insect hemocytes and their role in immunity. Insect Biochem. Mol. Biol. 32 (10), 1295–1309.
Lemos, B., Branco, A.T., Hartl, D.L., 2010. Epigenetic effects of polymorphic Y chromosomes modulate chromatin components, immune response, and sexual conflict. Proc. Natl. Acad. Sci. U.S.A. 107 (36), 15826–15831.
Lim, P.S., Li, J., Holloway, A.F., Rao, S., 2013. Epigenetic regulation of inducible gene expression in the immune system. Immunology 139, 141–150.
Lu, L.-F., Liston, A., 2009. MicroRNA in the immune system, microRNA as an immune system. Immunology 127 (3), 291–298.
Lyko, F., Ramsahoye, B.H., Jaenisch, R., 2000. DNA methylation in *Drosophila melanogaster.* Nature 408, 538–540.
Marhold, J., Rothe, N., Pauli, A., Mund, C., Kuehle, K., Brueckner, B., Lyko, F., 2004. Conservation of DNA methylation in dipteran insects. Insect Mol. Biol. 13 (2), 117–123.
Marks, P.A., Miller, T., Richon, V.M., 2003. Histone deacetylases. Curr. Opin. Pharmacol. 3, 344–351.
Milutinovic, B., Peuß, R., Ferro, K., Kurtz, J., 2016. Immune priming in arthropods: an update focusing on the red flour beetle. Zoology (Jena) 119, 254–261.
Mukherjee, K., Vilcinskas, A., 2014. Development and immunity-related microRNAs of the lepidopteran model host *Galleria mellonella.* BMC Genomics 15, 705.
Mukherjee, K., Vilcinskas, A., 2017. The entomopathogenic fungus *Metarhizium robertsii* communicates with the insect host *Galleria mellonella* during infection, *Virulence*, in press.
Mukherjee, K., Altincicek, B., Hain, T., Domann, E., Vilcinskas, A., Chakraborty, T., 2010. *Galleria mellonella* as a model system for studying *Listeria* pathogenesis. Appl. Environ. Microbiol. 76 (1), 310–317.

Mukherjee, K., Fischer, R., Vilcinskas, A., 2012. Histone acetylation mediates epigenetic regulation of transcriptional reprogramming in insects during metamorphosis, wounding and infection. Front. Zool. 9, 25.

Mukherjee, K., Hain, T., Fischer, R., Chakraborty, T., Vilcinskas, A., 2013. Brain infection and activation of neuronal repair mechanisms caused by the human pathogen *Listeria monocytogenes* in the lepidopteran model host *Galleria mellonella*. Virulence 4, 324–332.

Mukherjee, K., Twyman, R.M., Vilcinskas, A., 2015. Insects as models to study the epigenetic basis of disease. Prog. Biophys. Mol. Biol. 118, 69–78.

Mukherjee, K., Grizanova, E., Chetkova, E., Lehmann, R., Dubovskiy, I., Vilcinskas, A., 2017. Experimental evolution of resistance against *Bacillus thuringiensis* in the insect modelhost *Galleria mellonella* results in epigeneticmodifications, *Virulence*, in press.

Mylonakis, E., Podsiadlowski, L., Muhammed, M., Vilcinskas, A., 2016. Diversity, evolution and medical applications of insect antimicrobial peptides. Philos. Trans. R. Soc. B 371, 20150290.

Parsons, B., Foley, E., 2016. Cellular immune defense of *Drosophila melanogaster*. Dev. Comp. Immunol. 58, 95–101.

Peleg, S., Feller, C., Forne, I., Schiller, E., Sévin, D., Schauer, T., Regnard, C., Straub, T., Prestel, M., Klima, C., Schmitt Nogueira, M., Becker, L., Klopstock, T., Sauer, U., Becker, P.B., Imhof, A., Ladurner, A.G., 2016. Life span extension by targeting a link between metabolism and histone acetylation in Drosophila. EMBO Rep. 17 (3), 455–469.

Pigeault, R., Garnier, R., Rivero, A., Gandon, S., 2016. Evolution of transgenerational immunity in invertebrates. Proc. Biol. Sci. 283 (1839). pii: 20161136.

Rahnamaeian, M., Cytryńska, M., Zdybicka-Barabas, A., Dobslaff, K., Wiesner, J., Twyman, R.M., Zuchner, T., Sadd, B.M., Regoes, R.R., Schmid-Hempel, P., Vilcinskas, A., 2015. Insect antimicrobial peptides show potentiating functional interactions against Gram-negative bacteria. Proc. R. Soc. B 282, 20150293.

Regev, A., Lamb, M., Jablonka, E., 1998. The role of DNA methylation in invertebrates: developmental regulation of genome defense? Mol. Biol. Evol. 15, 880–891.

Reynolds, J.A., Bautista-Jimenez, R., Denlinger, D.L., 2016. Changes in histone acetylation as potential mediators of pupal diapause in the flesh fly, *Sarcophaga bullata*. Insect Biochem. Mol. Biol. 76, 29–37.

Sato, F., Tsuchiya, S., Meltzer, S., Shimizu, K., 2011. MicroRNAs and epigenetics. FEBS J. 278, 1598–1609.

Schmeck, B., Beermann, W., van Laak, V., Zahlten, J., Opitz, B., Witzenrath, M., Hocke, A.C., Chakraborty, T., Kracht, M., Rosseau, S., Suttorp, N., Hippenstiel, S., 2005. Intracellular bacteria differentially regulated endothelial cytokine release by MAPK-dependent histone modification. J. Immunol. 175 (5), 843–850.

Sharakhov, I.V., Sharakhova, M.S., 2015. Heterochromatin, histone modifications, and nuclear architecture in disease vectors. Curr. Opin. Insect Sci. 10, 110–117.

Skinner, M.K., 2015. Environmental epigenetics and a unified theory of the molecular aspects of evolution: a neo-Lamarckian concept that facilitates neo-Darwinian evolution. Genome Biol. Evol. 7 (5), 1296–1302.

Slonchak, A., Hussain, M., Torres, S., Asgari, S., Khromykh, A.A., 2014. Expression of mosquito microRNA Aae-miR-2940-5p is downregulated in response to West Nile virus infection to restrict viral replication. J. Virol. 88 (15), 8457–8467.

Song, K.H., Jung, M.K., Eum, J.H., Hwang, I.C., Han, S.S., 2008. Proteomic analysis of parasitized *Plutella xylostella* larvae plasma. J. Insect Physiol. 54, 1270–1280.

Steiner, H., 2004. Peptidoglycan recognition proteins: on and off switches for innate immunity. Immunol. Rev. 198, 83–96.

Tonk, M., Vilcinskas, A., 2017. The medical potential of antimicrobial peptides from insects. Curr. Top. Med. Chem. 17, 554–575.

Vandegehuchte, M.B., Janssen, C.R., 2014. Epigenetics in an ecotoxicological context. Mutat. Res. Genet. Toxicol. Environ. Mutagen. 764–765, 36–45.
Vilcinskas, A., 2013. Evolutionary plasticity of insect immunity. J. Insect Physiol. 59, 123–129.
Vilcinskas, A., 2016. The role of epigenetics in host–parasite coevolution: lessons from the model host insects *Galleria mellonella* and *Tribolium castaneum*. Zoology (Jena) 119, 273–280.
Vilcinskas, A., Matha, V., 1997. Effect of the entomopathogenic fungus *Beauveria bassiana* on humoral immune response of *Galleria mellonella* larvae (Lepidoptera: Pyralidae). Eur. J. Entomol. 94, 461–472.
Vogel, H., Altincicek, B., Glöckner, G., Vilcinskas, A., 2011. A comprehensive transcriptome and immune-gene repertoire of the lepidopteran model host *Galleria mellonella*. BMC Genomics 12, 308.
Will, T., Vilcinskas, A., 2013. Aphid proof plants: biotechnology-based approaches for aphid control. Adv. Biochem. Eng. Biotechnol. 136, 179–203.
Xiang, H., Li, X., Dai, F., Xu, X., Tan, A., Chen, L., Zhang, G., Ding, Y., Li, Q., Lian, J., Willden, A., Guo, Q., Xia, Q., Wang, J., Wang, W., 2013. Comparative methylomics between domesticated and wild silkworms implies possible epigenetic influences on silkworm domestication. BMC Genomics 14, 646.
Xiao, C., Rajewsky, K., 2009. MicroRNA control in the immune system: basic principles. Cell 136, 26–36.
Ye, Y.H., Wollfit, M., Huttley, G.A., Rancès, E., Caragata, E.P., Popovici, J., O'Neill, S.L., McGraw, E.A., 2013. Infection with a virulent strain of disrupts genome wide-patterns of cytosine methylation in the mosquito *Aedes aegypti*. PLoS One 8, e66482.
Zug, R., Hammerstein, P., 2012. Still a host of hosts for Wolbachia: analysis of recent data suggests that 40% of terrestrial arthropod species are infected. PLoS One 7, e38544.

FURTHER READING

Vilcinskas, A., Mukherjee, K., Vogel, H., 2013. Expansion of the antimicrobial peptide repertoire in the invasive ladybird *Harmonia axyridis*. Proc. R. Soc. B 280, 2012–2113.

CHAPTER SEVEN

From Molecules to Management: Mechanisms and Consequences of Locust Phase Polyphenism

Darron A. Cullen[a], Arianne J. Cease[b,c], Alexandre V. Latchininsky[d], Amir Ayali[e], Kevin Berry[f], Jerome Buhl[g], Rien De Keyser[h], Bert Foquet[i], Joleen C. Hadrich[j], Tom Matheson[h], Swidbert R. Ott[h], Mario A. Poot-Pech[k], Brian E. Robinson[l], Jonathan M. Smith[h], Hojun Song[i], Gregory A. Sword[i], Jozef Vanden Broeck[a], Rik Verdonck[a], Heleen Verlinden[a], Stephen M. Rogers[b,m]

[a]KU Leuven (University of Leuven), Zoological Institute, Leuven, Belgium
[b]Arizona State University, School of Life Sciences, Tempe, Arizona, United States
[c]Arizona State University, School of Sustainability, Tempe, Arizona, United States
[d]University of Wyoming, Ecosystem Science and Management, Laramie, Wyoming, United States
[e]Tel-Aviv University, Department of Zoology, Tel-Aviv, Israel
[f]University of Alaska Anchorage, Institute of Social and Economic Research, Anchorage, Alaska, United States
[g]University of Adelaide, School of Agriculture, Food and Wine, Adelaide, Australia
[h]University of Leicester, Department of Neuroscience, Psychology and Behaviour, Leicester, United Kingdom
[i]Texas A&M University, Department of Entomology, College Station, Texas, United States
[j]Colorado State University, Department of Agricultural & Resource Economics, Fort Collins, Colorado, United States
[k]Yucatán State Plant Protection Committee (Locust Program), Mérida, Yucatán, México
[l]McGill University, Department of Geography, Montreal, Quebec, Canada
[m]University of Cambridge, Department of Zoology, Cambridge, United Kingdom

Contents

Advances in Insect Physiology, Volume 53
ISSN 0065-2806
http://dx.doi.org/10.1016/bs.aiip.2017.06.002

Abstract

Locusts are grasshoppers (Orthoptera: Acrididae) that are characterised by their capacity for extreme population density-dependent polyphenism, transforming between a cryptic solitarious phase that avoids other locusts, and a swarming gregarious phase that aggregates and undergoes collective migration. The two phases differ in many aspects of behaviour, physiology and ecology, making locusts a useful model through which to investigate the phenotypic interface of molecular processes and environmental cues. This review summarises recent progress in understanding the mechanisms and consequences of locust phase change, from differential gene expression and epigenetic regulation through to neuronal plasticity and altered behaviour. The impact of techniques such as RNA interference and the sequencing of the first locust genome are discussed, and we consider the evidence from comparative analyses between related locust species for the possible evolution of locust-like phenotypic plasticity. Collective movement, and new ways of measuring the behaviour of both migrating bands in the field and individuals in the laboratory, are analysed. We also examine the environmental factors that affect phase change, along with the wider impact of land use and management strategies that may unwittingly create environments conducive to outbreaks. Finally, we consider the human costs of locust swarming behaviour, and use combined social, economic and environmental approaches to suggest potential ways forward for locust monitoring and management.

1. INTRODUCTION

1.1 Scope and Intent of this Review

This review arose out of two symposia held at the XXV International Congress of Entomology, Orlando, Florida (September 2016), which succeeded in bringing together researchers representing many of the groups around the world currently working on all aspects of locust phase polyphenism and its wider consequences. Most of the attendees at these symposia agreed to write synopses of recent progress in different areas of locust research, each according to their interests and expertise, which form the core of this chapter. We have three objectives in this review. First, to act as an update to the seminal review of Pener and Simpson (2009) and so the emphasis here is on papers published in the 8 years since that synthesis. We refer the reader to other reviews, such as Buhl and Rogers (2016), Ernst et al. (2015), Ariel and Ayali (2015), Wang and Kang (2014), Burrows et al. (2011), Verlinden et al. (2009), Pener and Yerushalmi (1998), Pener (1991) and the books by Uvarov (1966, 1977) for material covering different aspects of the entire history of locust phase change research.

Our second objective is to cover as broad a spectrum of current research interests in locust phase change as possible. As expected, we review the dramatic developments in our understanding of the molecular, epigenetic and physiological mechanisms that drive the phase change process, which have been made possible by the rapid progress of molecular and behavioural research techniques (Cullen, De Keyser, Ott, Rogers, Smith, Vanden Broeck, Verdonck and Verlinden). Perhaps the most prominent advance over recent years has been the publication of a draft genome sequence for the Migratory Locust, *Locusta migratoria* (Wang et al., 2014), bringing locust phase research into the genomic era for the first time. This genome, and other locust sequence data, has greatly facilitated gene-driven studies of locust physiology, many of which have used RNA interference (RNAi), which has become a powerful tool in locust research. RNAi is a method of post-transcriptional gene silencing, which employs double-stranded RNA (dsRNA) complementary to an RNA transcript of interest (typically protein-coding mRNA) to generate functional knockdowns via a highly conserved enzymatic pathway. RNAi has been used to great effect across a wide range of insects (Bellés, 2010; Cullen, 2012), and can cause a potent and specific systemic gene knockdown in locusts via a simple injection of dsRNA into the body cavity (Wynant et al., 2012). RNAi has been used to corroborate the dramatic role of the neuropeptide [His^7] corazonin in causing the black colouration and some morphometric characteristics of gregarious locusts (Sugahara et al., 2015, 2016), as well as being an important tool in investigating the molecular signalling pathways underlying behavioural phase change (see Sections 4 and 5). Intriguingly, crowding solitarious locusts leads to a rapid but transient increase (30 min to 1 h), in the expression of genes involved in mediating the RNAi response (Wynant et al., 2015). Other recent advances include automated video tracking and new ways of interpreting the resulting data, which have improved our methods for assaying locust behavioural phase (e.g. Cullen et al., 2010, 2012; Ma et al., 2011; Stettin and Ott, 2012) and underpin many of the physiological experiments discussed throughout this review.

We go on to survey how the altered lifestyles and ecology of solitarious and gregarious locusts necessitate alterations in brain structure, neural processing and learning capacity, to deal with the different requirements imposed by changes in their social and physical environments (Matheson and Rogers). The self-organisation of gregarious locusts into coherent migratory bands, potentially comprised of millions of individuals, all with a mostly uniform direction of marching is one of their most striking behaviours. Recent progress in the modelling and measuring of collective motion in locusts using new methodologies is presented (Ayali, Buhl and Sword).

We also cover work using comparative analyses of locusts and other acridid species showing only limited polyphenism, to give new insights into the evolution of full density-dependent phase change (Foquet and Song). The ecology of locusts, and the ways in which landscape, vegetation type and land use may influence the likelihood of locust outbreaks and plagues are also considered in some more detail than in Pener and Simpson (2009) (Cease and Latchininsky) with the Central American Locust, *Schistocerca piceifrons* being used as a specific case study (Poot-Pech and Song). Finally, we consider the costs and consequences of locust swarms and the damage they inflict from a socioeconomic perspective (Berry, Hadrich and Robinson). Much research on locusts, from the molecular to the ecological, is justified on the ground of the economic damage they inflict, yet this is often only loosely tied to quantitative analyses of the actual economic, and often surprising wider social, costs that they inflict. Here we include a brief survey of the priorities, methods, models and major findings of some of the social scientists working on human–locust interactions in the context of wider land-use strategies. As a result, this review aims to bridge the mechanisms and consequences of locust phase polyphenism from molecules to livelihoods.

The third and final objective is that this review will stand as a means to introduce the Global Locust Initiative (locust.asu.edu). This initiative's purpose is to enable innovative advances in locust research and management by promoting interdisciplinary systems- and solutions-oriented approaches. The initiative is based on three pillars: advancing fundamental research, creating and maintaining a global network, and developing local solutions to the global challenge of locust plagues. The initiative grew out of many conversations and collaborations between researchers, government and non-government agencies, and farmers' groups around the world. We will formally launch the Global Locust Initiative in 2018 and aim to keep it broadly inclusive and transparent.

1.2 Nomenclature, and a Note on Laboratory Strains

This review follows the standard convention of naming species by both their common and scientific names in the first instance, and abbreviated scientific name thereafter. Common names are sometimes reused to aid readability, and the reader is also referred to Table 1 for a full list of all accepted locust species, along with naming authorities and geographic range. Table 1 also gives the same details for all other species of Orthoptera named in the review.

Much of the work discussed in this review focusses on two species of locust reared in laboratory colonies: the Desert Locust, *Schistocerca gregaria*, and the Migratory Locust, *Locusta migratoria*. Different laboratory strains of the same species or subspecies can vary in many aspects of their physiology and behaviour, reflecting their region of origin, rearing conditions, and the number of generations in captivity. For further discussion see Berthier et al. (2010), who performed population genetic and inbreeding analyses across a range of laboratory locust strains.

Discussions of laboratory-reared *S. gregaria* will almost invariably refer to colonies derived from wild populations of the nominate phase-changing northern race, *S. gregaria gregaria* (Forskål, 1775), which has been brought into captivity from many regions of North Africa and the Middle East. By contrast, the southern race, *S. gregaria flaviventris* (Burmeister, 1838), does not undergo phase change (Chapuis et al., 2017) and is therefore highly unlikely to have been the founder subspecies for any laboratory colony used to successfully study this form of behavioural plasticity.

As the most widely distributed grasshopper species in the world, the situation for *L. migratoria* is more complex. Many subspecies of *L. migratoria* have been historically recognised, based primarily on their geographic range;

Table 1 Worldwide Locust Species

Scientific Name (Locusts)	Common Name	Geographic Range
Anacridium melanorhodon (Walker, 1870)	Sahelian Tree Locust	North and Central Africa, Middle East
Aiolopus simulatrix (Walker, 1870)	Sudan Plague Locust	North and Central Africa
Austracris guttulosa (Walker, 1870)	Spur-Throated Locust	Australia
Calliptamus italicus (Linnaeus, 1758)	Italian Locust	Europe and Central Asia
Ceracris kiangsu Tsai, 1929	Yellow-Spined Bamboo Locust	East Asia
Chortoicetes terminifera (Walker, 1870)	Australian Plague Locust	Australia
Dociostaurus maroccanus (Thunberg, 1815)	Moroccan Locust	North Africa, Europe, Caucasus
Gastrimargus musicus (Fabricius, 1775)	Yellow-Winged Locust	Australia
Gomphocerus sibiricus (Linnaeus, 1767)	Siberian Locust	Europe, Caucasus, North Asia
Locusta migratoria migratoria (Linnaeus, 1758)	Asian Migratory Locust	West and Central Asia, Eastern Europe
Locusta migratoria migratorioides (Reiche and Fairmaire, 1849)	African Migratory Locust	Mainland Africa and Atlantic Islands, Madagascar, South and East Asia, Australasia, Tibetan plateau
Locustana pardalina (Walker, 1870)	Brown Locust	Southern Africa
Nomadacris septemfasciata Audinet-Serville, 1883	Red Locust	Sub-Saharan and Southern Africa
Oedaleus asiaticus Bey-Bienko, 1941	Mongolian Locust	Northern Asia
Oedaleus senegalensis (Krauss, 1877)	Senegalese Locust	North Africa, Middle East
Patanga succincta (Johansson, 1763)	Bombay Locust	India, South-East Asia

Table 1 Worldwide Locust Species—cont'd

Scientific Name (Locusts)	**Common Name**	**Geographic Range**
Schistocerca cancellata (Serville, 1838)	South American Locust	South America
Schistocerca gregaria flaviventris (Burmeister, 1838)	Desert Locust (southern race)	Southern Africa (N.B., typically non-swarming)
Schistocerca gregaria gregaria (Forskål, 1775)	Desert Locust (northern race)	Southern Europe, North Africa, Middle East, Indian subcontinent
Schistocerca interrita Scudder, 1899	Peruvian Locust	Peru
Schistocerca piceifrons peruviana Lynch Arribalzaga, 1903	Central American Locust	Peru, Ecuador, Colombia, Venezuela, Panama, Trinidad and Tobago
Schistocerca piceifrons piceifrons (Walker, 1870)	Central American Locust	Mexico, northern Costa Rica
Scientific Name (Non-Locusts)	**Common Name**	**Geographic Range**
Aiolopus thalassinus (Fabricius, 1781)	Slender Green-Winged Grasshopper	Africa, Europe, Asia
Anabrus simplex Haldeman, 1852	Mormon Cricket	North America
Calliptamus barbarus (Costa, 1836)	Eurasian Pincer Grasshopper	North Africa, Europe, West and Central Asia
Calliptamus wattenwylianus Pantel, 1896	Western Pincer Grasshopper	North Africa, Europe
Camnula pellucida (Scudder, 1862)	Clear-Winged Grasshopper	North America
Chorthippus parallelus (Zetterstedt, 1821)	Meadow Grasshopper	Great Britain, Europe, Caucasus, West and Central Asia
Coscineuta virens (Thunberg, 1815)	Moruga Grasshopper	Trinidad and Tobago
Melanoplus differentialis (Thomas, 1865)	Differential Grasshopper	North America

Continued

Table 1 Worldwide Locust Species—cont'd

Scientific Name (Non-Locusts)	Common Name	Geographic Range
Melanoplus sanguinipes (Fabricius, 1798)	Migratory Grasshopper	North America
Oedaleus infernalis Saussure, 1884		Caucasus, Central and East Asia, Japan
Rhammatocerus schistocercoides (Rehn, 1906)	Mato Grosso Grasshopper	Brazil
Schistocerca americana (Drury, 1770)	American Bird Grasshopper	North America
Schistocerca ceratiola (Hubbell and Walker, 1928)	Florida Rosemary Grasshopper	Florida
Schistocerca damnifica (Saussure, 1861)	Mischievous Grasshopper	North America
Schistocerca lineata Scudder, 1889	Spotted Bird Grasshopper	North America
Schistocerca nitens caribbeana Dirsh, 1974		Caribbean
Schistocerca nitens nitens (Thunberg, 1815)	Grey Bird Grasshopper	North America, Mexico
Schistocerca serialis cubense (Saussure, 1861)		Florida Keys, Caribbean, Cuba, Jamaica

Other species of Orthoptera named in this review are also listed in the table.

this is reflected in the literature for both laboratory- and field-based studies. Ma et al. (2012) removed much of the uncertainty about *Locusta* subspecies in a comprehensive phylogeographical analysis, in which they sequenced 3 mitochondrial genes for 263 individuals across 53 worldwide sampling sites, along with 65 complete mitochondrial genomes. The authors tentatively collapsed all subspecies into just two, based on broadly northern and southern lineages: the nominate Asian Migratory Locust, *L. migratoria migratoria* (Linnaeus, 1758), and the African Migratory Locust, *L. migratoria migratorioides* (Reiche and Fairmaire, 1849). The latter race is of particular interest; it subsumed, perhaps unsurprisingly, the Malagasy Migratory Locust, *L. migratoria capito* Saussure, 1884 from Madagascar, but also the Oriental (*L. migratoria manilensis* Meyen, 1835) and Tibetan

(*L. migratoria tibetensis* Chen, 1963) subspecies (Ma et al., 2012). Throughout this review, we will use the recent reclassification of Ma et al. (2012) where a subspecies is given, but will also mention the defunct synonym given by the authors where applicable.

A further result of this reclassification is that some countries around central and east Asia are home to both of the currently accepted subspecies of *L. migratoria*. This is particularly obvious in China, which has wild populations of *L. migratoria migratoria* in the north and *L. migratoria migratorioides* in the south. Much of the molecular and physiological work discussed throughout this review was performed at the Chinese Academy of Sciences, Beijing. Their laboratory colony originates from the neighbouring northern provinces of Hebei and Shandong and, since they therefore belong to the nominate subspecies *L. migratoria migratoria*, the authors generally refer to their laboratory strain as simply *L. migratoria* L. (Professor Kang Le, personal communication to D. A. Cullen; Ma et al., 2012). In their highly anticipated genome description (Wang et al., 2014), the authors also used DNA from a male locust captured in the southern island province of Hainan, which would be *L. migratoria migratorioides* according to Ma et al. (2012). For all studies, and for all species or subspecies, the reader is advised to contact the relevant authors where taxonomic information or founder colony location is unclear.

1.3 Recent Locust Outbreaks

In the past decade, there have been widespread reports of locust upsurges around the world. The infamous Desert Locust, *S. gregaria*, can invade up to 29 million km^2 from Northern Africa to India, nearly 20% of the Earth's land area (Popov et al., 1991). The last major invasion of *S. gregaria* was in 2003–2005 when 13 million ha were treated with pesticides in 22 countries (Belayneh, 2005). The Food and Agriculture Organization of the United Nations (FAO) estimated that this plague caused crop losses of 80%–100% in afflicted regions, mostly in Sub-Saharan Africa (Brader et al., 2006), and cost over US$500 million to control (Belayneh, 2005). While there has not been an *S. gregaria* plague in the past 10 years, there have been numerous localised outbreaks that have led to swarm formation (FAO Locust Watch: http://www.fao.org/ag/locusts). In 2008, there were swarms in Ethiopia and Saudi Arabia. Widespread outbreaks occurred between 2012 and 2013 from Mauritania in the western region of this species' range across to Israel, Yemen and Saudi Arabia in the near east range. In Israel alone, control efforts by the Israeli Ministry of Agriculture were estimated to cost US$4 million, with more than 80% of this spent on nymphal band control (Israeli

Ministry of Agriculture and Nature and Parks Authority, personal communication to A. Ayali). Fortunately, damage to agriculture was very limited as the locusts were mostly confined to the Negev desert. The locust situation remained serious along the Red Sea coast and Horn of Africa through to March 2015. From 2015 to May 2017, the threat fluctuated in Yemen, Saudi Arabia, Sudan and Mauritania. As of May 2017, the risk and threat to crops have declined, but the situation is unclear in Yemen, where surveys cannot be conducted.

In east, central and southern Africa, several locust species are a periodic threat to food security including the Brown Locust, *Locustana pardalina*, the African Migratory Locust, *L. migratoria migratorioides*, and the Red Locust, *Nomadacris septemfasciata*. Vast *L. migratoria migratorioides* plagues covering parts of western, central and southern Africa were common historically in the late 1800s and the mid-1900s, but *N. septemfasciata* has likely posed the biggest threat in the past decade. In 2008–2009, Malawi, Mozambique, Tanzania and Zambia were hit with upsurges of *N. septemfasciata*. The FAO coordinated a campaign together with the International Red Locust Control Organization for Central and Southern Africa (IRLCO-CSA). This campaign was the first time biopesticides (the mycopesticide based on *Metarhizium acridum*) were used on a large scale against locusts in Africa (FAO Bulletin: http://www.fao.org/agriculture/crops/news-events-bulletins/detail/en/item/40241/icode/3/?no_cache=1). Without intervention, the FAO estimated that the invasion could have affected the food crops of around 15 million people (http://www.fao.org/news/story/en/item/21084/icode/). In January 2017, Zambia experienced outbreaks of both *L. migratoria migratorioides* and *N. septemfasciata*, which damaged over 10,000 ha of grassland and cropland. In February 2017, the director of the IRLCO-CSA said that US$4 million would be needed to tackle the problem (http://www.herald.co.zw/region-needs-4m-to-fight-locusts/). As of May 2017, the outbreaks are still a threat to food security across central and southern Africa.

In Madagascar, a plague of Malagasy Migratory Locust, *L. migratoria migratorioides* (previously known as *L. migratoria capito*), began in June 2012 and threatened the livelihoods and food security of 13 million people (60% of the country's population). During three locust campaigns in 2013–2016, the locusts were controlled across 2.3 million ha at a cost of US$37 million (see http://www.fao.org/emergencies/crisis/madagascar-locust/en/).

Kazakhstan, Russia and other countries of the Caucasus and Central Asia experienced high levels of infestations by the Asian Migratory Locust, *L. migratoria migratoria*, Italian Locust, *Calliptamus italicus*, and Moroccan

Locust, *Dociostaurus maroccanus* in the past decade. From 2010 to 2016, infested areas in this region varied from 7.1 to 11.2 million ha per year. Anti-locust treatments peaked in 2014 with 6.7 million ha out of which 4.7 million ha were treated in Kazakhstan, principally against *Ca. italicus* (see http://www.fao.org/ag/locusts-CCA/en/index.html).

Records of *L. migratoria* outbreaks in China reach back more than 1000 years (Stige et al., 2007), and they remain a threat to agriculture in present times. For example, from 2012 to 2016, *L. migratoria* outbreaks covered about 1.5 million ha (Long Zhang, unpublished data, personal communication to A. J. Cease). Both subspecies of *L. migratoria* are common in China, with several synonyms common in the literature (Ma et al., 2012; see Section 1.2). China has made great strides to increase biopesticide use in place of synthetic pesticides, specifically the fungus *Metarhizium acridum* (Bischoff et al., 2009) and the microsporidian *Paranosema locustae*. In 2004, around 5% of locust treatments used these biopesticides, but in 2017, they account for more than 30%. At over 100,000 ha treated, this is the largest application of biopesticides against locusts yet (Long Zhang and David Hunter, unpublished data, personal communication to A. J. Cease).

In eastern Asia, the Yellow-Spined Bamboo Locust, *Ceracris kiangsu*, is a periodic threat. In 2015, Laos was affected by an unprecedented outbreak of *Ce. kiangsu*, which led to food insecurity and loss of livelihood for more than 20,000 people by July 2015 (http://www.fao.org/laos/news/detail-events/en/c/414367/). Despite management efforts by the FAO and the Laos and Chinese governments, as of May 2017 *Ce. kiangsu* is still at plague proportions (https://www.laotiantimes.com/2017/05/05/locusts-swarm-northern-laos-wreaking-havoc-crops/).

In 2010, Australia experienced one of the worst plagues of the Australian Plague Locust, *Chortoicetes terminifera* in 40 years (Deveson, 2011). This plague was established following successful breeding in New South Wales and Queensland, and exacerbated by an intense La Niña in March 2010 that increased populations further (Deveson, 2013). At its peak, over 2000 landholders reported high-density bands; the plague subsided in early 2011 after control by the Australian Plague Locust Commission and state-level agencies (Deveson and Woodman, 2014). Since 2011, outbreaks have been reported across the country annually, with intense localised damage in some areas but no widespread plagues (Australian Government: http://www.agriculture.gov.au/pests-diseases-weeds/locusts).

The Americas are home to a few locust species that have had outbreaks recently including the Central American Locust, *S. piceifrons* (Poot-Pech, 2017; see Section 10.3), the South American Locust, *Schistocerca cancellata*, and the Moruga grasshopper, *Coscineuta virens*. *Co. virens* exhibits some characteristics of locust phase polyphenism and has periodic outbreaks in its restricted range in Trinidad (Song, 2011). There were outbreaks in 2010, then again starting in 2013 and continuing as of May 2017 (http://www.trinidadexpress.com/20170517/news/ll).

The South American Locust, *S. cancellata*, was the most destructive agricultural pest in Argentina during the late 1800s to the early 1900s. However, after an extensive 1954 control campaign involving 12,000 tons of pesticide (de Wysiecki and Lange, 2005), its numbers decreased considerably with smaller outbreaks in 1961, 1989 and 2010 (Barrientos Lozano, 2011; Hunter and Cosenzo, 1990). In the latter half of 2015, an upsurge level unprecedented in 60 years originated from the long-standing recession zone in NW Argentina and spread through a large region including Santiago del Estero, Catamarca, Tucuman, Salta, Chaco and Cordoba. The 2015 swarm caused fear in many rural areas because this locust was historically devastating; however, the swarms were controlled and there was little agricultural impact from this upsurge (Wilda Ramírez and Hector Medina, SENASA; Eduardo Trumper, INTA, Argentina, personal communication to A. J. Cease). In January 2017 in Bolivia, a 50-year hiatus in reported locust activity was interrupted by the appearance of a swarm near Santa Cruz. In February 2017, SENASAG estimated that more than 1000 ha of crops had been affected and President Morales declared a state of emergency, dedicating US$700,000 to control efforts (Luis Sanchez Shimura, SENASAG, personal communication to A. J. Cease; http://www.bbc.com/news/world-latin-america-38914922). The *S. cancellata* challenge is predicted to be ongoing, and Bolivia, Paraguay and Argentina are working together to develop a long-term research and management strategy.

2. EVOLUTION AND DIVERSITY OF SWARMING GRASSHOPPERS

Out of 6787 known species in the grasshopper family, the Acrididae (Cigliano et al., 2017), only 19 species are currently considered true locusts, which belong to at least 4 different subfamilies (Table 1; see also Pener and Simpson, 2009 (table 1 therein); Song, 2011). It is clear from this spread across the Acrididae that swarming locusts have evolved multiple times but, at the same time, this has been a relatively rare occurrence (i.e. only 0.28% of known acridids are locusts). So, how did swarming locusts evolve, and what are the factors promoting this transition? In order to address this question properly, it is important first to define what a locust is, and how they differ from other grasshoppers. Pener (1983) defined locusts as acridids that (i) form dense migrating groups as bands of nymphs and/or swarms of adults, and (ii) are polyphenic in the sense that individuals living on their own or at low population densities differ in many characteristics from those living in groups. How rigidly demarcated is this definition?

Most of what we know about locusts comes from studies of two model species: the Desert Locust, *S. gregaria*, and the Migratory Locust, *L. migratoria* (Pener and Simpson, 2009). However, not all locusts are similar to these two species because there are a large number of species-specific variations among 'non-model' locust species (Song, 2011). For example, the Australian Spur-Throated Locust, *Austracris guttulosa*, and the Bombay Locust, *Patanga succincta*, do not form nymphal bands (Song and Wenzel, 2008), while other locust species march as nymphs. Nymphal colouration of *C. terminifera* and *Ca. italicus* does not change in response to population density changes (Song, 2011), while density-dependent colour change is typically associated with true locusts. Moreover, there are some non-migratory grasshoppers that show a certain level of collective movement, such as *Melanoplus sanguinipes* and *Camnula pellucida* (Pfadt, 1988; Song, 2011), as well as density-dependent colour change such as many species in the subfamily Cyrtacanthacridinae (Song and Wenzel, 2008; Fig. 1). Furthermore, there are non-swarming species that are closely related to locusts, such as *Schistocerca americana*, which show density-dependent phenotypic plasticity reminiscent of true locusts (Gotham and Song, 2013; Sword, 2003) and exhibit a certain level of collective movement when the density is high enough (Kuitert and Connin, 1952). In other words, there is no clear distinction between grasshoppers and locusts when we begin to examine the biology of acridids other than *S. gregaria* and *L. migratoria*. Instead, there seems to be a gradient of traits that range between 'sedentary' or 'non-migratory' grasshoppers and swarming locusts. Thus, to understand the evolution of swarming locusts, we need to examine the evolution of individual components that make up locust phase polyphenism, especially density-dependent phenotypic plasticity and collective movement. Furthermore, we need to approach these two phenomena from both process (i.e. adaptive significance) and pattern (i.e. phylogenetic) perspectives. When examined in detail, and as set out in this review, even species that show the full suite of traits associated with full locust phase polyphenism, such as *S. gregaria* and *L. migratoria*, differ markedly in their ecology, circumstances in which they

Fig. 1 Diversity of nymphal colour patterns within the genus *Schistocerca*. (A) *S. lineata*, which specialises on *Ptelea trifoliata*, reared at high density. This species shows density-dependent aposematic colouration. (B) *S. nitens caribbeana* reared in isolation. This species does not show density-dependent colour plasticity but exhibits a green-brown polymorphism irrespective of rearing density. (C) *S. damnifica* reared at high density. Isolated nymphs are typically *green*. (D) *S. ceratiola* reared in isolation. This species does not change colour when crowded. (E) *S. americana* reared in isolation (*left*) and under crowded conditions (*right*). (F) *S. piceifrons* reared at high density. (G) *S. serialis cubense* reared at high density. This species shows a very reduced level of colour plasticity. (H) *S. nitens nitens* reared at high density. Like *S. nitens caribbeana*, this species does not show density-dependent colour plasticity but exhibits a green-brown polymorphism that does not appear to be linked to rearing density.

undergo phase transition, rate of transformation, and underlying physiological mechanisms.

A defining feature of locusts is the expression of density-dependent phenotypic plasticity (Simpson and Sword, 2009; Sword and Simpson, 2008). While we often include colour, behaviour, morphometric ratios and life history traits as major plastic reaction norms of locust phase polyphenism, it is now clear that not all of these reaction norms are plastic in all locust species (Song, 2011), and that different reaction norms of locust phase polyphenism have evolved separately from each other (Song, 2005; Song and Wenzel, 2008). In particular, density-dependent colour plasticity provides a good case study for how an individual reaction norm might have evolved. Initially, the function of density-dependent colour plasticity was unclear because the most obvious explanation—that the conspicuous colour of gregarious nymphs is aposematic—was not supported because vertebrate predators would readily feed on gregarious nymphs (Andiatsirevombola et al., 2016; Gillett and Gonta, 1978). A novel insight was provided by Sword (1999), who studied a Texas population of *Schistocerca lineata*, which feeds on the toxic Wafer Ash, *Ptelea trifoliata*. In this population, nymphs show density-dependent warning colouration, in which crowding induces bright yellow and black colour patterns in nymphs (Fig. 1A), and it was experimentally shown that vertebrate predators quickly learn to associate this colour pattern with unpalatability. Based on this finding, Sword et al. (2000) tested whether a similar phenomenon occurred in *S. gregaria* and confirmed that, at the onset of population increase, gregarious *S. gregaria* nymphs in Mauritania preferentially feed on a native toxic plant, Egyptian Henbane, *Hyoscyamus muticus*, which suggested that *S. gregaria* also expresses density-dependent warning colouration. More recently, Maeno et al. (2012) showed that solitarious nymphs of *S. gregaria*, which have a cryptic colouration ranging from green to brown, preferred to shelter on a spiny plant, *Fagonia arabica*, suggesting that solitarious nymphs use this plant as a physical protection against predators. These studies collectively suggest that density-dependent colour plasticity should be considered as an induced anti-predator strategy in both directions (Simpson and Sword, 2009; Sword, 2002). A recent study on the food preference of *S. piceifrons* showed that this species has a strong attraction to the odour of *Pisonia aculeata*, but has a feeding preference to two other plant species (Poot-Pech et al., 2016; see Section 10.3). Interestingly, *P. aculeata* is an evergreen thorn bush, which is used as a shelter or a refuge. So, it is possible that further studies on *S. piceifrons* might uncover a similar pattern of behaviour to that observed in *S. gregaria*.

True aposematic warning patterns (as opposed to mimetic patterns) should signal that the organisms bearing them are toxic or unpalatable, which may be conferred either from ingested secondary plant compounds or by synthesising defensive chemicals de novo. However, the feeding ecology of grasshoppers and locusts in relation to host plant toxicity is largely unknown for many species. It is also unclear why certain locust species do not show density-dependent colour plasticity (Song, 2011). From a phylogenetic standpoint, a plastic reaction norm in nymphal colouration should not be considered as an adaptation that has evolved de novo in locusts, but rather an exaptation. It is clear that density-dependent colour plasticity is a phylogenetically conserved trait, based on a broad survey within the

Acrididae (Song, 2011), and at least within the Cyrtacanthacridinae (Fig. 1; Song and Wenzel, 2008). Most species of *Schistocerca*, regardless of whether they are locusts or not, show this trait although the colouration induced by crowding does not appear to be classic warning colouration in many species (Song et al., 2017). More widely, it appears that many species within Cyrtacanthacrinae have the capacity to change colour in response to increased population density (Song, 2011; Song and Wenzel, 2008). Although it is still unclear how and why this plastic reaction norm evolved in the first place, its widespread expression suggests that it may not be costly to maintain or, at least, it is not strongly selected against. Thus, most *Schistocerca* species exhibit density-dependent colour plasticity not because it is adaptive in every species, but because their common ancestor already possessed this trait. A recent study showed that the *Ptelea*-feeding *S. lineata* is a genetically distinct population within the species that has a wide distribution in North America (Raszick and Song, 2016). This suggests that this population was able to co-opt a pre-existing colour plasticity with an unknown function towards density-dependent warning colouration, as the population evolved a feeding preference to a toxic host plant. It is probable that a similar process has happened multiple times in various locust species, at least within the Cyrtacanthacridinae. See Fig. 1 for some examples of colour polyphenism and polymorphism in the genus *Schistocerca*; the reader is referred to Pener and Simpson (2009) for detailed coverage of the physiological control of colour polyphenism.

There is some evidence that cannibalism may contribute to collective movement in nymphal bands of *S. gregaria* (Bazazi et al., 2008) and, in turn, nutritional state may influence the tendency to cannibalise. For example, Hansen et al. (2011) found that *C. terminifera* nymphs in a marching arena tended to cannibalise weak conspecifics, especially under conditions of protein deprivation (Hansen et al., 2011). Furthermore, individuals that cannibalised survived longer and thus moved farther than those individuals without the opportunity to cannibalise. In another study, Bazazi et al. (2011) found that while nutritional state did not influence the movement of solitarious nymphs, feeding gregarious locusts a high-carbohydrate, low-protein diet led them to walk both more rapidly and more frequently (see Section 9 for more discussion on the influence of diet on phase change). More recently, Guttal et al. (2012) developed an evolutionary model, examining how the threat of cannibalism could influence the evolution of local interactions and collective movement in *S. gregaria*. Their model predicted that solitarious nymphs would show a clear avoidance of other individuals, but an increase in density will lead to a clear change in behaviour: an escape-pursuit behaviour in which locusts move away from nearing individuals, while at the same time pursuing individuals that are moving away. When simple avoidance behaviour does not inhibit conspecific interactions anymore, exploiting areas left vacant by other individuals is the most effective way to reduce encounters with conspecifics. This model shows that cannibalism could be one of the main drivers for the evolution of density-dependent behavioural plasticity (see Section 8.1 for further discussion on marching behaviour and the possible role of cannibalism).

A recent study showed that two non-swarming species of *Schistocerca*, *S. americana* and *S. serialis cubense*, display density-dependent behavioural plasticity, although in a reduced form (Gotham and Song, 2013). This implies that plastic reaction norms in behaviour may still be retained in those species that are phylogenetically close to the locusts. In other non-migratory *Schistocerca* species however, crowding does not seem to induce any changes in behaviour, which suggests that behavioural plasticity has been lost in these species (Song et al., 2017).

Final insights into the evolution of swarming locusts come from population genetic studies of several locust species. Due to their migratory potential, locusts are generally expected to show a high degree of gene flow between populations (Chapuis et al., 2009, 2011). Indeed, recent studies were not able to find any major population genetic structures in several locust species, including solitarious populations of *S. gregaria* (Chapuis et al., 2014), *C. terminifera* (Chapuis et al., 2011), *Ce. kiangsu* (Fan et al., 2014), and the non-phase-changing but highly mobile pest species *Oedaleus infernalis* (Sun et al., 2015). Studies of the population genetic structures of *L. migratoria* showed that there are two major worldwide lineages, in contrast to a number of subspecies that previous taxonomic studies proposed (Ma et al., 2012; Tokuda et al., 2010). While each lineage covers a huge geographical area, the genetic diversity within seems to be minimal, a pattern that is consistent with other locust species. Another interesting finding comes from the comparison between swarming and non-swarming species. Blanchet et al. (2011) compared population genetic structure in three different species of the genus *Calliptamus*: the Italian Locust, *Ca. italicus*, even though it never swarms in the region where the study took place; *Ca. barbarus*, which has a similar wing length to *Ca. italicus*; and *Ca. wattenwylianus*, which has significantly shorter wings that result in a reduced mobility. As expected, *Ca. wattenwylianus* showed a clear population genetic structure, even within a single field, but population genetic structure was absent in the other two species. Although most of these studies have relied on a relatively low number of molecular markers, they are consistent in highlighting the general pattern that there are very high heterozygosity values and/or high variance within populations, both of which are indicative of a large population size (Chapuis et al., 2009). It seems reasonable to postulate that where populations are this large and homogeneous, locust species are unlikely to speciate further because the lack of genetic structures and ecological differentiations among different populations would make it very difficult to develop reproductive isolation within species. The circumstances allowing for sympatric speciation or ecological speciation taking place in locusts would be exceptionally rare. The only way locusts could give rise to a new species is probably a founder event to an area where future gene flow would be significantly reduced. One such example is the southern subspecies of the Desert Locust, *S. gregaria flaviventris*. A recent study shows that this subspecies, which is restricted to the southwestern region of Africa, is genetically and morphologically divergent from the nominate subspecies in northern Africa (Chapuis et al., 2016). What is more intriguing is that this southern population has a much-reduced expression of density-dependent phenotypic plasticity (Chapuis et al., 2017; Waloff and Pedgley, 1986), which means that there could have been a selective pressure that promoted the loss of plastic reaction norms. When environmental conditions do not promote increases in population density and the expression of gregarious phase is not adaptive, perhaps because resources are homogeneously plentiful, locusts may give rise to non-swarming grasshoppers. A more clear case can be found in the wider *Schistocerca* genus, which contains approximately 50 species distributed throughout the New World (Song et al., 2013). It has been hypothesised that the North African *S. gregaria* is the earliest diverging lineage within the genus, and that the diversity of the genus in the New World is a case of adaptive radiation followed by multiple westward transatlantic colonisations by a '*gregaria*-like' ancestor (Lorenz, 2009; Lovejoy et al., 2006; Song et al., 2013). The fact that most *Schistocerca* species are non-swarming suggests that the genetic capacity to swarm that must have been present in the ancestral lineage has subsequently been selected against, suggesting that the environmental conditions in the New World must have been more favourable to cryptic and solitary lifestyles. The ability to swarm has eventually been lost, though remnants of density-dependent colour plasticity have remained.

In conclusion, we argue that the evolution of swarming locusts is a result of the evolution of numerous reaction norms that have followed different evolutionary trajectories throughout the diversification of grasshoppers. While we continue to gain an in-depth understanding of the proximate mechanisms of locust phase polyphenism from the studies of *S. gregaria* and *L. migratoria*, we should broaden our views to embrace a comparative perspective; there is clearly more than one way of becoming a locust.

3. MEASURING PHASE-RELATED BEHAVIOUR

Laboratory-based experiments conducted since Pener and Simpson (2009) have largely continued to use variants of the assay system established by Roessingh et al. (1993) to quantify behavioural phase state. This assay seeks to condense the multi-dimensional nature of locust phase-related behaviour into a single metric of gregariousness (or solitariousness). Two broad categories of behaviour, relating to activity and attraction to other locusts, are particularly important in assessing phase state, both of which are much higher in long-term gregarious than in solitarious locusts. In setting up this assay in the laboratory a baseline of behavioural range and variation is established using observations of approximately 100 solitary-reared and 100 crowd-reared locusts, which are individually observed in a rectangular arena with a back-lit 'stimulus group' of 20–30 conspecific locusts behind a clear, perforated partition at one end and an empty back-lit chamber at the other end (discussed in detail in Cullen et al., 2012; Pener and Simpson, 2009: p. 164; Rogers et al., 2014; Simpson et al., 1999). During each 5–10 min assay the observer records various behavioural measurements for the individual locust, which can be broadly divided into three categories: first, behaviours related to overall activity (e.g. distance moved, walking speed, time spent climbing the arena walls, ground covered/path tortuosity); second, behaviours related to overall attraction to, or repulsion from, the stimulus group (e.g. average distance to the stimulus group, time spent at the stimulus and/or non-stimulus end(s), or some composite thereof); and third, other small behaviours are sometimes included, particularly grooming time/frequency, which is consistently found to be higher in gregarious individuals of *S. gregaria* (e.g. Anstey et al., 2009; Ott et al., 2012; Rogers et al., 2014). Most studies of this nature have used immature locusts to avoid the added complications of flight and sexual behaviour that come with adult insects, though there have been some exceptions; see Cullen et al. (2012; appendix table A1 therein) for a comprehensive overview of all studies up to that year.

Following data collection for the 200 or more individual locusts of a known rearing history, the behavioural data for all insects are then fed into a stepwise binary logistic regression procedure, which selects the variables that best separate the average behavioural profiles of the two extreme phases before combining and weighting these variables to give the logistic equation. This equation can then be used to deliver a single value of an individual locust's behavioural state. With the baseline behavioural variation of the two phases established, locusts of unknown phase state can then be tested in the arena assay with reference to these two model populations. When designated as P_{greg} (i.e. the probability of being in the gregarious phase) a returned value close to 1 means that the behaviour shown by the test locust is much more likely to be observed among long-term gregarious locusts, while a P_{greg} close to 0 means that the behaviour of the test animal is much more similar to that seen in long-term solitarious locusts (or vice versa where the inverse 'P_{sol}', i.e. probability of being in the solitarious phase, is used instead). Any subsequent test locust that has been subject to some experimental manipulation (e.g. specific sensory stimulation, pharmacological treatment, RNAi, nutritional perturbation) is then assayed in the same way, with a value of P_{greg} returned as a single composite predictor of that animal's behavioural phase state.

3.1 Recent Behavioural Studies

3.1.1 Behavioural Assays of Schistocerca gregaria

Notable studies that have implemented the Roessingh et al. (1993) assay system over the past decade include the neuropharmacological work of Anstey et al. (2009) and Ott et al. (2012), which, respectively, showed that serotonin (5-HT) and Protein Kinase A (PKA), a downstream effector of several biogenic amine receptors, are both necessary and (in the case of 5-HT) sufficient to trigger the transition towards gregarious behaviour in *S. gregaria*. The behavioural models in both of these studies correctly classified 90% of the locusts from the model-building population and featured almost identical variables for returning the final measurements of P_{greg}; walking speed, proportion of time at rest, and time spent in the 25% of the arena next to the stimulus group were used in both cases. The fourth variable in both models was grooming, but measured in slightly different ways in each case (total grooming time in Anstey et al., 2009, and frequency of grooming episodes in Ott et al., 2012). The specific weightings for each variable also differed marginally between the two studies, which is unsurprising given that the two models were built using locusts from different laboratory colonies (the former Oxford University *S. gregaria* colony for Anstey et al., 2009, and the Cambridge University *S. gregaria* colony for Ott et al., 2012). The fact that such similar behavioural models could be independently applied to distinct colonies of *S. gregaria* stands as a testament to both the robustness of the Roessingh et al. (1993) assay and the P_{greg} behavioural metric, and to the reliability with which *S. gregaria* will generally perform in the relatively unnatural setting of a rectangular arena-based assay.

The four individual behaviours retained in these models change in concert throughout the course of behavioural phase transition, as demonstrated more recently in a meta-analysis of *S. gregaria* behavioural transition by Rogers et al. (2014). This study deconstructed P_{greg} into its constituent behavioural components, using data collected across an 8-year period of research (Anstey et al., 2009; Rogers et al., 2003; Simpson et al., 2001), and supported previous findings that forced crowding, tactile stimulation of the hind femora, and the pharmacological application of serotonin all induce synchronous changes in the full suite of retained behavioural measurements including grooming and attraction to other locusts, and not merely changes in activity (as had been suggested by Tanaka and Nishide, 2013). Furthermore, and importantly, Rogers et al. (2014) also showed that no single behavioural variable retained in the model was the sole driver of P_{greg}, and no single variable changed disproportionately strongly relative to the others in the face of serotonergic driving of behavioural phase. Indeed, across the 196 solitarious and 196 gregarious locusts assayed throughout the three studies considered in the meta-analysis, not a single locust with a P_{greg} below 0.43 in the composite behavioural model approached the stimulus group of conspecific locusts and, conversely, only two locusts with a P_{greg} above 0.8 did not approach the stimulus group. This study serves to highlight the necessity of a multi-dimensional behavioural model, in which measures of both activity and group-related behaviours are important for producing a well-balanced predictor of phase state.

A different methodology for measuring behavioural phase change in *S. gregaria* was taken by Geva et al. (2010). Instead of separate crowding treatments and assay measurements, test solitarious and gregarious fourth-instar nymphs were placed in 20 cm diameter circular arenas and crowded with 10 long-term gregarious locusts and continuously filmed. From these videos, 5-min bins were extracted for quantitative behavioural analysis. Solitarious locusts, on being placed in a crowd from which they could not remove themselves, were highly aroused and exhibited high activity, jumping, and avoidance behaviour as they sought to escape. By contrast, gregarious locusts were much less active in this behavioural context. This situation is therefore the opposite to that seen during the Roessingh et al. (1993) style behavioural assays, in which stimulus and testing regimes are separated, and where solitarious locusts are much less active than gregarious locusts. The overall activity of solitarious locusts decreased rapidly within 30 min as they remained in the crowd. A logistic regression model using the activity of solitarious and gregarious during the first 5 min of crowding as the model populations was constructed. The model showed some convergence in P_{sol} over the course of 4 h of crowding, driven principally by a dramatic change in the solitarious locusts, but the gregarious locusts also changed from their baseline, presumably as they became more active following prolonged crowding. The authors interpret these data as demonstrating a behavioural phase change process and that habituation to the presence of other locusts is an important element driving behavioural gregarisation. Solitarious locusts crowded for 30 min, removed from the crowd for 24 h and then reintroduced for a further 30 min retained a memory of their prior exposure and were less active on reintroduction to the crowd than naïve locusts, or even locusts that had been continuously crowded for an hour. This long-lasting effect could be removed by treating the locusts with the protein-synthesis inhibitor anisomycin one hour after a 30 min crowding session; on reintroduction into the crowd, locusts displayed the high activity characteristic of naïve locusts.

It is unclear whether the behaviour highlighted by Geva et al. (2010) is the same kind of behavioural change seen using other experimental paradigms. At the neuronal level, several studies have shown that solitarious locusts undergo pronounced habituation to repeated visual looming stimuli (Matheson et al., 2004; Rogers et al., 2007, 2010) and to tactile stimulation of the head (Fuchs et al., 2003), whereas habituation is much more muted in gregarious locusts (see Section 7). Of itself therefore, habituation is not evidence of gregarisation. Other experiments suggest that gregarisation, or at least some aspects of it, is not very habituation-like. For example, locusts subjected to tactile stimulation of the hind femora for 4 h (Simpson et al., 2001; see Section 4) behave in a manner identical to long-term gregarious locusts even though they have *never* had any exposure to other locusts. Furthermore, the sensory modality used to engender the behavioural change, touch, is not available in the Roessingh style arena, in which the locusts can see and smell other locusts in the stimulus group but not touch them; there is a complete disjunction between the stimuli inducing gregarisation and those available during the testing of it. A general loss of arousal during prolonged tactile stimulation can be excluded because of the specific nature of the tactile stimulus needed to evoke gregarisation, and by the general increase of activity characteristic of the gregarious phase seen in this behavioural context. Nevertheless, the rapid behavioural habituation and loss of arousal seen by solitarious locusts when placed in a crowd may be an important means of capturing and retaining them within a group so that they are subjected to prolonged exposure to gregarising stimuli.

3.1.2 Automated Video Tracking of Chortoicetes terminifera

Despite the consistently robust behavioural models that the Roessingh et al. (1993) assay and logistic regression have returned for *S. gregaria*, the observer-dependent nature of this manual assay system has historically meant that each researcher has had to build their own model from an independent set of solitarious and gregarious locusts. This has inevitably made it difficult to directly compare the raw behavioural data (and their resultant P_{greg} values) between studies, particularly those conducted in different research groups, since they were each based on measurements by different human observers, and therefore potentially subject to inherent observer-based variation. This has arguably led to the unfortunate situation in which each study (or laboratory-linked cohort of studies) of behavioural phase using the Roessingh et al. (1993) assay (or indeed, any other observer-dependent system) has tended to stand alone as a discrete set of experiments with only a limited capacity for direct comparison between data sets. This applies to both the baseline behaviour of the model-building gregarious and solitarious locust populations, and the experimental/transitional animals between the two extremes.

Fortunately, these issues should be resolved with the increasing implementation of automated video tracking to the Roessingh et al. (1993) assay system. This development is among the most significant advances in methodology that has occurred in locust behavioural phase research in recent years, with the first study to do so (Gray et al., 2009) the only one covered by Pener and Simpson (2009). Gray et al. (2009) used the commercial software EthoVision® 3.1 (Noldus Information Technology, Wageningen, The Netherlands; Noldus et al., 2001) to address the hitherto unresolved question of whether the Australian Plague Locust, *C. terminifera*, expresses behavioural phase polyphenism. EthoVision, which was originally developed for studies in rodents, uses a proprietary video tracking algorithm to return a suite of preprogrammed behavioural measurements, independent from the operator and therefore repeatable. Crucially, these include measurements from the two main categories of phase-related behaviour; activity (including distance moved, velocity and several measures of path tortuosity, all relative to a ruler on the calibration image taken with the recording camera) and position relative to the stimulus group (as defined by zones drawn by the user on a calibration image within the graphical user interface). Gray et al. (2009) definitively showed that *C. terminifera* expresses strong behavioural phase polyphenism. The logistic regression model built from EthoVision output data of final-instar nymphs retained total distance moved (activity-related), average distance to the stimulus chamber (position to conspecifics-related) and total time climbing the arena walls (a mixture of activity and positional data) to correctly classify 97.4% of the 78 crowd-reared model insects, and 100% of the 78 solitary-reared locusts. This overall accuracy of 98.7% remains the highest yet published for a locust behavioural model, and demonstrates both the power of automated video tracking and the robust expression of behavioural phase polyphenism in *C. terminifera*. Automated video tracking also confers the added advantage that multiple behavioural arenas can be tracked at the same time (two in the case of *C. terminifera*), speeding the conduct of laboratory-based experiments.

Following on from Gray et al. (2009), further behavioural work in final-instar nymphs of *C. terminifera* used automated video tracking to determine the proximal stimuli that induce phase change in this species. Cullen et al. (2010) tested the effects of olfactory and visual stimuli (both in isolation and in combination) and the effects of tactile stimulation directed to a range of body parts on phase transition in *C. terminifera*, in an experimental design similar to those in previous studies of *S. gregaria* (Rogers et al., 2003; Simpson et al., 2001). Cullen et al. (2010) determined that tactile stimulation of the antennae is the key stimulus triggering behavioural phase change in *C. terminifera* (see Section 4 for further discussion). Furthermore, Cullen et al. (2010) extended the methodology for automated behavioural tracking in *C. terminifera* by employing an identical EthoVision-based assay to Gray et al. (2009) but using a two-step model building and validation process. In the first step, a behavioural model was built using tracking data for 70% of the available model population (amounting to 98 solitarious and 75 gregarious locusts in this subset). The most accurate logistic regression model correctly classified 93.9% of the solitarious and 90.7% of the gregarious insects, and retained four of the available behavioural variables incorporating both activity- and position-based data: mean angular velocity, mean velocity, time spent in the non-stimulus third of the arena, and time spent climbing the non-stimulus wall. In the second step, this model was validated by testing it against the remaining 30% of the available model population (25 solitarious and 49 gregarious locusts). The model correctly classified 96% of the solitarious and 97.9% of the gregarious locusts in the validation subset, giving an overall accuracy of 93.9% (not 92.5% as the authors erroneously conclude in their discussion). This simple 70:30 division of the available model population into separate model-building and model verification cohorts ensured that the model was not over-fitted to the behaviours of the 70% cohort from which the model was built but had general applicability, and guaranteed the robust downstream assaying of experimentally manipulated locusts in the same study. Cullen et al. (2012) performed further optimisation experiments for the automated video tracking of *C. terminifera* final-instar nymphs, with a focus on ascertaining the most biologically pertinent behavioural variables to include in the model-building process. They showed that climbing variables should be treated with caution, as they were essentially redundant 'edge effect' measurements of the times spent in the stimulus and non-stimulus zones. This study also determined the optimum assay duration, by building models from the data extracted from prerecorded behavioural tracks that were arrested after the first 2, 4, 6 or 8 min. The only robust model, which correctly classified 92.4% of the 251 model insects, was built from the streamlined suite of behavioural variables for the full 8-min tracks; all other models were substandard in some way. Finally, and perhaps surprisingly, Cullen et al. (2012) also showed that automated video tracking is capable of extracting behavioural data that were previously unnoticed, or otherwise unquantified. 'Mean angular velocity' varied significantly between gregarious and solitarious *C. terminifera*, and upon closer inspection of the raw video tracks the authors realised that this was due to a greater number of small turns and body movements in the solitarious animals, which were typically performed on the non-stimulus side of the tracking arena. This variable, which would have gone unnoticed without automated video tracking, was retained in their optimal behavioural model. Conversely, thus far, automated tracking has not been successful in picking up some of the discrete behavioural differences between solitarious and gregarious locusts that were a feature of some of the manual models. These behaviours include grooming, which is consistently higher in gregarious *S. gregaria* (Rogers et al., 2014), and peering, in which locusts move their head side to side to estimate distance from motion parallax cues (Collett, 1978). Although

EthoVision is currently the de facto standard for automated video tracking in locust phase research, a freely available open-source software alternative has been developed and used successfully in *S. gregaria* (Stettin, 2015; Stettin and Ott, 2012).

3.1.3 Automated Video Tracking of Locusta migratoria

The Roessingh et al. (1993) assay system augmented with automated video tracking by EthoVision has also been successfully employed across a program of phase-related research in *L. migratoria*, to unravel the molecular physiology of behavioural phase in fourth-instar nymphs (discussed at length in Section 5.2). The first study to perform automated video tracking in *L. migratoria* (Ma et al., 2011) implicated dopamine synthesis and release as the triggers for phase change in this species, and started by attempting to establish separate logistic regression models for each of the five larval instars. The stepwise model selection retained different subsets of behavioural predictor variables for different instars, but this does not necessarily indicate that behavioural phase differences differ between instars. Instead, the different sets of retained variables may at least in part be coincidental outcomes that reflect the instability of stepwise variable selection in the presence of correlations among predictors (Harrell, 2001). Logistic regression models for the first- and second-instar nymphs of *L. migratoria* failed to retain any positional variables and, in the case of the first-instar assay, the two retained variables were likely measurements of the same thing (total distance moved and frequency of movement). The authors describe these first two instars as exhibiting 'partial phase-specific behaviour', but basing this conclusion on stepwise regression alone is potentially problematic. These early instars of *L. migratoria* may indeed have lacked any phase-related differences in attraction to conspecifics. Alternatively, such differences may have merely eluded detection due to the particular variables and arena zones stipulated by the authors, or due to the shortcomings of automated stepwise model building. Models for the third, fourth and fifth instars successfully retained measurements of both activity and position relative to the stimulus group, but again with an apparent degree of redundancy. Alongside frequency of movement, the third-instar model retained entry frequency into the 25% of the arena near the stimulus zone and total duration close to walls, though the *particular* walls were not specified so it is possible that both variables were measurements of the locusts' movements near the stimulus group. Furthermore, entry frequency into the stimulus zone (or any other zone) is also highly likely to correlate with activity; a locust walking in rings around the outside of the arena will enter zones more frequently than a locust that has approached the stimulus zone and then remained there for the duration of the assay. The fourth-instar model, by contrast, only retained one position-related variable ('attraction index', a simple composite of the time spent in the 25% of the arena near the stimulus group, minus the time spent in the opposite non-stimulus end) but two similar measurements of activity (total distance moved and total duration of movement). Accuracies of these behavioural models were not given, though an almost identical version of the fourth-instar *L. migratoria* model was used in Guo et al. (2011), in which the authors implicate the olfactory-related *chemosensory protein* (*CSP*) and *takeout* genes in phase-related attraction/repulsion behaviour of this species. This model combined the same three behavioural variables (total distance moved, total duration of movement and attraction index) with almost identical weightings in the model, to return an average accuracy of 87.1% (correctly classifying 87.2% of 86 gregarious model locusts and 87% of 69 solitarious animals). Further evidence of variable redundancy is evident in Tong et al. (2015), who again use a model retaining the same three variables but also present plots for each of them separately (see fig. 4 therein). Their bar charts demonstrating the separation between solitarious and gregarious *L. migratoria* for total distance moved and total duration of movement are highly similar, suggesting that one variable is probably redundant. More recently, this canon of work in *L. migratoria* has settled on a common model in three studies investigating the role of biogenic amines in behavioural phase in this species (Guo et al., 2013; Ma et al., 2015; Xu et al., 2017). Again focussing on fourth-instar nymphs, this model achieved a slightly higher accuracy of 90.2% (89.2% of 100 solitarious model locusts and 91.2% of 100 gregarious animals) but also contains a potential redundancy among the two activity-related variables, total distance moved and frequency of movement, retained alongside attraction index.

The behavioural models for *L. migratoria* each contain at least one activity- and one position-related variable, thereby capturing some element of the multi-dimensional nature of locust behavioural phase. Nevertheless, there does appear to be some redundancy in each of the logistic regression models highlighted above. Total distance moved, total duration of movement, and frequency of movement are highly likely to be significantly correlated and, without access to the raw data (as is usual), it is impossible to determine the extent to which these separate measures of activity add some additional behavioural information into their respective models, if any. Furthermore, the accuracies of these models (where given) are not as uniformly high as they have been for those of *S. gregaria* (e.g. Roessingh et al., 1998; Rogers et al., 2003; Simpson et al., 2001) or more recently for *C. terminifera*, which were also obtained using automated video tracking (Cullen et al., 2010, 2012; Gray et al., 2009). Nevertheless, the recent behavioural models for *L. migratoria* obtained through automated tracking have still been a substantial step forward since, prior to the work discussed here, there were few studies in which *L. migratoria* were assayed in this type of arena at all. In one notable exception, Hoste et al. (2002) assessed the phase-related behaviour of an albino strain of *L. migratoria* deficient in the neuropeptide [His^7]-corazonin. Their model, which retained turn rate, single leg movement count and time in the non-stimulus third of the arena, only achieved an accuracy of 70.2% (a value of 50% would be obtained by random guessing).

3.2 Suggestions for Future Studies

We finish this section by suggesting a small number of recommendations with a view to obtaining some universally recognised standards in automated video tracking that capture the multi-dimensional nature of locust behavioural phase: (i) Wherever possible, try to ensure that models retain both activity- and position-related data. (ii) When selecting behavioural variables to input into the behavioural model (either logistic regression-based or otherwise), choose only those that have the potential to add some extra information to the model; therefore, avoid using redundant measurements of the same behaviour (e.g. total distance moved, walking duration, and number of walking bouts). Ensure that climbing does not correlate strongly with some other activity- or position-related variable, and that frequency of zone entry is not simply a redundant measurement of activity. (iii) Where possible, verify the accuracy of the model with a separate subset of behavioural data, obtained from animals with a known history. Perform formal tests of model fit wherever possible (see those discussed in Hosmer et al., 1997 for logistic regression). (iv) Finally, present the raw data for each of the variables retained in the model, and plot as separate graphs (e.g. in the supplementary data for the publication) to allow scrutiny of how specific behavioural changes and their intercorrelation lead to an altered P_{greg}. This final point has become increasingly important as more strains and species of locusts are assessed for behavioural phase change. The logistic regression approach is bounded by the variation shown by the model populations used to construct each model. It cannot meaningfully compare the variation size between solitarious and gregarious locusts across different strains or species.

For example, a 'strong' classification may be achieved between 'solitarious' and 'gregarious' populations with little absolute magnitude of difference between them if the variance of each model population is small. Conversely, model populations where the average difference is large may not be easily classified if the behavioural characteristics within the solitarious and gregarious groups all have large variances. Presently the only way to compare between different solitarious and gregarious populations across laboratories, strains or species is to look at the actual magnitude of behavioural difference using external measurement criteria (e.g. path length in cm), rather than the internal yardstick provided by P_{greg}. A further reason for presenting data for individual behaviours is that it helps assess whether an experimental treatment affects all aspects of locust behavioural phase, or just particular behavioural components. This has the potential to greatly assist in our understanding of the mechanistic basis of behavioural phase change, and its evolution in different species of locust.

4. PROXIMAL STIMULI INDUCING PHASE CHANGE

Prior to Pener and Simpson (2009), the proximal stimuli that induce behavioural phase change were far better understood in *S. gregaria* than in any other locust species. Roessingh et al. (1998) showed that solitarious *S. gregaria* nymphs underwent full behavioural gregarisation within 4 h when presented with combined visual and olfactory stimuli from a group of gregarious nymphs, but neither visual nor olfactory stimulation on its own was effective. The same study also indicated that mechanosensory stimulation was a potent gregarising stimulus, as solitarious nymphs gently buffeted with papier maché spheres for 4 h also subsequently exhibited fully gregarious behaviour. This mechanosensory stimulus was localised to the hind femora by Simpson et al. (2001), in which various body parts were each stroked with a paintbrush for 5 s each minute over a 4-h period, and yet further still by Rogers et al. (2003), who showed that stroking half of the femur produced a stronger shift in behavioural phase than that produced by only stroking one quarter. Importantly, Rogers et al. (2003) also showed that restraining the femur relative to the thorax did not lead to behavioural gregarisation with subsequent paintbrush tickling, demonstrating that both mechanosensory stimuli from the hind femora and proprioceptive stimuli from the proximal leg joints are required to induce behavioural gregarisation in *S. gregaria* through this pathway.

More recently, progress has been made identifying the proximal stimuli that induce behavioural gregarisation in *C. terminifera* and *L. migratoria*. Cullen et al. (2010) emulated the experimental design of Roessingh et al. (1998) and Simpson et al. (2001) in a behavioural study of *C. terminifera*. After 6 h of treatment, the only stimulus that triggered significant gregariousness in *C. terminifera* final-instar nymphs was tactile stimulation of either the head (including antennae) or the antennae in isolation. Tactile stimulation of the abdomen also produced a shift towards gregarious behaviour that was nearly significant, but the authors showed in a separate experiment that locusts tickled on the abdomen jumped over 17 × more frequently than those tickled on the antennae during their respective treatments (1151 vs 66 jumps between 10 locusts over 6 h, respectively). The authors therefore concluded that the apparent effect of abdominal mechanostimulation was likely due to these locusts self-stimulating their antennae as they jumped against the inside walls of their holding chambers (Cullen et al., 2010; fig. 3 therein). It is important to note that the only previous paper to investigate behavioural gregarisation in *C. terminifera* (Gray et al., 2009) used 72 h of crowding with conspecifics as its shortest period of stimulation. Behavioural gregarisation in *C. terminifera* was achieved after 6 h of tactile stimulation, which suggests that much shorter periods of crowding should also be sufficient to induce phase change in this species, though this has yet to be reported.

Equivalent experiments to identify potential gregarising stimuli in *L. migratoria* have yet to be reported. However, recent experiments in *L. migratoria* tentatively suggest a role of olfaction during the process of phase change. Guo et al. (2011) used microarrays and qPCR to demonstrate a strong differential expression in two chemosensory-related gene families between antennal tissues of the two phases. Several *CSP* transcripts showed a significantly higher expression in the antennae of gregarious nymphs than in the antennae of solitarious nymphs, and knockdown of one of these (*CSP3*) with RNAi caused a significant repulsion from the stimulus group in their behavioural assay, as measured by an attraction index (Guo et al., 2011; fig. 5 therein). As the authors point out, attraction to the stimulus group in the Roessingh et al. (1993) assay could be due to either visual or olfactory attraction, or both, so Guo et al. (2011) also performed a separate experiment to test for olfactory attraction/repulsion using a simple Y-maze choice assay, in which nymphs walking through a plastic tube could enter an arm drawing through clean air, or an arm drawing through volatiles from 30 conspecific nymphs. This assay confirmed the authors' findings from the arena assay; gregarious nymphs treated with RNAi against *CSP3* significantly reduced their preference for the volatile-containing arm, opting instead to follow the clean air. By contrast, one *takeout* gene showed a significantly higher expression in the antennae of solitarious than in gregarious nymphs, and RNAi of *takeout* in solitarious animals produced the opposite effect; they showed a significantly increased attraction to the stimulus group in the behavioural assay, and a significantly decreased aversion to locust volatiles in the Y-maze.

An important observation by Guo et al. (2011) is that, despite the clear effects of *CSP3*-RNAi on olfaction-led attraction behaviour in gregarious *L. migratoria*, and of *takeout*-RNAi on olfaction-led repulsion behaviour in their solitarious counterparts, neither treatment influenced activity levels as measured in the arena assay (see their fig. 5 and discussion). Furthermore, a time course of forced crowding in the early stages of this study showed that solitarious nymphs were still significantly behaviourally distinguishable from the gregarious model population after 64 h of crowding. The authors attribute this relatively slow shift in P_{greg} to an asynchronous time course of changes in the component behaviours of their model; attraction index increased steadily with forced crowding, while the two activity-related variables (total distance moved and total duration of movement) took 32 h of crowding before they significantly increased (see their fig. 1 for P_{greg} and supplementary fig. 1 for the component behaviours). The same trend is demonstrated in Ma et al. (2015), in which similar data for attraction index and Y-maze assays show a steady increase in olfactory-based attraction behaviour but without a corresponding change in activity until at least 32 h of crowding. The fact that these experimental *L. migratoria* were still distinguishable from the long-term model building populations in both Guo et al. (2011) and Ma et al. (2015) means that they were, by definition, still in the transitional stage between solitarious and gregarious. Coupled with the apparent importance of olfactory-led attraction during this period, it seems very likely that olfaction plays a key role in inducing phase change in *L. migratoria*. Nevertheless, visual and tactile stimuli should also be tested, both in isolation and in interaction with the olfactory stimulus, to unravel the full suite of stimuli capable of causing gregarisation in this species.

Stimuli inducing phase change in *C. terminifera* and *L. migratoria* are both potentially mediated via the antennae. Furthermore, both species are members of the family Oedipodinae and are therefore more closely related to each other than either species is to *S. gregaria* (a member of the Cyrtacanthacridinae), in which phase change is induced by quite different mechanisms (the olfactory route in this species is in interaction with a visual stimulus). It could be speculated that *L. migratoria* and *C. terminifera* share a common mechanism of behavioural phase change, which has been overlooked due to the different methods used to assess the two species. In the case of *C. terminifera*, nymphs of this species do not tend to engage in prolonged tactile interaction in laboratory culture (D. A. Cullen, unpublished observations), so the precise nature of tactile stimulation as demonstrated by Cullen et al. (2010), and how it relates to natural populations of *C. terminifera*, is yet to be determined. Even so, Cullen et al. (2010) also showed beyond doubt that olfactory stimuli did not induce behavioural phase change in this species; the olfactory stimulus (the volatiles from 50 conspecifics for 6 h) produced a median P_{greg} of 0.0638, the lowest of all the stimuli tested. It should also be noted that the stimuli that *induce* gregarisation need not necessarily be the same as the stimuli that *maintain* gregarious behaviour, and future experiments should also seek to identify the minimum stimuli required for maintaining gregarious behaviour in both *L. migratoria* and *C. terminifera*, with a view to making a comprehensive comparison between these two species and the more distantly related *S. gregaria*. As more species of acridid become identified as undergoing full or partial density-dependent behavioural polyphenism (see Section 2), it is possible that the range of mechanisms and sensory modalities that mediate gregarisation will expand, reflecting the independent evolution of phase change across several different lineages.

5. BIOGENIC AMINE SIGNALLING

The biogenic amines serotonin, dopamine, octopamine and tyramine are implicated in behavioural phase change in two well-studied locust species: *S. gregaria* and *L. migratoria*. In this section, we first introduce the different receptor classes found in insects and review the likely efficacy and specificity of different pharmaceutical agonists and antagonists when used on locusts. Second, we provide an overview of our current understanding of the role of biogenic amines and their receptors in locust phase transition.

5.1 Phylogeny and Pharmacology of Biogenic Amine Receptors

The biogenic amines are central to many cellular signalling networks underlying plasticity by inducing transient changes in intracellular second messengers. These regulate a wide range of kinase enzymes, which change target protein function by altering their activity, cellular location and/or interaction with other proteins. Biogenic amines thus lead to transient functional changes of target proteins and, in the longer-term, regulate transcription in the cell nucleus. They are closely associated with a wide variety of insect behaviours (Blenau and Baumann, 2001; Libersat and Pflueger, 2004; Verlinden et al., 2014).

In insects, biogenic amines can be divided into the indolamine serotonin (5-hydroxytryptamine or 5-HT; derived from tryptophan), the catecholamine dopamine (derived from tyrosine via L-DOPA), the monophenolic amines octopamine and tyramine (also derived from tyrosine) and the imidazolamine histamine (derived from histidine). Serotonin and dopamine occur universally across the animal kingdom, where they are involved in major physiological processes with many conserved functions. Octopamine and tyramine are the invertebrate counterparts of noradrenaline and adrenaline, respectively (Pflüger and Stevenson, 2005), but only occur in trace amounts in vertebrates. Histamine is also present in both vertebrates and invertebrates, but whereas most signalling in vertebrates takes place through G-protein-coupled receptors (GPCRs), in invertebrates all known histamine signalling is mediated by ligand-gated ion channels (Fleck et al., 2012; Kita et al., 2017). In general, most biogenic amine receptors are GPCRs, belonging to the rhodopsin GPCR family. However, several amines can also interact with specific ligand-gated ion channels.

5.1.1 Phylogeny of the Biogenic Amine Receptor Families

If all the sequences of biogenic amine GPCRs are placed together in a phylogenetic tree, it is clear that the different receptor classes for each of the biogenic amines have common origins, rather than being linked most closely by the biogenic amine ligands themselves (Hauser et al., 2006, 2008; Qi et al., 2014; Spielman et al., 2015; Wu et al., 2014). This may indicate that a hypothetical ancient signalling molecule had a range of receptor classes, each coupled with different downstream signalling pathways. Slightly different signalling molecules then evolved, either by evolutionary drift or by a selected for process, each with their own set of associated receptors. Most of the biogenic amine receptors can be activated by high doses of other biogenic amines, which supports this hypothesis. Another explanation for the observation that biogenic amine receptors do not cluster based on ligand binding could be that these receptors have undergone extensive functional convergent evolution, which seems less plausible. Whatever their exact origin, it is certain that the major classes of biogenic amine receptors arose before the divergence of protostomes and deuterostomes. Subsequent gene duplications and structural/functional differentiation occurred generating further subtypes within each main receptor class. These receptor subtypes are believed to have evolved independently in vertebrates and invertebrates (Peroutka and Howell, 1994; Tierney, 2001; Vleugels et al., 2015). During this evolutionary process, selection would have been based on functionally important receptor characteristics such as ligand binding and coupling to second messengers, not the conservation of recognition sites for synthetic pharmaceutical ligands (Vleugels et al., 2014). Hence, as yet there is no general pharmacological classification system available for invertebrate biogenic amine receptors, which needs careful consideration when designing experiments. Most pharmaceutical agonists and antagonists are classified purely on their effects on mammalian biogenic amine receptors, and there is no guarantee that such agents will affect insect receptors in a similar manner, if indeed at all.

Some general findings are listed below, although it should be noted, first, that not all chemicals were tested on all types of receptor; second, that cell assay-specific differences have been observed for some receptors; and third, that putative species-specific differences should also be considered. Furthermore, no direct measurement of the binding of pharmaceutical agents onto specific G proteins in insects has been performed so far; instead, indirect measures using secondary messenger molecules have been used.

5.1.2 Serotonin Receptors

In vertebrates, 5-HT receptors are divided into seven classes. All but 5-HT_3, a ligand-gated ion channel, are GPCRs. 5-HT_1 and 5-HT_5 receptors couple to $G_{i/o}$ proteins and thus reduce cAMP synthesis in the target cells. 5-HT_2 receptors preferentially couple to $G_{q/11}$ proteins, leading to increase of intracellular calcium levels. 5-HT_4, 5-HT_6 and 5-HT_7 receptors all bind to G_s proteins, increasing intracellular cAMP production (Nichols and Nichols, 2008).

Thus far, six types of 5-HT receptor belonging to four classes have been characterised in insects. Five of them, namely 5-HT_{1A}, 5-HT_{1B}, 5-HT_{2A}, 5-HT_{2B} and 5-HT_7, have similar conserved amino acid sequences and activate the same second-messenger systems as the equivalently numbered vertebrate classes (Colas et al., 1995; Röser et al., 2012; Saudou et al., 1992; Schlenstedt et al., 2006; Thamm et al., 2010; Troppmann et al., 2010; Vleugels et al., 2013, 2014; Witz et al., 1990). The sixth type is an invertebrate-specific 5-HT_8 receptor, which shows less sequence conservation to any other receptor subtype (Qi et al., 2014). It mediates increases in intracellular calcium levels and does not affect intracellular cAMP levels (Qi et al., 2014).

Agonists (excluding naturally occurring biogenic amines) able to activate all tested invertebrate 5-HT receptors when applied in the micromolar range are 5-methoxytryptamine, (±)-8-hydroxy-2-(di-propylamino) tetralin (8-OH-DPAT), 5-carboxamidotryptamine (5-CT) and possibly also α-methyl 5-hydroxytryptamine (α-Me5-HT), although this has not been tested on 5-HT_8 receptors. (±)-1-(2,5-Dimethoxy-4-iodophenyl)-2-aminopropane [(±)-DOI] is a potent hallucinogenic drug in humans that is a highly selective agonist of the mammalian 5-HT_2 receptor family. This also seems to be the case in insects, although it has only been tested on *Drosophila* 5-HT_{1A}- and 5-HT_2-type receptors (Colas et al., 1995; Johnson et al., 2009).

Prazosin, dihydroergocryptine, L-butaclamol and D-blutaclamol may be good antagonists for 5-HT_1 and 5-HT_7 receptors, while 5-HT_2 receptors can be antagonised by clozapine and cyproheptadine. WAY-100635 inhibits 5-HT_1, 5-HT_7 and 5-HT_8 receptors, while ketanserin inhibits 5-HT_2 and 5-HT_7 receptors (Colas et al., 1995; Qi et al., 2014; Thamm et al., 2013; Vleugels et al., 2014; for an overview see Vleugels et al., 2015). Methysergide is a non-selective 5-HT receptor antagonist, but was not tested on 5-HT_8. In addition, in the blowfly, *Calliphora vicina*, methysergide activates 5-HT_7 but inhibits 5-HT_{2A}-mediated responses (Röser et al., 2012). In the honey bee, *Apis mellifera*, methysergide displays high potency at 5-HT_{2A} but does not block 5-HT_{2B} (Thamm et al., 2013). The other non-selective 5-HT receptor antagonist methiothepin interestingly had agonistic properties on the 5-HT_8 receptor of the small white butterfly, *Pieris rapae* (Qi et al., 2014). Mianserin was shown to inhibit 5-HT_2 and 5-HT_7 receptors (Colas et al., 1995; Thamm et al., 2013; Vleugels et al., 2014; for an overview see Vleugels et al., 2015).

5.1.3 Dopamine Receptors

In vertebrates, five distinct dopamine GPCRs mediate all known functions of dopamine. They can be divided into two major groups based on their structural, pharmacological and biochemical properties. The first group, called D1-class dopamine receptors (D1 and D5, which may also be referred to as D1A and D1B since they are presumably closely related paralogues that arose from a gene duplication event), couple to $G_{\alpha s}$ and thus increase intracellular cAMP levels. The second group, called D2-class dopamine receptors (D2, D3 and D4), mediate cAMP reduction by interaction with $G_{\alpha i}$ (Beaulieu and Gainetdinov, 2011; Callier et al., 2003; Cardinaud et al., 1998).

Because of the conservation of common functional and structural properties, the three types of dopamine receptors found in most invertebrates are generally categorised using the same classification system as vertebrates. Members of the Dop1 group are closely related to the vertebrate D1 receptor class and thus increase intracellular cAMP levels. Members of the Dop2/INDR (invertebrate-specific dopamine receptor) group also show more sequence similarity with D1-like receptors than with D2-like receptors, and also increase intracellular cAMP levels (Blenau and Erber, 1998; Hill et al., 2016; Humphries et al., 2003; Mustard et al., 2003; Ohta et al., 2009). However, the Dop3 class is more closely related to the vertebrate D2 class, and acts to reduce cAMP levels (Beggs and Mercer, 2009; Beggs et al., 2005, 2011; Missale et al., 1998; Mustard et al., 2005; Verlinden et al., 2015). The *Tribolium castaneum* Dop3 receptor was reported to have constitutive effects on cAMP levels and upon ligand binding the receptor also increased intracellular calcium levels (Verlinden et al., 2015). An additional receptor, the Dop/Ecd, has been characterised in *Drosophila melanogaster.* This receptor increases intracellular cAMP levels, but does not seem to be closely related to other dopamine receptors. This receptor interacts with both dopamine and ecdysone (Evans et al., 2014; Srivastava et al., 2005).

The agonist 6-amino-5,6,7,8-tetrahydronaphthalene-2,3-diol hydrobromide (6,7-ADTN) activates Dop1, Dop2 and Dop3 receptor types (Mustard et al., 2003; Sugamori et al., 1995; Troppmann et al., 2014; Verlinden et al., 2015). Bromocriptine seems to be a potent and effective agonist for insect Dop3 receptors (Verlinden et al., 2015), with only minor or no agonist activity on Dop2 receptors (Reale et al., 1997; Troppmann et al., 2014). It may thus be a good candidate to pharmacologically distinguish D2-like receptors from D1-like receptors and INDRs in insects. A general agonist of 5-HT receptors, 5-methoxytryptamine, also induced considerable Dop3 receptor activation (Verlinden et al., 2015).

Flupenthixol, an antagonist known to inhibit mammalian dopamine (D1–D5), serotonin (5-HT2), adrenaline (α1) and histamine (H1) receptors, is a potent antagonist of insect Dop1, Dop2 and Dop/Ecd receptors (Blenau et al., 1998; Han et al., 1996; Hearn et al., 2002; Mustard et al., 2003, 2005; Reale et al., 1997; Srivastava et al., 2005). Flupenthixol also has antagonistic properties on Dop3 receptors when applied in high concentrations

(Hearn et al., 2002; Verlinden et al., 2015). The two structurally related compounds chlorpromazine and chlorprothixene antagonise the same mammalian receptors as flupenthixol and also Dop1, Dop2 and Dop3 receptors (Hill et al., 2016; Sugamori et al., 1995; Verlinden et al., 2015).

Butaclamol is a potent general dopamine receptor antagonist, inhibiting all insect dopamine receptors it was tested on. 7-Chloro-8-hydroxy-3-methyl-1-phenyl-2,3,4,5-tetrahydro-1H-3-benzazepine (SCH23390) seems to be a selective antagonist for Dop1 and Dop2 receptors (Hearn et al., 2002; Srivastava et al., 2005; Verlinden et al., 2015), while spiperone additionally antagonises the Dop/Ecd receptor (Srivastava et al., 2005). Mianserin and methysergide are potent antagonists for Dop3 (Verlinden et al., 2015).

5.1.4 Octopamine and Tyramine Receptors

As indicated before, octopamine and tyramine are the invertebrate counterparts of norepinephrine and epinephrine. The vertebrate compounds interact through α- and β-adrenergic receptors, which show considerate homology with the receptors for octopamine and tyramine.

Insect/invertebrate octopamine and tyramine receptors can be grouped into five classes: (i) α1-adrenergic-like octopamine receptors (Octα1R); (ii) β-adrenergic-like octopamine receptors (OctβR); (iii) the recently characterised α2-adrenergic-like octopamine receptors (Octα2R); (iv) octopamine/tyramine or tyramine receptors (TyrRI); and (v) tyramine receptors (TyrRII) (Beggs et al., 2011; Evans and Maqueira, 2005; Verlinden et al., 2010a; Wu et al., 2014). The Octα1R class shows sequence similarity with vertebrate α1-adrenergic receptors. These receptors have higher affinity for octopamine than for tyramine and induce an increase in intracellular calcium levels, and also a small increase in levels of intracellular cAMP (Balfanz et al., 2005; Grohmann et al., 2003; Wu et al., 2014). The OctβR class has sequence similarities with vertebrate β-adrenergic receptors and specifically increases intracellular cAMP levels in response to octopamine (Hauser et al., 2006; Maqueira et al., 2005; Wu et al., 2012a). The Octα2R class shows sequence similarities with vertebrate α2-adrenergic receptor and was only recently characterised in the rice stem borer, *Chilo suppressalis*. Upon activation, they decrease intracellular cAMP levels and increase intracellular calcium levels. Octopamine is more potent in activating this receptor as compared to tyramine (Wu et al., 2014). The TyrRI class similarly shows structural and pharmacological similarities with vertebrate α2-adrenergic receptors. In most studies, these receptors have a preference for tyramine over octopamine in inducing a decrease in intracellular cAMP levels (Blenau et al., 2000;

Ohta et al., 2003; Poels et al., 2001; Rotte et al., 2009; Saudou et al., 1990; Vanden Broeck et al., 1995; Wu et al., 2013), whereas in some preparations of the *D. melanogaster* receptor octopamine has shown to be equally or more potent as tyramine in increasing intracellular calcium release (Reale et al., 1997; Robb et al., 1994). The TyrRII class is specific for tyramine and leads to an increase in intracellular calcium levels (Bayliss et al., 2013; Cazzamali et al., 2005; Huang et al., 2009; Reim et al., 2017; Wu et al., 2015).

Naphazoline and clonidine are very potent agonists of OctβR and Octα2R (Chen et al., 2010a; Maqueira et al., 2005; Wu et al., 2012a, 2014). Naphazoline also activates TyrRI and TyrRII receptors (Huang et al., 2009; Vanden Broeck et al., 1995). Phentolamine is capable of activating many octopamine and tyramine receptors. However, since it does not activate the receptors equally efficiently (partial agonist) and it competes over the same binding place, its presence reduces the activation of the receptor by other agonists (Wu et al., 2014). Demethylchlordimeform activates the Octα1R and OctβR of *Bombyx mori* (Chen et al., 2010a; Ohtani et al., 2006). Chlordimeform does not activate the TyrRI or TyrRII of *C. suppressalis* (Wu et al., 2013, 2015) and only induces small activity in the Octα1R of *Bombyx* (Ohtani et al., 2006); in high doses it has even been reported to have antagonistic effects (Nathanson and Hunnicutt, 1981).

Epinastine is a strong antagonist of Octα2R (Wu et al., 2014) and a weak antagonist of Octα1R (Huang et al., 2010), with no effect on OctβR (Wu et al., 2012a). Chlorpromazine and mianserin are potent antagonists of OctβR, with no effect on Octα2R (Wu et al., 2014). Cyproheptadine and metoclopramide are antagonists of Octα1R, OctβR and TyrRI. Yohimbine is an antagonist of TyrRI and TyrRII (Rotte et al., 2009; Saudou et al., 1990; Wu et al., 2013), which also inhibits Octα1R (Bischof and Enan, 2004; Ohtani et al., 2006; for an overview see Verlinden et al., 2010a).

5.2 Involvement of Biogenic Amines and Their Receptors in Locust Phase Transition

5.2.1 *Serotonin*

In *S. gregaria*, gregarising stimuli lead to a rapid ninefold transient increase in serotonin content in the thoracic ganglia and a slower fourfold increase in both the brain and optic lobes over a 24-h period (Rogers et al., 2004). Later, pharmacological manipulations provided evidence that serotonin is both necessary and sufficient to initiate gregarisation (Anstey et al., 2009).

Injections of serotonin or serotonin receptor agonists (a mix of α-Me-HT and 5-CT) induced some degree of behavioural gregarisation in isolated locusts. Injections of the serotonin precursor 5-hydroxytryptophan had little effect on their own, but greatly accelerated the rate of behavioural change after brief exposure to gregarising stimuli compared to controls. Conversely, α-methyltryptophan (AMTP), an inhibitor of tyrosine hydroxylase, or a mix of serotonin receptor antagonists (ketanserin and methiothepin) prevented behavioural gregarisation in briefly crowded locusts (Anstey et al., 2009). More recently, immunofluorescence staining has identified serotonergic neurons in *S. gregaria* nymphs that respond to gregarising stimuli (Rogers and Ott, 2015) and has qualitatively supported the earlier finding (Rogers et al., 2004) that solitarious phase nymph ganglia contain 20%–35% more serotonin than those of gregarious nymphs. The cAMP-dependent PKA, an intracellular effector molecule closely linked to learning and other forms of neuronal plasticity, is also required for behavioural gregarisation (Ott et al., 2012). PKA is a downstream effector of many biogenic amine receptors, including 5-HT_7, a likely target of serotonergic signalling during gregarisation in *S. gregaria*. Despite these several lines of evidence, the role of serotonin in behavioural phase change in *S. gregaria* has been challenged. Tanaka and Nishide (2013) found that injections of serotonin into the thoracic haemocoel failed to induce a gregarious-like attraction to a stimulus group, although it may be that serotonin injections are only effective with direct application onto the thoracic ganglia (Anstey et al., 2009; Rogers et al., 2014). A further analysis of gregarisation in *S. gregaria*, which looked at the individual behaviours used to construct the logistic regression model, suggested that behaviours related to both activity and attraction to other locusts were indeed affected by serotonin (Rogers et al., 2014; see also Section 3).

Solitarisation in *S. gregaria* has been less thoroughly analysed at the behavioural level than gregarisation, and there have been few studies looking for possible neurochemical mechanisms. The most thorough behavioural analysis suggests that after a period of rapid behavioural change in the first few hours of isolation, locusts remain in a state intermediate between gregarious and solitarious values for a period of at least 96 h (Roessingh and Simpson, 1994). It is unclear whether the initial rapid loss of activity following isolation is genuine solitarisation or represents a loss of arousal following the removal of stimulation from other locusts.

More recently, Alessi et al. (2014) analysed changes in titres of biogenic amines following isolation of long-term gregarious *S. gregaria*. Their behavioural analysis suggested that there was an exceptionally rapid loss of gregarious behaviour starting with a change in attraction/repulsion to other locusts within 1 h of isolation, followed by greatly decreased activity within the next 24 h. The arena design and behavioural assay differed in some regards from the Roessingh-type design used by many other researchers (see Section 3), and there was no long-term (multi-generational) isolated solitarious group used to benchmark fully solitarious behaviour, so its correspondence to the earlier study (Roessingh and Simpson, 1994) is unclear. Alessi et al. (2014) found that the amount of serotonin in the thoracic ganglia increased following as little as 1 h of isolation and was elevated for at least 72 h post-isolation, consistent with the earlier finding by Rogers et al. (2004). They also found, however, that the amount of serotonin in the brain decreased significantly in contrast to the earlier study, which found a substantial increase, particularly in the optic lobes. It is clear that further work is required to unravel this complex issue.

In contrast to *S. gregaria*, levels of serotonin in the brains of *L. migratoria* did not differ between long-term solitarious and gregarious fourth-instar nymphs (Ma et al., 2011). Crowding solitarious nymphs for up to 32 h did not induce any change in brain serotonin titres (Guo et al., 2013), but it should be noted that time sequences of behavioural change in *L. migratoria* following crowding of isolated-reared nymphs from the same lab (Guo et al., 2011; Ma et al., 2011, 2015) suggested that gregarisation is an extremely slow process in this species, and locusts were only partially gregarised even after 64 h of crowding (median P_{greg} 0.2–0.3 compared to median values >0.8 for fully behaviourally gregarious locusts, and <0.2 for fully solitarious locusts from the model populations). There is some evidence for serotonin's role as an initiator of gregarisation in *L. migratoria* from injections into the thoracic haemocoel with serotonin receptor agonists and antagonists (Ma et al., 2011). Injections of the 5-HT_2 receptor agonist (±)-1-(2,5-dimethoxy-4-iodophenyl)-2-aminopropane hydrochloride [(±)-DOI] caused an acute shift towards gregarious behaviour when injected into the head capsule of solitarious locusts, but by contrast, locusts treated with the same agonist whilst crowded for 32 h showed somewhat slower behavioural change than saline-injected controls (Guo et al., 2013).

Isolation of gregarious nymphs led to a progressive increase in serotonin titres in the brain until by 32 h the amounts were threefold greater (Guo et al., 2013). Here again, the correlation with the change in behaviour was not particularly strong: *L. migratoria* appears to undergo a very rapid behavioural solitarisation, with gregarious locusts isolated for just 1 h showing a substantial loss of gregarious behaviour, and locusts isolated for 4 h or longer indistinguishable from long-term solitarious locusts (Guo et al., 2011). Serotonin had no acute effect on the behavioural phase state of gregarious locusts when injected, but there was evidence that it acted to accelerate the rate of solitarisation when combined with brief periods of isolation (Guo et al., 2013). The long-lasting increase in serotonin on isolation, well beyond the full expression of fully solitarious behaviour, suggests that it may have a role in the further consolidation of phase state. It is currently unknown how long the elevated serotonin titre lasts and what its full functional significance is.

This suggests that serotonergic signalling in *L. migratoria* is linked to both behavioural solitarisation and gregarisation. There are several potential explanations for the seemingly contradictory effects of serotonin on behavioural phase change in *L. migratoria*. First, specific serotonin receptor classes could carry out different functions in solitarisation or gregarisation. This is supported by the finding that mRNA expression of only the 5-HT_2 receptor increased during behavioural gregarisation, and injecting a specific 5-HT_2 receptor agonist directly into the brain of solitarious nymphs induced gregarisation (Guo et al., 2013). However, RNAi against the 5-HT_2 receptor did not influence phase state or behavioural gregarisation relative to controls (Guo et al., 2013). Second, there could be time- and tissue-specific effects of serotonin on behavioural phase change, as brain serotonin levels in *L. migratoria* change in response to solitarisation, but systemic injections primarily affected gregarisation (Ma et al., 2011). It is possible that the role of serotonin is to facilitate behavioural/neuronal plasticity during both gregarisation and solitarisation, but does not have any direct effect on the expression of gregarious or solitarious behaviours themselves. Full phase change is accompanied by extensive changes in gene expression (Badisco et al., 2011a,b; Guo et al., 2011; Kang et al., 2004), which means that signalling systems are acting on different underlying physiologies in the two phases and may therefore have different effects. Further work is required to characterise potentially distinct neuromodulatory roles of serotonin in different regions of the nervous system.

5.2.2 Dopamine

Dopamine plays important roles in social interactions in a range of insects (Penick et al., 2014; Rillich and Stevenson, 2014; Vergoz et al., 2009), so it is plausible that it would also play a role in behavioural phase change. Crowding solitarious *S. gregaria* nymphs transiently increased dopamine content in the brain and thoracic ganglia, but the time course was slower than that of behavioural gregarisation (Rogers et al., 2004). Upon isolating *S. gregaria*, brain dopamine titres also increased transiently (Alessi et al., 2014; Rogers et al., 2004), suggesting that there may be a role for dopamine in the solitarisation process. Alessi et al. (2014) found an increase within one hour of isolation which was maintained for at least 72h, while Rogers et al. (2004) only examined a single 24h isolation time point. Injecting gregarious and solitarious *S. gregaria* with dopamine increased the likelihood of test locusts avoiding other locusts in a behavioural assay, whereas fluphenazine, an antagonist of vertebrate D1- and D2-class receptors, increased their propensity to approach other locusts (Alessi et al., 2014).

Ma et al. (2011) studied the catecholamine pathway in *L. migratoria* and reported lower amounts of dopamine in solitarious nymphs compared with gregarious nymphs. Brain dopamine titres in solitarious nymphs transiently increased above long-term gregarious values upon crowding, which corresponded with an upregulation of the Dop1 receptor (Guo et al., 2015) and an increased expression of genes in the dopamine synthesis pathway (Ma et al., 2011, but see comments on the time course of behavioural change in *L. migratoria* above). Gregarious locusts displayed more solitarious-like behaviour when (i) dopamine biosynthesis was interfered with using the tyrosine hydroxylase inhibitor AMTP; (ii) dopamine release was prevented by reserpine, an irreversible blocker of the vesicular monoamine transporter (which affects all monoamines); (iii) dopamine receptor signalling was inhibited by SCH23390, an antagonist of Dop1 and Dop2 receptors (Ma et al., 2011); and (iv) following RNAi-mediated knockdown of Dop1 (Guo et al., 2015). SKF38393, an agonist of vertebrate D1-class receptors, induced an acute change towards gregarious behaviour in solitarious nymphs, and also accelerated the shift towards gregarious behaviour upon crowding, and was reported to affect both activity and attraction towards conspecifics (Guo et al., 2015). Dopamine acting via Dop1 thus seems to play a role in gregarisation of *L. migratoria*.

In gregarious *L. migratoria* nymphs, there was a drop in dopamine titre in the brain after only 1 h of isolation, which mirrored an upregulation of the Dop2 receptor (Guo et al., 2015). Solitarious nymphs injected with a vertebrate D2-class receptor agonist *R*(−)-2,10,11-trihydroxy-*N*-propylnoraporphine hydrobromide hydrate [*R*(−)-TNPA], and left uncrowded, did not display any behavioural change. Gregarious locusts treated with the Dop2 antagonist *S*(−)-sulpiride and subsequently isolated were much more gregarious in behaviour after 1 h than control animals (i.e. the rate of solitarisation had been slowed down). A similar result was seen when RNAi was performed against the Dop2 receptor (Guo et al., 2015). Dop2 signalling could thus be needed for solitarisation in both *L. migratoria* and *S. gregaria.*

5.2.3 Octopamine and Tyramine

There has been relatively little attention to the role of octopamine and its precursor tyramine in the behavioural phase change of *S. gregaria*. Tyramine titres across the whole nervous system decreased upon isolating gregarious nymphs, but there was no noticeable change in octopamine (Rogers et al., 2004). Conversely, in a separate study of gregarious adults, isolation corresponded with a fall in octopamine titres in the metathoracic ganglia after 1 h of isolation and levels stayed low until 24 h of isolation, but there was no change in tyramine levels (Alessi et al., 2014). In *S. gregaria*, Octα1R and OctβR transcript levels are significantly higher in long-term gregarious animals as compared to long-term solitarious animals. The transcript levels of OctβR in the central nervous system (CNS) are already significantly increased after 30 min of crowding (Verlinden et al., 2010b).

In *L. migratoria*, early studies reported elevated levels of octopamine in the brains of solitarious animals (Fuzeau-Braesch and David, 1978; Fuzeau-Braesch and Nicolas, 1981). Ma et al. (2015) made a weakly supported claim for lower titres of octopamine in solitarious nymphs compared to long-term gregarious nymphs (see their fig. 2A; four of the six measurements for gregarious locusts are similar to those of solitarious locusts). In the same study, isolation of gregarious nymphs was accompanied by an increase in the amount of tyramine according to the duration of isolation. Locusts in these experiments were placed in a Y-maze and given the choice of odour volatiles from a group of gregarious locusts or a clean air supply following crowding or isolation treatments. In gregarious locusts that had been isolated the amount of tyramine in the brain was positively correlated with the percentage of locusts choosing the clean-air arm of the Y-maze (Ma et al., 2015; see fig. 2C therein). There was no clear trend in the expression of a tyramine receptor during isolation. The involvement of tyramine signalling in the early stage of solitarisation was supported by a dramatic shift towards solitarious-like behaviour in gregarious locusts given systemic injections of tyramine, and RNAi against TyrR1 which, whilst not affecting gregarious behaviour, appeared to protect against the loss of gregarious behaviour when locusts were isolated.

Expression of the octopamine receptor Octα1R showed no change during isolation. Ma et al. (2015) also reported that the decrease in brain octopamine levels during isolation corresponded with a loss of attraction to locust odours. However, this interpretation is only possible because of two extreme outliers in the long-term gregarious group in the data (see their supplementary fig. 4A, in which the data points for gregarious and solitarious octopamine titres are duplicated from fig. 2A in the main text).

Upon crowding, tyramine-β-hydroxylase, the enzyme which converts tyramine into octopamine, may be upregulated. There was also evidence for an increase in the expression levels of the Octα1R during crowding, correlated with the increased attraction to conspecifics, but expression in long-term gregarious and solitarious locusts appears more similar. There was no correlation between total brain octopamine concentration and attraction to locust volatiles in a Y-maze. Systemic injections of octopamine into solitarious locusts promoted a small increase in gregarious-like behaviours and some alteration in odour preferences in a Y-maze towards preference for conspecific odours. RNAi against the Octα1R had no effect on solitarious behaviour in a behavioural assay, but treated locusts failed to show any behavioural change following 32h of crowding, unlike saline treated controls (Ma et al., 2015). There may therefore be a role for octopamine in *L. migratoria* during gregarisation and in mediating the attraction to locust odour in solitarious nymphs during crowding.

5.2.4 Conclusions

Given the complex ways that biogenic amines act on a wide range of behaviours and processes (Andretic et al., 2005; Buhl et al., 2008; Goosey and Candy, 1980; Leitch et al., 2003), it is challenging to identify key mechanisms that control phase change, as these amines are not uniquely linked to single functions, interact with a range of receptors, and likely have a mix of mutually synergistic and antagonistic relationships with each other in modulating plasticity and behaviour. Many studies have utilised pharmacological or molecular interventions that may be expected to have very wide-ranging and systemic effects beyond behavioural phase change. A major challenge is to identify and zoom in on the specific neural circuits and systems mediating behavioural phase change. Very strong effects have been reported for different individual biogenic amine and receptor combinations on both gregarisation and solitarisation within the same species.

There is a need to improve our understanding of how these separate elements work together to produce the phenotypic change characteristic of natural behavioural phase change. There is considerable variation in the time course of phase change both within and across locust species, further stressing the importance of taking a holistic and multi-disciplinary approach to uncovering the underlying mechanisms.

6. GENOMICS, TRANSCRIPTOMICS AND EPIGENETICS

Transcriptomic studies in a range of insect species have shown that differential gene expression is an important mechanism underlying phenotypic plasticity; for example, in the determination of caste in eusocial insects (Corona et al., 2016; Oppenheim et al., 2015). Since locusts can produce two dramatically different phenotypes from a single genome, some gene products must either be differentially expressed or have differential activity or function between phases. Most quantitative transcriptomic studies in locusts have focussed on differences between long-term phases (in *L. migratoria* and *S. gregaria*). Many of the earlier studies (Chen et al., 2010b; Kang et al., 2004; Wei et al., 2009) lacked substantive biological replication, and although different transcript families and life stages have been targeted, our understanding of the transcriptomic underpinnings of phase polyphenism is still fragmentary.

6.1 Availability of Sequence Information

The availability of sequence information has long been a limiting factor in the quest for key molecular regulators of locust phase change, but substantial progress has been made in recent years. The first large resources of orthopteran transcriptome sequences were databases of Sanger-sequenced expressed sequence tag (EST) clones for *L. migratoria* (Kang et al., 2004; Ma et al., 2006) and *S. gregaria* (Badisco et al., 2011a). Although contig numbers were similar for both species (approximately 12,000), there was only limited sequence overlap between both sets, which may be explained by the use of RNA extracts from different tissues or life stages and/or by EST clones not typically covering the entire length of mRNA transcripts. These databases also provided annotation information using classifications frameworks such as Clusters of Orthologous Groups (COG), Gene Ontology (GO), the Kyoto Encyclopedia of Genes and Genomes (KEGG) and InterPro, but about half the sequences remain unannotated. Microarrays have also been designed for *L. migratoria* and *S. gregaria* to study various aspects of phase polyphenism (e.g. Badisco et al., 2011b; Guo et al., 2011; Ma et al., 2011).

With the advent of next-generation sequencing (NGS) technologies, especially Illumina sequencing, the number of available insect transcriptomes is growing rapidly, with the ever-broadening range of taxa including several orthopteran species (Oppenheim et al., 2015). To date, the most extensively covered orthopteran species is *L. migratoria*, with transcriptome projects based on different tissues, life stages and transcript families (Chen et al., 2010b; Ma et al., 2011; Wei et al., 2009), and a recently published reference genome with 17,307 predicted genes (Wang et al., 2014), the first published genome for any locust species.

6.2 Transcriptomic Differences Between Long-Term Phases

The first quantitative transcriptomic study in *L. migratoria* compared the transcriptomes of solitarious and gregarious fifth-instar nymphs, using cDNA derived from either the head, hind legs or midgut (Kang et al., 2004). A remarkable observation was the higher expression of juvenile hormone binding proteins (JHPH superfamily of proteins: juvenile hormone binding protein, hexamerins, prophenoloxidase and hemocyanins) in the heads of gregarious nymphs. More general patterns of tissue-specific differential expression were also observed, and the authors claimed a total of 532 transcripts as being putative differentially expressed genes (DEGs). However, since this study lacked biological replication, it is hard to estimate the real extent of differential gene expression.

The first publication to use NGS in a study of locusts (Wei et al., 2009) focussed on the small RNA fraction of the transcriptome of *L. migratoria*. Small RNAs belonging to 50 known gene families were found, and 185 new, potentially locust-specific microRNA (miRNA) transcripts were predicted. This study claimed substantial differences in miRNA expression between long-term phases. However, these claims were based on the comparison of two samples that seemed to differ substantially in their content of tRNA, rRNA and unannotated small RNA, suggesting that technical differences between both cDNA libraries may have existed that could equally well explain the differences in miRNA. Shortly after this, Chen et al. (2010b) published an updated *L. migratoria* transcriptome using Illumina sequencing based on whole-body RNA extracts of gregarious and solitarious animals of different ages. This transcriptome contained 72,977 contigs larger than 100 base pairs, with 11,490 predicted protein-coding genes. During early larval stages, transcriptomic differences became increasingly pronounced, culminating in the most substantial difference recorded at the fourth-instar stage. In older nymphs and adults, the differences were smaller again. Since only one replicate was used per age cohort, it was still very hard to quantify the extent of differential gene expression, but a core set of 242 genes differed between both phases throughout development and adult life. These transcripts were suggested to be candidate *phase marker genes*. Most of these genes were involved in metabolism and biosynthesis, but a number of signal transduction components were also differentially expressed. Ma et al. (2011) focussed on the differences between long-term phases of *L. migratoria* throughout development using a microarray. They also found that differences between both phases peaked at the fourth-instar stage, supporting the findings of Chen et al. (2010b) and suggesting that the fourth instar is a key developmental stage for the determination of phase state. Among the 594 DEGs at the fourth-instar stage, the most common functional categories were general metabolism, molecular transport, the production of cuticle protein and chemosensory transduction. The molecular pathway of catecholamine biosynthesis and metabolism ranked as the most enriched pathway in genes upregulated in gregarious locusts. The important role of this pathway was further confirmed with quantitative PCR and functional experiments (see Section 5.2).

The first transcriptomic comparison between long-term phases of *S. gregaria* was a microarray study by Badisco et al. (2011b). The brain transcriptome of adult solitarious and gregarious locusts was compared, and 214 transcripts were found to be differentially expressed. In gregarious brains, DEGs were enriched in stress response genes including heat-shock proteins, a group of genes that had already been shown to be differentially expressed in *L. migratoria* (Wang et al., 2007). Conversely, solitarious locusts seemed to invest more in oxidative stress resistance, which may facilitate their increased longevity compared to their gregarious counterparts (Maeno et al., 2013). Much as in *L. migratoria*, long-term phases also differed in metabolism, sensory perception, signal transduction and some developmental pathways (despite the maturity of the subjects).

6.3 The Transcriptome During Phase Transition

The first transcriptomics study focussing on the time course of behavioural change was a microarray study in *L. migratoria* by Guo et al. (2011). 794

transcripts were shown to be differentially expressed at some point during a 64-h isolation of gregarious fourth-instar nymphs, and 1103 were differentially expressed at some point during a 64-h crowding of solitarious fourth-instar nymphs. Solitarisation appears to be a relatively fast process in *L. migratoria*, which was reflected in large fluctuations in the patterns of differential expression. On the other hand, gregarisation seems to take much longer (see Section 5.2), which was reflected by more stable trends in differential expression. In both time courses, insect cuticle proteins, pheromone binding, odorant binding and members of the JHPH family were most enriched among DEGs. Two groups of antenna-specific DEGs were also functionally validated; several *chemosensory proteins* and *takeout* appear to play a role in olfaction-led group attraction and repulsion by gregarious and solitarious locusts, respectively (see Section 3.2).

A transcriptomic study was also published alongside the *L. migratoria* reference genome (Wang et al., 2014), using a similar 64-h time course. The authors stated that 4893 genes were differentially expressed at some point over this time course (i.e. 28.3% of all predicted genes in their genome), but this was based on only one replicate per time point. Furthermore, a list of these putative DEGs, or any details concerning their annotations or functions, was not given.

6.4 Mobile Elements

Orthopterans have a wide range of genome sizes and include some of the largest known animal genomes (Camacho et al., 2015; Wang et al., 2014; Westerman et al., 1987). One possible driver of genome size expansion is colonisation by mobile DNA elements. Jiang et al. (2012) screened the transcriptome of *L. migratoria* for retrotransposons and found evidence for transcription of retroelements from approximately 100 loci in the genome, a figure comparable to observations in *D. melanogaster* and the malarial mosquito *Anopheles gambiae*. Locusts did, however, show a higher proportion of non-long terminal repeat (non-LTR) retroelements (73 non-LTR vs 32 LTR retroelements) with a greater diversity (members of 9 families vs 4 or 5 in dipterans) and with higher numbers of divergent copies. They also had a higher number of intact open reading frames (ORFs), suggesting that retrotransposition is still an active process. Based on observed differences in retroelement transcription between both phases in fifth instar and adult animals (but not in earlier life stages), the authors suggested that somatic retrotransposition may play a role in the regulation of phase-related phenotypic plasticity.

The reference genome of *L. migratoria* (Wang et al., 2014) estimated the haploid genome size to be 6.3 Gb in this species. Large introns and repetitive elements (both intergenic and in introns) were the apparent reasons behind this large genome size, covering about 60% of the total assembly. These elements mainly consisted of DNA transposons and long interspersed elements (LINEs). The latter was the most prevalent repeat family (about 4% of the genome) and contained many full-length copies with intact ORFs. This may be an indication of active transposition, which is in accordance with the earlier observed transcription of non-LTR retrotransposons (Jiang et al., 2012). Transposon diversity was high, with no single transposon dominating, indicating a general susceptibility to transposition rather than an aggressive colonisation by a single element.

No other locust genomes have been sequenced so far. Based on Feulgen image analysis densitometry and C-banding techniques, the haploid genome size of *S. gregaria* was estimated to be around 8.55 Gb (Camacho et al., 2015). Though transposons were not the focus of this study, non-LTR transposons were also identified as an important part of repetitive DNA. In a microarray study comparing both phases in *S. gregaria*, one reverse transcriptase had a higher expression in the solitarious phase (Badisco et al., 2011b).

6.5 DNA Methylation in Locusts

Together with chromatin modifications and non-coding RNAs, cytosine methylation of DNA has emerged as a key epigenetic mechanism in regulating gene expression to tailor phenotypes to the environment (Angers et al., 2010; Dias et al., 2015; Isles, 2015; Szyf et al., 2008). In social insects, DNA methylation has been implicated in the differential gene expression that underpins caste polyphenism and the division of labour (Yan et al., 2015). It is therefore reasonable to expect that DNA methylation has a role in phase polyphenism and phase-related behavioural plasticity.

In mammals, DNA methylation is accomplished by two classes of DNA (cytosine-5)-methyltransferases, DNMT1 and DNMT3. It is commonly assumed that DNMT3 is responsible for establishing methylation patterns de novo, thus interfacing the 'methylome' with the environment, whereas DNMT1 maintains the established pattern during cell divisions. A further enzyme, DNMT2, was so named due to its pronounced sequence similarity with DNMT1, but does not appreciably methylate DNA in mammals. Instead, DNMT2 very efficiently methylates aspartate tRNA

(Goll et al., 2006) although a dual role in both tRNA and DNA methylation remains a possibility in some organisms (Jeltsch et al., 2016).

In locusts, the evidence for the existence of the *Dnmt1* and *Dnmt2* genes is uncontroversial. Two *Dnmt1* paralogues and one *Dnmt2* are present in the genome of *L. migratoria* (Robinson et al., 2011, 2016; Wang et al., 2014). In *S. gregaria*, where whole-genome information is not yet available, current evidence indicates the presence of at least one *Dnmt1* (Falckenhayn et al., 2013) and one *Dnmt2* (Boerjan et al., 2011). By contrast, the search for DNMT3 has long been unsuccessful in both species. cDNA fragments related to *Dnmt3* are absent in the published *S. gregaria* (Badisco et al., 2011a) and *L. migratoria* (Ma et al., 2006) EST databases (Robinson et al., 2011). Analysis of *S. gregaria* transcriptome-scale shotgun cDNA sequences gave 75 BLAST hits (*E*-values <0.1) with *A. mellifera* DNMT1a as query, but no hits with *A. mellifera* DNMT3 (Falckenhayn et al., 2013). Subsequently, the genome announcement of Wang et al. (2014) reported the presence of a *Dnmt3* gene in *L. migratoria*, but did not present evidence in the paper. We have verified the presence of a hypothetical DNMT3-like protein, LOCMI17543-m1, in the current version of the LocustBase Official Gene Set (OGS CDS V2.4.1) associated with Wang et al. (2014). LOCMI17543-m1 shows only weak sequence conservation with other DNMT3s (about 30% over a query cover of approximately 95% in *tblastn* searches against the NCBI *nt* nucleotide collection database), but all top-scoring hits are *Dnmt3b* representatives and the three canonical domains present in mammalian DNMT3b (PWWP, ADDz_Dnmt3 and Dcm) can be identified by Conserved Domain Database BLAST (S. R. Ott, unpublished). Interestingly, however, BLAST-searching a *L. migratoria* CNS transcriptome (41,179 sequences; Zhang et al., 2012) with LOCMI17543-m1 as query gave no hit (S. R. Ott, unpublished). Robinson et al. (2016) likewise reported little evidence for DNMT3 expression in publicly available *L. migratoria* transcriptome data sets.

Possible explanations for the elusiveness of *Dnmt3* transcripts in locust EST and transcriptome data are either first, that expression is much lower than for *Dnmt1* and *Dnmt2*, or second, that the *Dnmt3*-like sequence in the *L. migratoria* genome (Wang et al., 2014) is no longer a functional gene. Locusts would not be alone among insects with functionally important DNA methylation to lack DNMT3 (Bewick et al., 2016). In these insect species, its function is possibly taken over by DNMT1.

There is some evidence that DNMTs are differentially expressed between phases. In *L. migratoria*, *Dnmt1* is more strongly expressed in the eggs and ovaries of solitarious locusts, and *Dnmt2* in nervous tissue of solitarious males (Benjamini–Hochberg corrected, $\alpha=0.05$; Robinson et al., 2016). In *S. gregaria*, Boerjan et al. (2011) reported differential expression of *Dnmt1* and *Dnmt2* in several of the 15 different tissues investigated, including in brain, testes and muscle (based on $N=3$ batches of pooled tissues per phase, not corrected for multiple comparisons).

Irrespective of the uncertainty surrounding the presence of DNMT3, locust genomes are extensively methylated by insect standards. Boerjan et al. (2011) quantified the ratio of methylated to total cytosines (mC/C) in CNS-derived DNA in *S. gregaria* by mass spectrometry and reported values of 1.6% and 1.9% in the metathoracic ganglion and brain, respectively. Comparable values were subsequently obtained by genome-scale bisulphite sequencing (Falckenhayn et al., 2013), which also revealed that >90% of mC residues occurred in a CpG dinucleotide context, as expected from other animals. These mC/C ratios are over 10× higher than in honeybees (0.11%; Lyko et al., 2010). Because methylcytosine shows hypermutability to thymidine, the ratio of observed to expected CpG sites ($CpG_{O/E}$) allows inference about the extent of historic DNA methylation in the germline. Assessing $CpG_{O/E}$ across many genes yields unambiguously bimodal distributions in both *L. migratoria* (Robinson et al., 2011; Wang et al., 2014) and *S. gregaria* (Falckenhayn et al., 2013), with one mode close to 1.0 and a second mode below 0.5, the latter indicating a large population of genes with a germline history of extensive methylation. In *L. migratoria*, this bimodal pattern was also observed in a subset of genes that have previously been found to be differentially expressed between phases (Robinson et al., 2011). Genes with historical germline hypermethylation (low $CpG_{O/E}$) were significantly associated with the GO terms *metabolism*, *protein transport*, *intracellular protein transport*, *biosynthesis* and *electron transport*. However, genes with a $CpG_{O/E}$ close to 1.0 may still undergo phase-related changes in methylation during the lifetime of a locust. If, for example, a behavioural phase change were underpinned by hypermethylation of particular genes specifically in the CNS, this would leave no trace in the germline.

Genome-wide analyses of methylation patterns with single-gene resolution have proved difficult in locusts due to the large size of their genomes (Camacho et al., 2015; Wang et al., 2014). Using shotgun bisulphite sequencing of genomic DNA from the CNS of gregarious *S. gregaria*, Falckenhayn et al. (2013) obtained 180 million sequence reads corresponding to 32 Gb of DNA sequence. However, the absence of a reference genome precluded the assembling into a genome-wide methylation map, and the absence of data from solitarious phase DNA precluded the identification of phase differences. Using the *S. gregaria* EST database of Badisco et al. (2011a) as a reference, the mC/C ratio in genic sequences was found to be over 3%, markedly higher than the genomic ratio. This suggested that methylation might preferentially target exons, as had been previously found in other insects. Contig methylation densities showed an extremely bimodal distribution, with about 20% of contigs fully methylated (mCpG/CpG > 0.95) and around 45% entirely unmethylated. This large proportion of densely methylated exons was in marked contrast with previously analysed insect species, but may not be unique to locusts (Bewick et al., 2016).

Wang et al. (2014) analysed the methylation of brain DNA in solitarious and gregarious fourth-instar females of *L. migratoria* by reduced representation bisulphite sequencing. The overall genomic mC/C ratio was 1.6%, comparable with that found in *S. gregaria*. This figure rose to 11% for CpG sites across the genome, and to 25.5% for gene-body CpGs. Strikingly, however, the exonic mCpG/CpG ratio was only 12.8%, less than 2 × the intergenic ratio of 7.3% and only marginally higher than the genomic ratio of 11%. The high gene-body mCpG/CpG ratio of 25.5% instead reflected an intronic ratio of 31% (all values measured from fig. 2B of Wang et al., 2014). The intronic ratio is thus almost 2.5 × higher than the exonic ratio. This contrasts sharply with honeybees where the mCpG/CpG ratio is about 40 × higher in exonic than in intronic DNA (Lyko et al., 2010). The ratios reported in *L. migratoria* also contrast with the available evidence in *S. gregaria*, where methylation appears to preferentially target exons, with 40% of exonic sequences showing methylation ratios greater than 65% (Falckenhayn et al., 2013). The available data in the two species are not directly comparable, however, and a direct characterisation of exonic and intronic methylation densities in *S. gregaria* is needed to resolve whether locust species differ fundamentally in this respect. In both *L. migratoria* and *S. gregaria*, methylation occurs in a significant fraction of repetitive DNA, including transposons and rDNA repeats (Falckenhayn et al., 2013; Robinson et al., 2011, 2016). In *L. migratoria*, the average mCpG/CpG ratio in DNA repeats is 15.5% (Wang et al., 2014). The supplementary material of Wang et al. (2014) lists 90 genes that showed differential methylation in the brains of solitarious and gregarious nymphs. Many of these genes were also differentially expressed according to a 'false discovery rate of the significant level of the differential expression'. A sizeable proportion of the differentially methylated genes were functionally classified as coding for *kinases*, other *signal transduction proteins* and proteins implicated in *microtubule cytoskeleton organisation*. This implies long-term differences in neuronal signalling and microtubule dynamics between phases, an interesting finding that must, however, not be confused with evidence for a role of these genes in the *process* of phase change.

In an alternative approach, Mallon et al. (2016) used methylation-sensitive amplified fragment length polymorphisms (MS-AFLP) to compare DNA methylation patterns in the CNS (brain and thoracic ganglia) of long-term solitarious, long-term gregarious and recently crowded (24 h) solitarious adults in *S. gregaria*. MS-AFLP is a gel-based fingerprinting technique that can provide an index of 'epigenetic differentiation' (Φ_{ST}) similar to the F_{ST} index of genetic differentiation in population genetics. A major limitation of MS-AFLP is that it screens only a limited number of anonymous loci. The methylation fingerprints of solitarious and gregarious *S. gregaria* showed significant differentiation across 282 methylation-susceptible loci. However, crowding for 24 h resulted in a methylation fingerprint that was differentiated from both solitarious and long-term gregarious locusts. Furthermore, in a principal coordinate analysis, the fingerprints of these recently gregarised locusts occupied a position distinctly away from the axis defined by the two long-term phases. This suggests that methylation patterns seen after 24 h of crowding are not simply partial realisations of the endpoints, but reflect a distinct transitional epigenetic state. This finding adds to previous evidence for the existence of mechanistically distinct transitional phase states at different levels from the molecular to the behavioural (Rogers and Ott, 2015; Rogers et al., 2004; Simões et al., 2013).

In summary, locusts have more densely methylated genomes than most other insects analysed to date (Bewick et al., 2016). One might be tempted to speculate whether this directly reflects their capacity for pronounced phenotypic plasticity, but this simplistic idea is easily refuted: the Meadow Grasshopper, *Chorthippus parallelus*, is incapable of phase change and has a genomic mC/C ratio of 4% (Lechner et al., 2013); and the German Cockroach, *Blattella germanica*, has genomic and exonic mCpG/CpG ratios of 12% and 40%, respectively (Bewick et al., 2016). Similarly, the notion that DNA methylation in insects is required for caste polyphenism and the evolution of sociality is not supported by recent analyses (Bewick et al., 2016; Standage et al., 2016). Diptera show no appreciable DNA methylation, demonstrating furthermore that the extreme *developmental* plasticity manifest in holometabolous insects can be achieved entirely through other regulatory mechanisms. At this point, there is clear evidence for phase-related differences in DNA methylation, but we are a long way from understanding the mechanistic role of DNA methylation in phase change. The results of Mallon et al. (2016) indicate that further progress will require analyses of changes in DNA methylation across different time points during phase transitions, rather than just comparisons of endpoints.

7. NEUROPHYSIOLOGICAL CONSEQUENCES OF PHASE CHANGE

Although a great deal of effort has been focussed on the mechanisms and plasticity underlying phase transition, analyses of the endpoints of this process—long-term solitarious and gregarious animals—have also revealed important insights through comparative analyses of how altered neuronal function relates to changes in behaviour. Phase transition requires extensive changes in the ways in which sensory information is gathered, processed and used to shape behaviour. Understanding these differences at the cellular and circuit levels will permit a deeper understanding of how processes ranging from short-term modification of existing neural structures through to the epigenetic regulation of developmental gene expression give rise to markedly different behavioural phenotypes.

Blackburn et al. (2010) related differences in the walking behaviour of adult solitarious and gregarious *S. gregaria* to the activity of one of the motor neurons controlling hind leg movements. Solitarious locusts generally walk slowly with the body held low to the ground as part of their behaviourally cryptic phenotype, but gregarious locusts walk rapidly with the body held high. Gregarious locusts take larger steps than solitarious locusts, with the femoro-tibial joint undergoing larger excursions and being kept more extended throughout the entire step cycle. The metathoracic slow extensor tibiae (SETi) motor neurone of solitarious locusts fired tonically at 5–14 spikes s^{-1} in quiescent animals. By contrast, the SETi of gregarious locusts showed little tonic activity (0–3 spikes s^{-1}). Imposed flexion of the femoro-tibial joint elicited strong transient spiking activity in the SETi of both phases, resisting the imposed movement. The spiking rate of SETi in solitarious locusts during these resistance reflexes was twice as great as that of gregarious locusts regardless of femoro-tibial joint angle or excursion, indicating a higher gain of the sensory-motor feedback loop from the main femoro-tibial joint proprioceptor. The extensor tibiae muscle of solitarious locusts produces less force than that of gregarious locusts at low stimulation frequencies (Blackburn, 2005), but the high gain of the sensory-motor feedback loop allows for fine-grained control of muscle force amplitude and thus movements (Cruse, 1985; Cruse and Schmitz, 1983). This in turn may be linked to the occurrence of pronounced behavioural catalepsis (Rogers and Simpson, 2014) in solitarious locusts, similar to that seen in other cryptic insects such as phasmids.

7.1 Social Complexity and Brain Size

The social interactions shown by gregarious locusts lack the behavioural complexity of eusocial Hymenoptera and termites. Nevertheless, the organised collective motion of locust swarms is, like many complex social behaviours, maintained by positive and negative behavioural feedback loops arising entirely from local interactions between near neighbours, which elicit simple behavioural responses that constitute rules of interaction (Buhl and Rogers, 2016; Buhl et al., 2006, 2011, 2012). In *C. terminifera*, for example, alignment of walking individuals only occurs if locusts are closer than 13.5 cm (Buhl et al., 2011). Above a critical population density, these local interactions on the scale of a few centimetres can generate highly ordered marching behaviour within groups of locusts that extend for hundreds of metres (Buhl et al., 2006). This does not require sophisticated large-scale coordination, but comes about solely as a result of local self-organisation (Camazine et al., 2003).

The inherent sensory complexity of a locust's immediate environment greatly increases when locusts become gregarious and form dense, constantly moving groups that demand dynamic responses. Gregarisation requires that locusts adopt new behaviours, and subjects them to different kinds of risk. This includes intense competition for resources such as food and roosting sites and, at its most extreme, this competition can result in an increased threat of cannibalism (Bazazi et al., 2008, 2011; Hansen et al., 2011).

The demands that group living places on behaviour and the nervous system may be indicated by the 30% larger brain of gregarious compared to solitarious *S. gregaria*, despite gregarious locusts being typically smaller in body mass (Ott and Rogers, 2010). This increased volume is not uniformly distributed: neuropiles associated with higher-order processing and integration of sensory information are disproportionately affected. For example, gregarious locusts have smaller eyes than solitarious locusts (Dirsh, 1953; Rogers et al., 2010) and the first visual neuropile in the optic lobes, the lamina, is little different between phases. However, the second visual neuropile, the medulla, is disproportionately larger in gregarious locusts than in solitarious locusts and the third visual neuropile, the lobula, is disproportionately larger again. Likewise, in the olfactory system, gregarious locusts have fewer olfactory sensilla than solitarious locusts (*L. migratoria*: Greenwood and Chapman, 1984; *S. gregaria*: Ochieng' et al., 1998). The antennal lobe, which is the primary olfactory neuropile, is 10% smaller in gregarious locusts than in solitarious locusts of matching brain size. The primary calyx of the mushroom bodies, a higher-order olfactory and multi-modal integration centre, is conversely 50% *larger* in gregarious compared to solitarious locusts. Phase-related differences in brain region proportions seem to arise from two processes: first, allometric growth as overall brain size increases; and second, a mosaic of step changes in size of some neuropiles, for example, the central complex. These differences in brain region size were found between adult long-term (several generation) solitarious and gregarious locusts. It is unclear at present whether regional size differences are the result of neuronal proliferation or arise from the growth of existing arborisations. It is also unknown how phase change initiated during post-embryonic development affects the subsequent growth and development of the brain as the locust adopts new life history strategies.

7.2 Learning Abilities of Solitarious and Gregarious Locusts

Locusts, in common with many insect species, have a considerable capacity for learning (Simões et al., 2011, 2012, 2013, 2016), which is one of the most basic and universal manifestations of phenotypic plasticity. Although the ability to learn confers fitness advantages (Dukas, 2008a,b), it can also incur substantial costs (Mery and Kawecki, 2004, 2005). Consequently, the learning abilities of different species are often fine-tuned to their specific ecological and life history requirements and show wide variation in performance (Dukas, 2008a; Ngumbi et al., 2012). It is therefore perhaps unsurprising that solitarious and gregarious *S. gregaria*, with their very different life histories, also differ in their learning abilities as shown in a series of papers by Simões et al. (2011, 2012, 2013, 2016).

7.2.1 Appetitive and Aversive Olfactory Learning in Gregarious Locusts

Simões et al. (2011) first developed a paradigm using restrained *S. gregaria* that demonstrated classical Pavlovian associative conditioning for appetitive stimuli (Simões et al., 2011). This assay was based on the palp opening reflex (POR) in which locusts that are presented with a food stimulus (in this case a synthetic food) extend their maxillary and labial palps. This simple reflex can be used to measure memory acquisition. Locusts can learn to associate a food stimulus (unconditioned stimulus) with a simultaneously presented odour (conditioned stimulus). Thus, the frequency of the POR in response to presentation of an odour alone was substantially higher in locusts previously trained with paired presentations of both the odour and a food stimulus than in locusts trained with the odour alone, or with both stimuli presented separately and hence unpaired. These appetitive memories lasted for at least 24 h and were transferable to a different behavioural context: locusts trained on a particular odour in Pavlovian food-conditioning trials showed a subsequent preference for that odour when walking in a Y-maze under operant conditions.

These same authors subsequently demonstrated that gregarious locusts can undergo aversive conditioning using a similar Y-maze paradigm (Simões et al., 2012). The innate olfactory preference of locusts was altered by pairing a naively preferred odour with a synthetic food to which noxious nicotine hydrogen tartrate (NHT) had been added. The concentration of NHT was sufficient to cause a variety of malaise symptoms (Bernays and Lee, 1988) without causing long-term damage. Only a single trial pairing was needed to produce robust aversive memories that lasted for at least 24h, and the strength of the aversion was dependent on the concentration of NHT in the food. The aversion was not immediately apparent in subsequent behavioural tests in the Y-maze however, but increased in strength over the course of 4h, suggesting that a post-ingestive mechanism was responsible for the memory formation. This was confirmed by injecting NHT directly into the haemolymph and pairing this with odour presentation. An olfactory association with the post-ingestive effects of NHT could be formed with a gap of up to 30min between injection and subsequent odour presentation, but required ingestion of neutral food simultaneously with exposure to the odour.

7.2.2 Phase State Affects Aversive but Not Appetitive Conditioning

Naïve solitarious and gregarious *S. gregaria* have a similar 70:30 baseline preference for vanilla odour over lemon odour in a Y-maze (Simões et al., 2013). Locusts of both phases were trained by coupling a single presentation of this preferred vanilla odour with a synthetic non-nutritious food containing 10% NHT. Just 10min later the locusts were given the choice between the vanilla odour and a novel lemon odour. Only 34% of solitarious locusts now chose vanilla, clearly indicating that they had acquired a learned aversion, whereas 59% of gregarious locusts still retained a preference for the vanilla odour. The rapidity of the learned aversion in solitarious locusts suggests that it is not mediated by the post-ingestive process acting in gregarious locusts (Simões et al., 2012). By 4h after training both solitarious and gregarious locusts avoided the vanilla odour stimulus. Solitarious locusts therefore form aversive olfactory memory associations much more rapidly than gregarious locusts.

By contrast, recently gregarised locusts, which had been crowded for 24h prior to testing, completely failed to learn to make any aversive association between the vanilla odour and toxic food. No difference in preference from naïve locusts was seen 10min, 4h or 24h post-training. Solitarious locusts crowded for more than 4h are in most respects behaviourally indistinguishable from long-term gregarious locusts (Roessingh and Simpson, 1994), but Simões et al. (2012) show that their learning performance differs markedly from that of both long-term gregarious and long-term solitarious locusts.

Simões et al. (2012) also went on to show that, unlike the case with aversive learning, behavioural phase state did not affect the acquisition or retention of appetitive memories. Long-term solitarious, long-term gregarious and recently gregarised locusts all displayed the same ability to form appetitive memories following four training trials pairing an odour with a food stimulus, and performed similarly in the Y-maze after 10min, 4h and 24h post-training.

Learnt memories that are acquired prior to undergoing gregarisation are not affected by the phase-change process. Solitarious locusts that were first trained to acquire either appetitive or aversive memories retained them when they were subsequently gregarised by 24-h crowding. So although newly gregarious locusts are impeded from forming new aversive memories by behavioural phase change, the mechanism of memory recall is not compromised.

7.2.3 Learning Using a Substance That Changes From Innately Aversive to Appetitive During Phase Change

How are memories affected when a stimulus is changed from being intrinsically aversive to appetitive by behavioural phase change? Hyoscyamine is an alkaloid found in Black Henbane (*Hyoscyamus niger*: Solanaceae), which occurs in the natural environment of *S. gregaria*. Solitarious individuals avoid ingesting this plant in their diet, but recently gregarious locusts actively choose to eat it (Despland and Simpson, 2005a,b). Hyoscyamine can therefore act as either an appetitive or aversive stimulus depending on a locust's phase. When artificial food containing hyoscyamine was used as an unconditioned stimulus and paired with an odour as a conditioned stimulus, solitarious locusts developed strong aversive memories to the odour, but neither long-term nor recently gregarised locusts did so (Simões et al., 2013). Hyoscyamine is therefore a negative reinforcer for solitarious locusts but neutral (and potentially positive) for gregarious locusts.

This potentially raises a problem for recently-gregarised locusts that may have formed negative associations whilst still solitarious, as memories acquired prior to gregarisation are maintained throughout the gregarisation process. How do recently gregarised locusts remove a memory to an odour induced by aversion to hyoscyamine, that is, no longer useful and may prevent them eating foods that are now appetitive because the toxin they contain is now needed for chemical defence? As with NHT, the hyoscyamine-induced aversive memory is robust to gregarisation. Simões et al. (2013) showed that a single further exposure to the hyoscyamine-odour pairing post-gregarisation rapidly overwrites the previous negative association and results in a positive association instead. The blockade on aversive memory formation during gregarisation may help to prevent reinforcement of such pre-existing memories, while the intact appetitive learning pathway would facilitate acquisition of the new positive association.

7.3 Phase-Related Differences in Daily Rhythms of Activity

Gregarious *S. gregaria* and *L. migratoria* are generally day-active (diurnal), whereas solitarious animals are most active just after dusk (crepuscular: Ely et al., 2011; Michel, 1970; Roffey and Popov, 1968; Steedman, 1990; Uvarov, 1977), but this is not always the case.

7.3.1 Solitarious Locust Behaviour

In the field, solitarious adult *S. gregaria* roost in trees at night and behave cryptically during the day. They move to the ground to feed soon after dawn, shelter in low vegetation during the heat of the day and return to the ground in the late afternoon (Maeno et al., 2016). In contrast, solitarious nymphs tend to remain on their host plant throughout the entire 24-h period (Roffey and Popov, 1968; K.O. Maeno unpublished observation, cited in Maeno et al., 2016). This makes sense as solitarious nymphs tend to be camouflaged against plants, whereas solitarious adults are better camouflaged against desert sand. The diurnal switching of microhabitats is influenced by light, temperature, humidity and specific plant species. During the day, solitarious adults only fly when disturbed (Steedman, 1990). In warm weather they may start flying spontaneously after dusk and remain active for 1–2 h, sometimes covering large distances in relatively large numbers (Roffey, 1963; Roffey and Popov, 1968). Flights begin in response to the evening decrease in solar radiation and cease when temperatures fall below 25°C. In the lab, they show increased walking, scanning (side-to-side movements of the head and thorax) and propensity to spontaneously take flight in the first 2 h after lights-off (Ely et al., 2011). Solitarious *L. migratoria* also fly mostly at night (Uvarov, 1977). In Mauritania solitarious adult *S. gregaria* probably feed little at night because they roost in unpalatable trees, but nymphs may feed during both night and day because they rarely move from their preferred host plant (Maeno et al., 2016).

7.3.2 Gregarious Locust Behaviour

In the field, gregarious *S. gregaria* nymphs typically roost overnight, descend to the ground before dawn (almost certainly in response to increasing light intensity), become active and march as temperatures increase, then roost again in the hottest part of the day. They then march again in the afternoon, form static groups at around sunset, and then either roost or march for several hours before roosting for the night (Ellis and Ashall, 1957). On some occasions nymphs march throughout the night. Feeding occurs at all times of the day and night, but rarely at dawn. Basking at right angles to the sun occurs relatively infrequently but generally in the morning when temperatures are low: more commonly nymphs simply cluster in warm sunny spots (Ellis and Ashall, 1957). Swarms of adults tend to take off and fly in the late morning, settle in the late afternoon (Roffey, 1963), and roost again overnight. The majority of these behavioural patterns are driven primarily by fluctuations in environmental temperature. The diurnal rhythms are, however, relatively weak: Cloudsley-Thompson (1977) states that *S. gregaria* are not rhythmic in the field and that their rhythms in the lab are 'feeble'. They are nevertheless described as true circadian rhythms that persist under constant dark, and which can be entrained by light and temperature. Distant approaching stimuli tend to elicit jumps during the day but hiding responses in the evening (H. Wilps, personal communication cited in Hassenstein and Hustert, 1999; but it is unclear whether this relates to solitarious or gregarious animals). In *L. migratoria*, entrainment to a 12:12 h light:dark cycle induces a diurnal rhythm in amount of locomotion, which dies away again after a few days in constant dark (Chapman, 1954; Edney, 1937; Uvarov, 1977). The diel activity of adults differs from day to day and between individuals, but generally resembles that of nymphs. There has been no thorough analysis of diurnal patterns of activity using modern automated arena assays.

In the lab, gregarious nymphs of *L. migratoria* feed at all times, but less so in the 4h before lights-on (Simpson, 1982). It is not known whether this rhythm persists under constant light conditions.

7.3.3 Diel Rhythms in Emergence From the Substratum

In *S. gregaria*, although eggs may be laid at any time of the day or night (Ellis and Ashall, 1957; Popov, 1958), hatchlings emerge from the substratum at around dawn in both the field (gregarious phase, from at least 0.5h before to 4h after dawn, Ellis and Ashall, 1957) and the lab (gregarious phase, Padgham, 1981). In the lab, hatchlings from eggs laid by solitarious females also emerge from the substratum in the 4h prior to dawn (Shand, 2016; see Section 7.3.5).

7.3.4 Diel Rhythms in Hatching From the Egg

In both the field and the lab, hatchlings from a single egg pod tend to hatch together, often within minutes (Ellis and Ashall, 1957; Nishide and Tanaka, 2016; Shand, 2016). This synchronicity is achieved through light, temperature or vibrational cues (Nishide and Tanaka, 2016; Nishide et al., 2015a,b; Padgham, 1981; Shand, 2016). When incubated in the lab under constant light with a 12:12h thermoperiod (20:30°C), *S. gregaria* eggs hatch during the cryoperiod, whereas those of *L. migratoria* hatch during the thermoperiod (Nishide et al., 2015a). When diel thermal and luminance cues are both removed more than 3 days before hatching, hatching becomes arrhythmic (Nishide et al., 2015a). However, if the cues are removed only 1–2 days prior to hatching, eggs tend to hatch at around the expected time over the next 1–2 days, indicating the persistence of a weak circadian rhythm entrained to the thermocycle. There is a similar but even weaker effect in *L. migratoria* (Nishide et al., 2015a). Under constant temperature but with a 12:12h light:dark cycle, separated eggs of *S. gregaria* incubated on the substratum surface hatch during the dark phase, whereas those of *L. migratoria* hatch during the light phase. The precise time of hatching is modulated by the ambient (constant) temperature, with higher temperatures leading to earlier hatching (Nishide et al., 2015b). Under these lab conditions, eggs buried in sand hatch arrhythmically, so the authors conclude that light may not be an important zeitgeber (external cue that entrains a circadian rhythm) in the field; but light intensities in the incubator (200lx) were 5000-fold lower than intense daylight (10^5lx).

In the cricket *Gryllus bimaculatus*, the timing of egg hatching is influenced by an internal circadian clock that is active by about mid embryogenesis and which thereafter entrains to photoperiod (Tomioka et al., 1991). The photoperiodic conditions to which the parents were exposed can also influence the timing of hatching. So far, the data for locusts are less clear: there is a suggestion of weak circadian control of hatching in *S. gregaria*, but no such evidence for *L. migratoria*. Further experiments with natural zeitgebers would be useful.

7.3.5 Phase Differences in Hatching and Emergence

In the lab, the eggs of solitarious *S. gregaria* hatch approximately 2 h later than those of gregarious animals when maintained under constant light with a 12:12 h thermoperiod and placed separately on or just under the surface of the substratum. No such difference is observed in *L. migratoria* (Nishide et al., 2015a). Under these conditions, however, hatching is very spread out during the cryophase (*S. gregaria*) or thermophase (*L. migratoria*), so it is difficult to relate these observations to the normal situation in the field where eggs hatch synchronously at around dawn. It is not known whether similar phase differences are observed for separated eggs under a 12:12 h photoperiod with constant temperature.

Emergence of *S. gregaria* nymphs from natural egg pods laid in sand in the lab and incubated under coincident 12:12 h thermo- and photoperiods differs between the phases (Shand, 2016; Shand et al., 2015). Those of solitarious parents emerge on average 40 min earlier than those of gregarious parents (57 vs 17 min prior to dawn). Importantly, eggs of solitarious parents placed individually under a 1–2-mm thick layer of sand and incubated under the same 12:12 h thermo- and photoperiod *hatched* 42 min *later* than similarly treated eggs of gregarious parents (2:12 vs 2:54 h prior to lights on; Shand, 2016), which is consistent with Nishide et al. (2015a). It thus appears that under natural conditions, eggs of solitarious *S. gregaria* may hatch later but emerge from the substratum earlier than those of gregarious animals.

7.4 Neurophysiological Correlates of Diel Patterns of Activity

Pener and Simpson (2009) reviewed the relatively few studies of neural differences between the phases available at that time. Matheson et al. (2004) and Rogers et al. (2007) demonstrated phase-related differences in the descending contralateral movement detector (DCMD) visuo-motor pathway, while Fuchs et al. (2003) demonstrated differences in the tritocerebral commissure giant (TCG) sensory-motor pathway involved in flight in *S. gregaria*. The tritocerebral commissure dwarf (TCD) is a GABAergic interneuron in the same sensory-motor pathway that is sensitive to ambient illumination intensity (Tyrer et al., 1988). The spontaneous activity of the TCD was greater under dark compared to light conditions in both phases, but the dark-related activity was 5 × higher in gregarious locusts than in solitarious locusts, suggesting that there is far greater synaptic inhibition onto the flight circuitry under dark conditions in gregarious locusts (Fuchs et al., 2003).

Rogers et al. (2010) have built on the earlier DCMD work by showing that solitarious *S. gregaria* have larger eyes located differently on the head and greater binocular overlap than gregarious animals. The receptive field of DCMD to looming stimuli is broadly similar in shape and sensitivity in the two phases, but there is a small caudolateral region of enhanced sensitivity in gregarious animals. Gregarious locusts habituate less to repeated looming stimuli along the eye equator, as previously reported, but there is no difference in the rate of habituation in other parts of the receptive field. Rogers et al. (2010) suggest that the different DCMD receptive field properties may reflect the need for diurnal gregarious and nocturnal solitarious locusts to escape from different types of predators. Gaten et al. (2012) show that the responsiveness of DCMD to looming objects varies in a true circadian rhythm, and that the pattern differs in solitarious and gregarious *S. gregaria*. The strength of response varies sinusoidally with a period close to 24 h under constant light in solitarious locusts, but in gregarious locusts the 24-h rhythm is modified by secondary ultradian rhythms. The strongest responses occur at around expected dusk in solitarious locusts, but some 6 h earlier in gregarious locusts, thus matching the times of day at which locusts of each type are most active. Shand (2016) shows that overall eye sensitivity measured using an electroretinogram varies in a very similar diurnal pattern in solitarious and gregarious *S. gregaria*, with little or no evidence for a true circadian rhythm. The peak of sensitivity is after dusk in both phases, which means that the robust difference in time of peak sensitivity in DCMD reported by Gaten et al. (2012) must originate downstream of the photoreceptors.

Gordon et al. (2014) show that adult solitarious *S. gregaria* are more sensitive than gregarious animals to high-frequency sounds, and relate this to the need for nocturnal solitarious locusts to avoid bat predators. The work does not, however, attempt to determine if there are diel changes in auditory sensitivity.

A key area where research is lacking is in an understanding of the functioning of central clock neurons in any locust.

7.5 Circadian Clock Genes in *Schistocerca gregaria*

Clock genes regulate not only the function of neurons that comprise the central circadian clock, but also the functions of cells and tissues throughout the body (reviewed for insects by, e.g., Tomioka and Matsumoto, 2010). Tobback et al. (2011) produced partial sequences for the core clock genes *clock*, *period* and *timeless* from an existing EST database for *S. gregaria*. Both gregarious and solitarious locusts show the same diel pattern of expression of these three genes in animals maintained under a 14:10 h light:dark cycle. The authors describe this as a 'circadian' rhythm, but since animals were not examined under constant dark, this term is not justified. *Clock* is highly expressed in the brain at lights-on, whereas *period* and *timeless* both have double peaks at 12 and 18 h later, with a dip at 15 h. This strange pattern may be related to the 14-h photoperiod (Tobback et al., 2011). The similarity of the diel patterns seen in solitarious and gregarious locusts suggests that observed differences in diel patterns of behaviour are regulated downstream of the molecular clock.

Tobback et al. (2011) used RNAi to knock down expression of *period* and *timeless* in gregarious female *S. gregaria*, but this resulted in reduced expression (mRNA levels) only in ovary, and not in brain. The knockdown in ovary resulted in females laying fewer eggs. RNAi of *clock* was lethal in females prior to sexual maturity. Tobback et al. (2012) performed RNAi against *period* and *timeless* in gregarious male *S. gregaria*, which resulted in reduced expression only in testes and accessory glands, and not in brain. Untreated females mated with RNAi-treated males produced fewer fertilised eggs, possibly because of disrupted diel patterns of sperm transfer in the male reproductive organs.

There has been no comparative study of the molecular biology of the circadian clock in any species of locust.

8. COLLECTIVE BEHAVIOUR

8.1 Laboratory Studies of Locust Collective Behaviour

Animal collective behaviour, specifically the emergence of synchronisation among moving groups of animals, is an important ecological phenomenon (Sumpter, 2010). The vastness of locust swarms, and their severe impact on both the natural environment and human agriculture, make them a particularly important example.

In recent years, there have been several new attempts to describe or capture locust collective behaviour using theoretical modelling, as part of the long-lasting endeavour to understand the locust phenomenon. A recent review on locust collective motion by Ariel and Ayali (2015) focussed mostly on theoretical and modelling efforts to explain the emergence and maintenance of order in marching locust nymphs. While flying swarms of adults may attract general attention, the marching bands of nymphs, beyond constituting a critical stage in the development of locust aggregations, also demonstrate remarkable synchronisation and coordination. Some modelling studies are closely associated with biological data and observations, which aim to specifically explain locust behaviour, while others offer a more abstract mathematical or physics-based perspective (Ariel and Ayali, 2015). This dualism can be found in many locust studies and reflects the open question of how unique an example of collective movement is locust swarming, and what general lessons are to be learned from it?

8.1.1 Locust Swarming and Coordinated Movement

Aggregation, the formation of dense groups of nymphs, is a major behavioural characteristic of gregarious locusts that appears soon after hatching. It should be noted, however, that locust aggregation per se differs from coordinated movement. The latter, referred to as marching behaviour, also appears a few days after hatching, and continues to constitute a dominant part of the nymphs' behaviour until adulthood, when they begin to fly.

Collective marching is affected by environmental conditions (e.g. temperature, rain, wind and topography; Amichay et al., 2016; Uvarov, 1977), but it is a self-emergent process, induced by local interactions among the locusts within the group. The nature of these local interactions is a major focus of research.

Undisturbed nymphs within marching bands will advance by means of persistent walking, intermittently interrupted by pauses (Ariel et al., 2014; Ellis and Ashall, 1957; Uvarov, 1977). The speed of walking varies, as do the durations of the walking or pausing bouts. Locusts at the edges of the swarm will alter their behaviour, which maintains the integrity of the band: e.g. locusts at the front will slow down or turn more often, reducing their net displacement in comparison to those behind (Uvarov, 1977). The major factors that affect the general direction of marching are still unclear. There may well be fundamental species-related differences in the interaction of the marching band and the environment (as well as in other aspects of the collective behaviour), reflected in differences in the overall shape of the bands (Hunter et al., 2008; Sumpter, 2010; Uvarov, 1977).

Relatively few studies have focussed on the local dynamics within flying swarms due to obvious technical limitations, though see Edelstein-Keshet et al. (1998) for a mathematical approach. Flying swarms are assumed to migrate predominantly downwind (Cooter, 1989). Interestingly, and somewhat similar to the case of the nymphs at the edge of the marching band, locusts flying out of the migrating swarm will turn back to rejoin it.

8.1.2 Laboratory Studies of Locust Collective Movement

Recent progress in our understanding of the rules governing animal collective motion has been made possible by the development of new methods for generating coordinated motion under controlled lab conditions and reliably tracking individuals within collectively moving groups. Several studies have revisited the pioneering work of Ellis and colleagues (e.g. Ellis, 1959, 1963a,b), investigating marching behaviour in closed circular arenas. Lab-based manipulations of locust density are now accompanied by automated video tracking and innovative statistical methods, which have yielded important insights into large-scale characteristics of locust swarms and the rules governing collective motion, as well as into the local animal–animal interactions or the details of locust motion within the crowd.

Buhl et al. (2006, 2011) demonstrated that locust collective movement is highly dependent on group density and is mediated by combinations of social interactions such as avoidance, alignment and attraction. These findings spurred later modelling studies. Ariel et al. (2014) reconfirmed that a dominant feature characterising locust behaviour is their intermittent, pause-and-go, walking pattern. This finding, previously reported by Bazazi et al. (2012) for a single animal, was shown to persist in all individuals within a marching band. It was suggested that the coordinated marching behaviour of the locust swarms is manifested by the repeated decisions taken by individual animals to initiate or resume walking. These recurring decisions can be explained by sensory stimuli conveyed by the optical flux in the vicinity of the individual, due to the behaviour of others in the crowd (Ariel et al., 2014), supported by the locusts' high propensity for walking. Recent technological advances in achieving high accuracy and reliable tracking of all individuals in the arena promise to further augment our knowledge by way of carefully planned laboratory experiments.

8.1.3 Modelling Locust Collective Behaviour

Locust collective movement occurs across different spatial and temporal scales—from the individual (cm and minutes) all the way to swarms comprising millions of insects (km and hours). Mathematical models of these phenomena tend to assume a homogeneous population, in which the influence of conspecifics on each individual is subject to some local average (or weighted average). While these assumptions may be somewhat unrealistic from a biological perspective, recent models have nevertheless predicted some important aspects of locust behaviour.

Most theoretical work on locust collective movement has established either continuous-based models, which define broad patterns of interactions across the entire group, or individual-based models, which focus on the local interactions of single locusts with their neighbours. Yates et al. (2009) and Escudero et al. (2010) used continuous models to determine that the mean time between directional changes increased exponentially with group density. Escudero et al. (2010) also established that direction switches are an inherent property of collective movement, and that their precise timing is unpredictable. This work was expanded upon more recently by Dyson et al. (2015), who combined empirical data from Buhl et al. (2006) with an individual-based model to show that interactions between individuals in marching bands of *S. gregaria* occur with at least two neighbours simultaneously.

Examples of individual-based models include that of Buhl et al. (2006), which was an SPP (self-propelled particles) model adapted from Czirók et al. (1999; based in turn on Vicsek et al., 1995). The main feature of such models derives from the understanding that the macroscopic state of the system can be classified into one of two well-defined phases (in the sense of statistical physics): order and disorder. Switching between the phases requires an external change in the system parameters such as concentration or the amount of noise. This finding introduced the possibility that order is closely dependent on the number of individuals, as demonstrated in the laboratory experimental arena.

A specific class of two-dimensional SPP model is termed Escape and Pursuit (E&P; Romanczuk et al., 2009). This modelling approach, directly motivated by experimental work (Bazazi et al., 2008, 2011), postulated that marching locusts are continuously escaping from conspecifics behind them while at the same time pursuing others in front of them. The strength of this model is the suggested direct link between the local individuals' interactions and the coordinated motion of the swarm. However, this modelling approach was questioned by Ariel and Ayali (2015), who note that most observations of locust behaviour are not consistent with aggressive interactions between healthy individuals within the marching band (cannibalism is far more common where the cannibalised individual is injured or dead). This observation is not necessarily inconsistent with a role for cannibalism in shaping these collective behaviours, precisely *because* the strategies locusts might have evolved to avoid direct contact with one another are successful, and reduce the risk of antagonistic interactions to such a level that it is outweighed by the group-related benefits of reduced predation risk, and the possibility of future cannibalism-derived nutrients (Hansen et al., 2011). This interesting debate would benefit from field studies, of which Buhl et al. (2012) is presently the only recent example. Investigating a swarm of *C. terminifera*, the authors reported a tendency for locusts to interact with all their surrounding neighbours. There was also evidence of a short-range repulsion force, balanced by a clustering force, similar to what has been observed in a repulsion/alignment/attraction-based SPP model. This study rejected the E&P mechanism for *C. terminifera* where marching had already been established (as opposed to during its onset) but, given that this is only one study in one species, further fieldwork would clearly be highly beneficial. The extent to which cannibalism is observed (or not) in extant populations does not discount a role for this behaviour in the evolution of collective movement (Guttal et al., 2012 and see details in Section 2).

An alternative approach to the E&P model was suggested by Ariel et al. (2014). Further placing the emphasis on the individual animal and following their own experimental observations of locusts marching in an experimental arena, they base their model on a set of discrete decisions—when to stop, walk, or turn, rather than a continuous process. Here the calculated local averages (as well as the order parameter) consider only the walking individuals. This model effectively captures the basic behaviours of locusts in a marching band, in addition to offering testable predictions.

8.2 The Need for Large-Scale Movement Data

Laboratory experiments (Ariel et al., 2014; Buhl et al., 2006; Ellis, 1951, 1953, 1964a,b) and minimal collective movement models (Ariel et al., 2014; Buhl et al., 2006; Vicsek et al., 1995) have helped establish that local interactions between individuals, for example, alignment with nearest neighbours, are the fundamental elements underpinning locust collective movement. Collective movement is a central and defining part of the behavioural repertoire of gregarious locusts, but key questions remain concerning the benefits of collective movement to locusts, and how it originally evolved. Current models have barely started addressing these questions. Benefits to animal groups include protection from predation by a dilution effect, or through enhanced collective escape responses (Foster and Treherne, 1981; Hamilton, 1971); improved navigation capabilities such as gradient climbing (Sumpter, 2010); and increased migration survival due to the potential for cannibalism (Hansen et al., 2011; though see discussion in Section 8.1.3). All these functions are either directly dependent on group structure or likely to be strongly influenced by variations of density within the group, its structure and dynamics. Yet, current models are unable to account for the large-scale group structures observed empirically (Sumpter, 2010). Locust migratory bands exhibit a wide range of patterns from 'columnar formations' (Uvarov, 1977), such as the intertwined narrow and dense stretching streams of *Locustana pardalina* (Uvarov, 1977), to 'frontal formations', such as wide crescent-shaped frontal formations followed by looser comet-like tails of *C. terminifera* (Buhl et al., 2011; Clark, 1949; Hunter et al., 2008; Uvarov, 1977). Some species, such as *D. maroccanus*, can alternate between both types of formations (Uvarov, 1977). There is a need to address locust collective marching at the large scale at which nymphal bands occur under natural conditions. While modelling techniques such as parallel computation using general-purpose computing on graphics processing units make it theoretically possible to simulate nymphal bands at sufficiently large scales (e.g. Erra et al., 2009; Torney et al., 2011), gathering the relevant quantitative empirical data from the field remains a major challenge.

8.2.1 Studying Nymphal Band Patterns and Their Movement Over Their Lifetime

Quantifying and mapping locust nymphal bands in the field is time consuming; detailed data are still extremely rare, and limited to a few species (e.g. in *S. gregaria*: Ashall and Ellis, 1962; Ellis and Ashall, 1957; *N. septemfasciata*: Dean, 1967; the outbreaking Gomphocerine Mato Grosso Grasshopper, *Rhammatocerus schistocercoides*: Lecoq and Balanca, 1998; Lecoq et al., 1999 see also Uvarov, 1977, for figures, photos and descriptions of bands in other species). While there are not enough data to determine which factors lead to the large variation of nymphal band patterns observed both within and across species, these studies reveal some key features specific to locust nymphal bands compared to other collectively moving animals. Locust bands tend to move at much slower speeds than the individuals within it. This is partially explained by individual locusts tending to move in short bursts of locomotor activity (in the field: Ellis and Ashall, 1957; Uvarov, 1977; observed in laboratory experiments by: Ariel et al., 2014; Buhl et al., 2006; Ellis, 1953) but also because nymphal bands are a product of both aggregation and marching. Nymphs exhibit aggregation behaviour in three main contexts. First, they form dense clusters overnight or when the temperature is too cold to march, which occur in vegetation or on the ground. Second, they climb up vegetation and aggregate when the ground temperature becomes too hot (perching). Third, they aggregate either on the ground or in vegetation when they reach the front of their band (Uvarov, 1977). The alternation of marching and inactivity at the individual level is also likely to be influenced by nutritional state. Locust feeding activity is regulated by internal rhythms (Simpson, 1981) and they need to weigh their own nutritional state against the need to move and remain within the group. An in-depth study of the link between individual behaviour and emergent nymphal band patterns will require the ability to systematically track and quantify patterns and movement at both individual and group levels, which still poses significant technical challenges.

8.2.2 Remote Piloted Aircrafts and Nymphal Bands and Vegetation Photogrammetry

Even with the use of global positioning system (GPS) units, ground surveys remain too time consuming to map small nymphal bands several times per day, let alone large ones. Bands can be visible during aerial survey operations, but such surveys are both expensive and not always capable of detecting smaller bands when relying on conventional manned aircraft operated at safe altitudes above 100 m (Hunter et al., 2008). Unmanned aerial vehicles (UAVs; also known as 'drones' or 'remotely piloted aircraft systems') promise to revolutionise surveying because they are both cheap to operate and typically fly at low altitudes, below 100 m. While their application to the surveying and studying of locust nymphal bands is at an early stage of development, UAVs are increasingly used in ecological and conservation contexts (Koh and Wich, 2012; Marvin et al., 2016). Their use ranges from monitoring areas using live video feedback; to mapping areas of study using high-resolution camera images (either conventional RGB or thermal, multi-spectral, hyperspectral) which can be stitched into larger pictures using photogrammetry; to remotely detecting sensors carried by animals. All of these applications could be adapted for use on locust nymphal bands.

Both main types of UAV, multi-rotor and fixed-wing aircraft, are suitable for surveying or mapping locust nymphal bands. For locust work, the main immediate limitation often comes from existing aviation regulations that insist that UAV operators keep their aircraft within visual line of sight, which effectively imposes a flight radius of approximately 500 m. Within such ranges, most UAVs are easily capable of overflying a locust band at altitudes ranging from 50 to 100 m, and recording overlapping pictures suitable for photogrammetry at resolutions of up to a few cm per pixel, within 20 min. Where flying beyond visual line of sight is allowed, fixed wing UAVs have the advantage of having a much longer flight time (up to an hour) and a larger range (up to 25 km in Koh and Wich, 2012), but the operational and training requirements are greater.

8.2.3 Individual Movement Quantification and Tracking in the Field

Tracking the movement of individuals remains challenging. Some studies have quantified individual movement over relatively small scales, starting as early as the 1960s with the pioneering work by Stower (1963), and more detailed recent papers have managed to quantify some movement characteristics and infer interaction rules (Buhl et al., 2012). For larger spatial and temporal scales, insect-borne sensors, including very small ones, are becoming increasingly accessible. The most promising tracking method uses harmonic radar transponders, such as those used to track individual bees (Lihoreau et al., 2012; Riley et al., 2005) or Mormon Crickets, *Anabrus simplex* (Sword et al., 2005). Recent progress in transponder miniaturisation is bringing these technologies within a size and weight range that would allow them to be fitted onto final-instar nymphs of *C. terminifera* (Tahir and Brooker, 2015). The radars required to track these sensors used to be too large to be frequently transported, but frequency-modulated continuous wave radars operating in the millimetre wave band will soon be small enough to be carried as UAV payloads (Tahir and Brooker, 2015).

8.2.4 Conclusions

Existing quantitative field data are still extremely scarce, and much more are needed to understand the emergence of the large-scale patterns produced by locust nymphal bands. Recent advances in both the development of UAV and miniature sensors will allow the gathering of data at unprecedented spatial and temporal scales, which should continue to be combined with laboratory, modelling and theoretical approaches. Furthermore, and given that collective behaviour is a crucial aspect of the locust phenotype, these varying approaches should be employed across a range of locust species, with a view to offering new perspectives for evolutionary and phylogenetic studies (see Section 2).

9. ENVIRONMENTAL STIMULI AFFECTING PHASE CHANGE

Locust phase change can be induced to a greater or lesser extent through some combination of visual, olfactory and tactile stimuli from other locusts. As discussed in Section 4, there is some evidence that olfaction plays a role in behavioural phase change of *L. migratoria*, and it also causes behavioural phase change in *S. gregaria* in combination with visual stimuli from other locusts (Roessingh et al., 1998). Tactile stimulation of the hind femur is also a potent gregarising factor in *S. gregaria* (Simpson et al., 2001), while tactile stimulation of the antennae has the same effect in *C. terminifera* (Cullen et al., 2010). It therefore follows that any environmental variables that bring locusts into close proximity, and especially those that increase direct contact between individuals, will generally promote gregarisation.

At a small scale, clumped resources promote interactions and subsequent gregarisation among *S. gregaria* (Babah and Sword, 2004; Bouaïchi et al., 1996; Collett et al., 1998; Despland and Simpson, 2000; Despland et al., 2000). In contrast, dispersed resources tend to keep locusts spread apart from each other, thereby maintaining them in the solitarious phase. Using survey data from routine monitoring from 2003 to 2011 in Mauritania, Cissé et al. (2013) used a statistical model to show that sparse and dry vegetation lowered the threshold of gregarisation relative to dense and green vegetation, presumably by increasing locust interactions. By analysing gregarious behaviour in circular arenas in the field, Cissé et al. (2015) corroborated those results showing that when at similar densities for a given $100\,\mathrm{m}^2$, locusts collected from areas with a lower percent vegetation cover tended to be more gregarious. However, other vegetation parameters such as greenness and distance between plants were found to be poor predictors of phase status. In a subsequent study, Cissé et al. (2016) used field assessments of locust phase state (predominantly clumping behaviour) collected in 2012 and 2013 to test the predictions of the Cissé et al. (2013) statistical model. The model predicted the phase of field populations of *S. gregaria* with only a 6% error rate. Using an occupancy model, Veran et al. (2015) showed that there was a higher probability of scattered populations leading to gregarious outbreaks of *C. terminifera* in desert areas as compared to grasslands. This pattern may be due to resources being more clumped in the desert. Despland and Simpson (2000) found an interaction between food quality and distribution; *S. gregaria* have an increased amount of direct contact, and are therefore more likely to gregarise, if they have to regularly switch among food sources that have complementary nutrient contents. Bazazi et al. (2011) found that crowd-reared *S. gregaria* nymphs fed low-protein diets for 2 days had an increased walking speed, which likely translates to increased interactions and the formation of migratory bands.

The effect of resource distribution on gregarisation likely differs by scale. In contrast to studies focussed at a small-scale, a large-scale landscape study using satellite imagery showed that *S. gregaria* outbreaks tended to occur when vegetation was more fragmented as opposed to clumped (Despland et al., 2004). These studies focussed at large and small scales illustrate that resource abundance, quality and distribution are all important factors influencing the likelihood that locusts will gregarise. Whether these same mechanisms apply to species beyond *S. gregaria* and *C. terminifera*, and why it might be functionally beneficial for locusts to undergo behavioural phase change under these different conditions, are still open questions.

Limited experiments have explicitly tested the major evolutionary drivers of locust phase change, but it is likely a complex set of interacting factors arising from living in a dynamic and unpredictable environment that differ to some extent among species (see Section 2). At the outbreak scale of localised populations (prior to a larger upsurge or plague), the benefits of some phase-related traits have been clearly demonstrated above a certain population density threshold. For example, behavioural phase change that leads to aggregations when at high density is likely an anti-predator strategy. The appearance of aposematic colouration only at high population density fits with warning colour theory, which predicts this strategy to be beneficial above a critical density of individuals expressing similar colouration (Sword, 2002). High population density-induced colour changes have been shown to signal toxicity (Sword, 1999). In *S. lineata*, high-density animals feed on toxic Wafer Ash, *P. trifoliata*, which causes them to gain gut-content mediated toxicity to vertebrate predators (Sword, 1999, 2001). Similarly, solitarious phase *S. gregaria* avoided foods containing the alkaloid hyoscyamine, while gregarious locusts did not (Despland and Simpson, 2005a). When solitarious locusts were subsequently crowded, they began to accept food containing hyoscyamine. Even in cases where insects lack aposematic colouration or plant-derived toxicity, there can be anti-predator advantages to being a part of a crowd. Sword et al. (2005) tracked Mormon Crickets, *A. simplex* in the field and found that those marching in a crowd had lower levels of predation than those not within the crowd. Crowd-reared locusts tend to have a faster development rate (Pélissié et al., 2016), which likely decreases opportunity for predation during the nymphal stages.

Once at high population density, locusts increase activity levels and may migrate *en masse* due to a combination of factors that may include the threat of cannibalism (Bazazi et al., 2008; Guttal et al., 2012), decreased disease transmission (Altizer et al., 2011) and/or seeking better resources. Locust phase change is often (but not always) coupled with migratory swarms (Key, 1950; Sword, 2003). Thus, it has been hypothesised that high population density increases intraspecific competition for food, and migration towards further resources by gregarious phenotypes at least partially mitigates the effects of this competition (Johnson, 1969). While solitarious locusts may migrate (e.g. Farrow, 1990), it is gregarious locusts that are most commonly associated with physiologically challenging long-distance flight. Indeed, gregarious locusts have characteristics generally associated with increased migratory potential. The forewing length:hind femur length ratio is higher in gregarious than solitarious adults of *S. gregaria*, *N. septemfasciata* and *Locustana pardalina* (Uvarov, 1966 and references therein; see also Nolte, 1967). Presumably, increased values are correlated with decreased wing loading (wing area/body mass) and higher flight capacity, although experimental tests are limited. In addition to changes in morphometry that enhance migratory capacity, gregarious locusts tend to have a proportionally higher resting metabolic rate (Applebaum and Heifetz, 1999; Butler and Innes, 1936; Cease et al., 2010; Heifetz, 1997; Uvarov, 1966), perhaps due to the elevated maintenance cost of larger flight muscles (Nespolo et al., 2008). Gregarious locusts also tend to have increased lipid stores for long-distance flight, and a delayed onset of fecundity (Applebaum and Heifetz, 1999; Hunter et al., 1981; Pener and Yerushalmi, 1998; Pener et al., 1997; Uvarov, 1966), which supports a general model of increased investment in flight-related physiology and morphology at the expense of reproductive capacity. Even as hatchlings, gregarious *S. gregaria* have higher body lipid contents than solitarious hatchlings (Maeno et al., 2013). The increased energy reserves made gregarious hatchlings less susceptible to starvation and would presumably also increase marching capacity. Density-dependent polyphenism has been documented even in acridids that do not form agriculturally damaging swarms. For example, the Slender Green-Winged Grasshopper, *Aiolopus thalassinus*, has never been recorded to march or swarm, but crowd-reared individuals develop more quickly and have increased activity levels, haemolymph lipids and carbohydrates, metabolic rate, and food consumption compared to those of solitary-reared insects (Heifetz and Applebaum, 1995). Such examples support the concept that different reaction norms of locust phase polyphenism have evolved separately from each other (Song, 2005; Song and Wenzel, 2008; see Section 2).

Because locust phase polyphenism is often, but not always, coupled with migration (Pener, 1991), it has been hypothesised that poor resources should promote phase change (Johnson, 1969). Nutrition is influential in regulating migratory plasticity in other insects (e.g. Denno et al., 1985; Dixon, 1998; Müller et al., 2001) and, in the short term (2 days), Bazazi et al. (2011) found that *S. gregaria* fed low-protein diets increased walking speed in gregarious nymphs. In a longer term study, Maeno and Tanaka (2011) found that *S. gregaria* reared on wilted grass had higher wing: leg length ratios, but decreased performance, relative to locusts reared on fresh cut grass. However, nymphs were reared individually and more experiments are required to determine if gregarious, crowd-reared *S. gregaria* respond in the same way.

In contrast to the hypothesis that poor resources should enhance development of migratory characteristics (Johnson, 1969), new evidence suggests that nutritionally optimal diets may be required to promote increased migratory capacity, at least in some species. Cease et al. (2017) found that Mongolian Locusts, *Oedaleus asiaticus*, reared at high population density and fed low-nitrogen grass (nutritionally optimal for this species) had increased wing area and decreased wing loading relative to locusts fed high-nitrogen grass (which decreases growth and survival in this species; Cease et al., 2012). The authors then reared locusts on three synthetic diets differing only in the protein: carbohydrate ratio for the duration of the fifth instar and measured flight behaviour in the adults. Locusts fed their nutritionally optimal dietary ratio of 1:2 protein:carbohydrate flew for longer than locusts fed imbalanced diets with lower or higher protein:carbohydrate ratios. Faki (2004) found that *S. gregaria* reared on no-choice diets of their preferred host plants *Pennisetum typhoideum* or *Heliotropium undulatum* tended to have morphometrics more closely aligned with gregarious locusts as compared to *S. gregaria* reared on a less-preferred and presumably less nutritionally optimal *Aerva javanica*. These studies suggest that a locust's nutritional state, and wider nutritional landscape, might directly influence its reaction norm for migratory capacity in response to population density, in addition to the described indirect effects of resource variation on phase change (Despland and Simpson, 2000; Despland et al., 2000). Other environmental factors may also modulate the effect of density by altering the threshold at which an individual locust responds to crowding stimuli. There is some evidence to suggest that the microsporidian *P. locustae* might suppress phase change in

L. migratoria migratorioides, although it is unknown if this was a direct effect or if it was behaviourally mediated by slowing locusts down and therefore decreasing their interactions with conspecifics (Fu et al., 2010; species given as *L. migratoria manilensis*).

These studies focussed on the environmental stimuli affecting phase change suggest a pattern whereby conditions that favour population growth may also favour development of gregarious phenotypes. Population growth often leads to increased effective density, which is the primary trigger of phase change. Therefore, it seems plausible that conditions promoting population growth might not suppress, but instead promote, gregarisation. Moreover, access to high-quality food resources may be necessary to support the energetic demands of the gregarious phase. Non-optimal diets may make it difficult for locusts to achieve the physiological condition necessary for successful migration and may preclude migration *en masse* even when faced with intraspecific competition under crowded conditions. This area of locust research supports recent calls for a renewed study of migratory and dispersal reaction norms (Clobert et al., 2009) to aid in developing an understanding of the regulation and evolution of migratory and dispersal behaviours (Burgess et al., 2015).

10. ECOLOGY AND LANDSCAPE-LEVEL PROCESSES

10.1 Remote Sensing

The immense, often transcontinental scale of potential locust habitats requires fast and efficient monitoring dictated by meteorological conditions and locust life cycles. Locust surveying is a daunting and challenging task as locust habitats often cover vast, remote and transborder areas with difficult and insecure access. Traditionally, it is performed by ground survey teams, who collect locust information and assess ecosystem parameters with particular attention to vegetation. Such ground-based surveying is often incapable of providing fast and reliable information on spatio-temporal distribution of potential locust habitats. Remote sensing in the form of satellite imaging appears to be an appropriate tool for addressing the challenges of scale and time in locust monitoring.

The first attempts of using information from satellites in locust monitoring were implemented by Pedgley (1974) and Hielkema (1981). Subsequently, Tucker et al. (1985) recognised the potential of the remotely sensed data for *S. gregaria* surveying and forecasting. Satellite images, particularly from Landsat, were used to map vegetation in the gregarisation areas of

S. gregaria in West Africa (Ghaout, 1990; Hielkema et al., 1986; Louveaux et al., 1990). In Australia, satellite data were first used by McCulloch and Hunter (1983) to inventorise *C. terminifera* habitats. The authors classified Landsat images and found that locust presence was confined to three ground cover classes: stony downs, stony plains, and high-elevation floodplains. Further development of remote sensing tools in Australia was done by Bryceson and colleagues, who demonstrated that it is possible to detect even small areas of vegetative growth after rains with Landsat data (Bryceson, 1984; Bryceson and Wright, 1986). Subsequently, Landsat images were instrumental in identifying oviposition beds (Bryceson, 1989) and the source areas from which plagues of *C. terminifera* developed (Bryceson, 1990, 1991). Remote sensing data are currently incorporated in a decision support system used by the Australian Plague Locust Commission (APLC) in its routine forecasting and management activities (Deveson, 2013; Deveson and Hunter, 2002).

Early detection of potential *S. gregaria* habitats became possible with the development of the Normalised Difference Vegetation Index (NDVI), a remote sensing technique which allows observers to distinguish areas of green vegetation from bare soil. Cherlet and Di Gregorio (1993) attempted to calibrate the NDVI delivered by the Advanced Very High Resolution Radiometer (AVHRR) satellite of the National Oceanic and Atmospheric Administration of the United States (NOAA), using extensive ground-collected field data from Niger. Voss and Dreiser (1994) used the NDVI to detect vegetated areas in Mauritania. Dreiser (1994) used Landsat data to produce *S. gregaria* habitat maps for certain areas of Sudan, Mali and Mauritania. Landsat data, which have a 16-day temporal resolution, were deemed adequate to assess the *potential* suitability of *S. gregaria* habitats. However, the identification of the *actual* habitats required higher temporal resolution imagery like that provided by the NOAA AVHRR satellite, in combination with meteorological data from Meteosat and other similar satellites (Voss and Dreiser, 1997).

Various remote sensing techniques have been used on a regular basis for Desert locust monitoring by Food and Agriculture Organization's Desert Locust Information Service (FAO-DLIS) in Rome since the 1990s. Early detection of habitat conditions favourable to locust outbreaks is a key element in locust preventive strategy, which relies on finding *S. gregaria* populations before they can expand into full-scale plagues (Magor et al., 2008). Satellite imagery is used to characterise changes in vegetation cover and soil moisture in order to direct ground survey teams to areas with the highest probability of conditions conducive to *S. gregaria* breeding. The first satellite imagery used in FAO-DLIS early warning systems was Meteosat cloud imagery, which made it possible to pinpoint rainfall areas (Cressman, 1996). A few years later, the NDVI delivered by NOAA AVHRR sensors with 7.6 km spatial resolution improved our understanding of locust habitats. Since then, the spatial resolution of satellite imagery has continued to improve, along with the treatment algorithms, which has greatly aided the identification of potential and actual *S. gregaria* habitats from West Africa to India through the estimation of rainfall and vegetation cover (Cressman, 2013). Since the 2000s, the FAO-DLIS has used Moderate Resolution Imaging Spectroradiometer (MODIS) images at a spatial resolution of 250 m, consisting of a 16-day composite image of NDVI values. Starting from the 2010s, the FAO-DLIS has incorporated dynamic greenness maps derived from the NDVI index from MODIS data (Pekel et al., 2011; Renier et al., 2015). In addition, daily, 10-day and monthly rainfall estimates are produced on a 0.25° × 0.25° latitude/longitude grid (Dinku et al., 2010). These various remote sensing data are incorporated into a Geographic Information System (GIS) called RAMSES (Reconnaissance And Monitoring System of the Environment of *Schistocerca*) allowing national locust information services to manage, query, display, analyse and map field and other data (Cherlet, 1993; Cressman, 2013). MODIS-derived NDVI is also currently used in Australia for locust forecasting (Deveson, 2013) although some specialists question its predictive capacity (Weiss, 2016).

Green vegetation emerging after rains in the desert is known to play a crucial role in the initial concentration and gregarisation of *S. gregaria* (Despland, 2003). Furthermore, the fractal dimension of vegetated areas, which measures its degree of spatial fragmentation, was demonstrated to be positively related to the gregarisation potential of *S. gregaria* populations (Despland et al., 2000). The authors used NDVI pixels of 7.6 km over a vegetation threshold in good and bad reproduction periods, across an area of approximately 500 km × 1300 km. However, Piou et al. (2013) did not find fractal dimension to be a good indicator of locust presence when analysed at a finer spatial resolution of 250 m and across a smaller area. Tratalos and Cheke (2006) also argued that NDVI data derived from satellite images covering large areas (from 4 to 8 km spatial resolution) were not a good predictor of *S. gregaria* presence. However, Piou et al. (2013) did find that NDVI at 250 m resolution can be a good predictor of *S. gregaria* occurrence. These studies support the potential of satellite imaging to identify habitats susceptible to locust invasions, which is critical for early warning and preventive management (Cressman, 2013). Predicting presence of locusts can be improved if remote sensing information and vegetation indices are coupled with historic survey data (Deveson, 2013; Latchininsky et al., 2016; Piou et al., 2013; Waldner et al., 2015).

With regard to *L. migratoria migratoria*, satellite images have been used to map spatio-temporal distribution of its preferred vegetation, the common reed *Phragmites australis*, in Central Asia (Latchininsky et al., 2007; Löw et al., 2013; Navratil and Wilps, 2013; Propastin, 2012, 2013; Sivanpillai and Latchininsky, 2007, 2008; Sivanpillai et al., 2006). Improved mapping of reed habitats has allowed for more targeted locust surveying and risk assessment (Latchininsky et al., 2016; Löw et al., 2016).

Satellite data have also been used for *L. migratoria migratorioides* habitat monitoring and post-outbreak damage assessment on vegetation in China (species given as *L. migratoria manilensis*; Ji et al., 2004; Ma et al., 2005; Tian et al., 2008; Zha et al., 2005). An attempt to apply Landsat data to estimate the risk of infestation by *L. migratoria migratorioides* in Madagascar was made by Franc et al. (2005; species given as *L. migratoria capito*).

Application of remote sensing data to other locust species is scarce. Franc et al. (2008) used SPOT satellite information to evaluate the impact of habitat fragmentation on population dynamics of *N. septemfasciata* in Madagascar. They showed that the SPOT images, combined with a digital elevation model, made it possible to identify three new migration pathways of *N. septemfasciata*, which were made possible by deforestation between 1986 and 2004. These pathways provided access to previously inaccessible oviposition zones and enabled locusts to breed, concentrate and gregarise (Franc and Duranton, 2012; Latchininsky et al., 2016). Gornyy et al. (2011) and Tronin et al. (2014) attempted to estimate *Ca. italicus* outbreak potential in Eurasia using satellite information but encountered difficulties due to great habitat heterogeneity. Similarly, the patchiness and mosaic character of the habitats of *D. maroccanus* made it difficult to characterise its distribution in Central Asia based on satellite data (Smith, 2012).

Currently available satellites are unable to directly detect locust swarms or nymphal bands. An attempt by Kibasa (2006) to detect actual locust aggregations with the high-resolution SPOT 5 satellite data failed because it was not possible to distinguish locusts from the background soil.

A different remote sensing technique, so-called 'active remote sensing' in the form of Vertically Looking RADAR (VLR), was first used to monitor *S. gregaria* flights over the Sahara in the late 1960s (Roffey, 1969). VLR provided a tool for measuring aerial density, orientation, direction and speed of flight of solitarious locusts (Schaefer, 1969, 1976). VLR data could be a useful complement to routine locust surveys (Chapman et al., 2003; Riley and Reynolds, 1997; Smith et al., 1993). Although some specialists considered the use of VLR devices for monitoring locust swarm migrations impractical because of complex data analysis (Reynolds, 1988; Riley, 1989), this technique is currently used by the APLC in Australia for monitoring migrations of *C. terminifera* swarms (Deveson et al., 2005; Drake and Wang, 2013; Drake et al., 2001).

A detailed history of remote sensing in locust monitoring and management is presented in Latchininsky and Sivanpillai (2010) and Latchininsky (2013). As soon as data from satellites became available for non-military purposes, remote sensing became a hope for locust specialists worldwide. Over the past three decades, satellite imagery has been increasingly applied to locust monitoring and has become a routine operational tool for two locust species, *S. gregaria* and *C. terminifera* (Cressman, 2013; Deveson, 2013). Based on satellite imagery, locust management teams target specific, high-risk gregarisation areas, which results in significantly reduced monitoring and management costs. This also contributes to a change in the paradigm of locust control from a curative to a preventive mode. Satellite data and associated GIS appear to be powerful instruments in the arsenal of locust managers, especially if accompanied by a thorough ground knowledge (Latchininsky and Sivanpillai, 2010). With advances in our understanding of the ecology of different locust species and the increased availability of satellites and processing software, remote sensing is gradually becoming an efficient tool for locust forecasting and management. Finally, the rapid development of remote sensing platforms, such as drones, is opening intriguing possibilities for various applications in locust research and management (see Section 8.2).

10.2 Climate Change

Climate change became obvious in the last few decades of the 20th century. One of the most powerful manifestations of climate change is global warming. Over the past 100 years the average temperature of the atmosphere near the Earth's surface has increased by 0.74°C (Houghton et al., 2001), and by the end of the 21st century, the Earth's temperature is expected to rise between 1.8°C and 4.6°C (IPPC, 2007).

Insects are ectotherms; their body temperature is strongly influenced by ambient air temperature, and their life cycles are usually short. Therefore, even small temperature variations can have a strong impact on population dynamics (Musolin and Saulich, 2012). Temperature increase, if it occurs within an optimal temperature range of a given species, is usually very beneficial for insects (Robinet and Roques, 2010). In acridids, increased temperature can trigger a variety of responses, including a shift in spatial distribution, earlier hatching and faster development.

Despite the expected benefits of increased global temperatures for warm-climate species like locusts, warmer temperatures may also incur detrimental effects counteracting these benefits, for example, a lack of oviposition sites. Recent studies of historic Chinese records have suggested a correlation between climatic cycles and locust plagues, though the trend seems to vary with the spatiotemporal scale under scrutiny. Using the 1000-year locust record compiled by Ma (1958), Stige et al. (2007) found that the mean abundance of *L. migratoria migratorioides* (named therein as *L. migratoria manilensis*) was negatively associated with temperature but positively associated with the frequencies of floods and droughts, both of which may have led to the creation of increased locust oviposition sites through either a waterbody receding, or the dampening of previously dry ground. The same authors (Zhang et al., 2009) used wavelet analysis to further analyse reported oscillations of locust population and climate, and suggested that periodic cooling events at 160- to 170-year intervals enhanced the likelihood of temperature-linked floods and droughts, the latter of which in particular led to a higher frequency of locust outbreaks. The implication of this work is that rising temperatures could reduce the likelihood of locust outbreaks in China, but the authors used locust data covering only the east of China and temperature data for the whole country, which could have led to regional inconsistencies. Indeed, analysis of the same long-term data set by Yu et al. (2009) drew the opposite conclusion; using a general circulation model and wavelet analysis, the authors found a positive association between locust abundance and temperature at both inter-annual and inter-decadal scales. These contradictory results were partly reconciled by Tian et al. (2011), who reconstructed an even longer time series of locust outbreaks spanning 1910 years (AD 2–1911) based on over 8000 historic documents. Across an annual scale (AD 1512–1911) and decadal scale (AD 1000–1900), there were more locusts under dry and cold conditions, and when locusts were more abundant in the previous year or decade. Furthermore, using a 200-year moving window the authors showed that the correlation between locust abundance and precipitation was consistently negative between AD 500 and 1900, except for a few decades around AD 1200. By contrast, at a decadal resolution, the negative correlation between locust abundance and temperature per se was not as consistent, and was most convincing when the average temperature for the whole of China was used (and less strong at the regional and hemisphere scale). It is therefore clear that global or continent-wide trends may be too coarse a scale to allow accurate predictions, and that local and regional circumstances must be considered for effective locust forecasting and control.

10.2.1 Changes in Distribution Area

Based on different climate simulation models, locust and grasshopper species of the Northern hemisphere could expand their distribution areas northwards. Olfert et al. (2011) predicted that *M. sanguinipes* is likely to inhabit more northerly Canadian latitudes than before. The distribution area of *Ca. italicus* is already shifting to the north in Russia due to climate warming (Popova et al., 2016). Likewise, *D. maroccanus* has also undergone a northward expansion of its distribution limit in Turkmenistan and southern Kazakhstan (Latchininsky, 2017a). A corollary of northward expansion is colonisation of higher altitudes of mountains. Historically, *D. maroccanus* inhabits semi-deserts in foothills of the Mediterranean (sensu lato) region alongside spring ephemeral vegetation, which develops at altitudes between 400 and 1000 m and includes *Poa bulbosa, Alyssum desertorum, Carex pachystylis* and *Medicago falcata* (Latchininsky and Launois-Luong, 1992). However, in recent years, climate warming has caused a gradual shift of vegetation belts to higher altitudes (Lenoir et al., 2008). As a result, in the 21st century, *D. maroccanus* breeding habitats have moved to higher altitudes, with their vertical distribution limit shifting approximately 300 m higher than before. In Central Asia, *D. maroccanus* is now frequently found at altitudes of up to 1800 m above sea level (Latchininsky et al., 2015).

10.2.2 Earlier Hatching and Faster Development

An important result of these changes in phenology may be an increased number of annual generations (Nufio et al., 2010; Wu et al., 2012b). While most acridids from temperate Eurasia and North America are univoltine (having only one generation per year), with usually overwintering diapausing eggs (Latchininsky, 2017b), some species are recorded as having more than one annual generation. For example, late-season hatching of a second generation of the normally univoltine *L. migratoria migratoria* became frequent in Uzbekistan and southern Russia in the 21st century (Latchininsky et al., 2015; Stamo et al., 2013). In North America, similar increases in annual generation number are reported for *M. sanguinipes* in Arizona (Jech, 2015). Even the embryonic development of grasshoppers, which takes place in the soil, occurs faster with higher ambient temperatures. In Alaska, grasshoppers *Melanoplus* spp. used to spend two winters as eggs in the soil, which meant a 2-year long life cycle. However, a 3°C temperature increase in recent decades caused eggs to hatch after the first winter, resulting in a shift to a univoltine life cycle (Fielding and Defoliart, 2010).

In summary, locust outbreaks are generally expected to become more frequent and severe under warming climatic conditions (Popova, 2014). Climate warming, which causes shifts in locust distribution, phenology and voltinism, is a serious challenge for both locust monitoring and management.

10.3 Locust Ecology Case Study: The Central American Locust, *Schistocerca piceifrons*

The Central American Locust, *S. piceifrons piceifrons*, is an important pest species that has caused enormous damage to agriculture in Central America (Barrientos Lozano et al., 1992; Contreras and Magaña, 2013; Harvey, 1983). This species is widely distributed across the southern half of México to northern Costa Rica, with perpetual gregarious zones in Yucatán, and frequent outbreak areas in Veracruz, San Luis Potosí, Tamaulipas and Tabasco in México. Within México, this locust is known to impact 4.9 million ha distributed throughout nine states, causing enormous agricultural and economic damage and making it one of the most important pest species in México and Central America. This locust is a highly polyphagous species known to damage numerous crop plants, including maize, soybeans, wheat, sesame, cotton, sorghum, banana, sugarcane and various fruits. Outbreaks of *S. piceifrons* are often associated with the dry season and high levels of starvation (Contreras and Galindo, 2013). Such outbreaks have been considered a possible explanation for the fall of the Mayan culture (Flores, 2011). This species has two generations per year, the first in summer and the second in winter. The first generation is shorter than the second one, which reaches maturity in the following spring (Hernández-Zul et al., 2013).

S. piceifrons has two subspecies: *S. piceifrons piceifrons* from México to northern Costa Rica and *S. piceifrons peruviana* in Peru, Ecuador, Colombia,

Venezuela, Panama and Trinidad and Tobago (Barrientos Lozano et al., 1992; Harvey, 1983). *S. piceifrons* is a tropical locust and its geographic distribution coincides with areas with suitable vegetation and favourable weather conditions for breeding and development (Contreras and Magaña, 2013). Within these areas, there are permanent gregarious zones where the vegetation and environmental conditions promote feeding and breeding of *S. piceifrons*, which leads to outbreaks and subsequent transformation to the gregarious phase (Barrientos Lozano et al., 1992; Harvey, 1983).

10.3.1 Vegetation

Vegetation plays a major role in providing food and shelter for locusts. Studies on *S. gregaria* have shown that high population density is correlated with abundance of a specific group of plants, like the dicotyledon *Heliotropium arbainense* and the monocotyledons *Pennisetum typhodium*, *Aristida pungens* and *Panicum turgidum* (Babah and Sword, 2004; Sword et al., 2010; Woldewahid et al., 2004, 2007). In addition, there is also evidence that food distribution and nutritional quality can influence phase change (see Section 9). When the distribution of vegetation is patchy, locusts experience higher levels of crowding and become more gregarious (Despland et al., 2000). The change in the threshold of gregarisation occurs according to the cover and status of the vegetation, in which low cover and dry vegetation lead to a low-density threshold of gregarisation, probably due to an increased probability of individuals mechanically stimulating each other (Cissé et al., 2013). Unlike *S. gregaria*, *S. piceifrons* is attracted more generally by odour of any plants that can provide refuge and protection from natural enemies. For example, *Pisonia aculeate* is a large shrub that attracts *S. piceifrons* and promotes aggregation because it stays green during the dry season and can be patchily distributed (Poot-Pech et al., 2016).

10.3.2 Land Use

Another important attribute of the gregarious zone is the pattern of land use. The original habitat of *S. piceifrons* was patchy grassland located in the same zones as dry tropical forest. However, a major challenge in Latin America is the intensification of land use, and the principal habitat of *S. piceifrons* today has been created by human activity (Harvey, 1983). Between 2000 and 2010, forests were significantly reduced due to changes in agricultural or livestock activities, specifically in Yucatán State (the principal gregarisation area of México and Central America). In 1981, Yucatán had 394,236 ha of grassland; by 2014, this had increased to 579,634 ha. This increase of grassy habitats by over 185,000 ha within a 30-year time span served to increase the breeding zone of *S. piceifrons* (Poot-Pech, 2016). In Yucatán, the principal grass that provides suitable conditions for development of *S. piceifrons* is *Panicum maximum* (Poot-Pech et al., 2017). During the dry season, it is common practise among ranches in the gregarious zone to remove dried grasses by burning. While this practice promotes grass regrowth in the rainy season, it also causes aggregation of *S. piceifrons* by creating patches of bare soil, which are favoured areas for ovipositing (M. A. Poot-Pech, unpublished observations). In addition to creating favourable oviposition sites, excessive livestock grazing may also promote locust outbreaks by creating a nutritionally optimal niche for locusts (e.g. Cease et al., 2012).

10.3.3 Weather

Characteristics of climate and soil contribute greatly to the fluctuation of population density of various insect species (Vinatier et al., 2011). In the case of locusts, temperature and soil moisture are highly correlated with egg laying and, in turn, increases in locust populations (Hernández-Zul et al., 2013; Van der Werf et al., 2005; Xian-Lei et al., 2007). Weather at a regional scale, particularly the distribution of precipitation, is also a critical factor that shapes the population density of locusts (de Wysiecki et al., 2011). Monthly averages for precipitation and temperature vary throughout the year in the breeding area, can be seen from the data compiled for the Municipality of Buctzotz since 1961–2015 (Table 2), which is reported to be a major gregarious zone (Contreras and Magaña, 2013).

Precipitation is higher during the first generation than the second generation, and therefore locusts have access to sufficient food to complete their development as well as adequate soil moisture to oviposit for the second generation. In the first generation the temperature is the highest, and thus this generation of *S. piceifrons* is shorter than the second generation (Hernández-Zul et al., 2013). In the second generation, the period between January and May is the recession period, where population density is low (Barrientos Lozano et al., 1992). *S. piceifrons* start to mate when the precipitation increases in September (first generation) and May (second generation) (Harvey, 1983).

Understanding the influence of biotic and abiotic factors, such as vegetation and weather, on population dynamics of *S. piceifrons* is critical not only for predicting outbreaks but also for effective field surveys (Sword et al., 2010; Symmons, 1992; Van der Werf et al., 2005). Further research in these areas will therefore offer some important tools to incorporate into management programs for this problematic species.

11. HUMAN–LOCUST INTERACTIONS

Locusts have played a visible role throughout human history, with some of our earliest records noting outbreaks that had massive and destructive impacts on human populations and livelihoods. The early Chinese history classic, the *Old Book of Tang*, notes that in 785 CE: 'The sky was shaded by the flying locusts. Hardly any vegetation and fur of animal remained after the place was visited by the plague of locusts. Roads were paved with starved people. Those survived steamed the locusts to eat as food' (Liu, 1975). Similarly, the book of Exodus (10:15 RSV) describes a particularly brutal locust plague: 'They covered the face of the whole land so that the land was darkened, and they ate all the plants in the land and all the fruit of the trees which the hail had left; not a green thing remained, neither tree nor plant of the field, through all the land of Egypt'. These early records reveal the longstanding connections between humans and locusts. While these are likely records of some of the most extreme events, they also reveal locusts to have been a common occurrence in human society throughout much of our history.

Today, of course, locusts still play a role in many communities, predominantly through impacts on agriculture. In lesser developed regions, where human–locust relationships are still tightly linked, we often find the populations that are most vulnerable to locust outbreaks. West Africa, for example, is highly susceptible to locust outbreaks or even persistent low-level crop degradation, and the impacts can be devastating for many rural dwellers (Brader et al., 2006). These impacts can have ripple effects in other areas of life. For example, outbreaks in Mali have been linked with decreased educational attainment for the children of affected families (de Vreyer et al., 2015). Low levels of capital assets in these areas make pre-emptive biological control difficult, and international agencies typically provide support only when outbreaks have reached 'emergency' status. Therefore, farmers in these kinds of situations can only *react* to locust problems when they occur. In industrialised regions, prevention and planning are often incorporated into agricultural management strategies, primarily through the use of insecticides to control locusts when even moderate densities begin to appear (Hunter, 2004). Still, prevention and control of locusts are far from solved, and many mysteries and challenges remain. Here we briefly review some perspectives from the social sciences, particularly from the viewpoint of agricultural and

Table 2 Life History of *S. piceifrons* in Buctzotz, Mexico

	Month											
Life Stage	**Jan.**	**Feb.**	**Mar.**	**Apr.**	**May**	**Jun.**	**Jul.**	**Aug.**	**Sep.**	**Oct.**	**Nov.**	**Dec.**
Nymph						×	×			×	×	×
Adult	×	×	×	×	×			×	×			×
Swarming	×	×						×	×			×
Mating					×	×			×			
Oviposition					×	×			×			
Generation	Second generation					First generation				Second generation		
Mean precipitation, cm	45.4	36.4	24.8	26.3	82.3	160.5	169.9	185.1	211.0	125.4	49.4	42.4
Mean temperature, °C	22.5	23.5	25.7	27.8	28.7	28.5	28.0	28.0	27.7	26.0	24.2	22.8

environmental economics, that may lend some insight into the nature of what might and might not work for better managing future outbreaks.

In general, we want to identify management levers that would allow individuals or policy makers to affect the likelihood of locust outbreaks occurring. The impact of locusts is greatest at the level of the farm; however, damage is heterogeneous and widespread across landscapes. The result is that the scope of intervention is diffuse and shared among many stakeholders. The mismatch of these scales creates an obstacle to sustainable locust management (Lockwood et al., 2001; Péloquin, 2014). Locusts are a continental-level problem that requires coordinated responses across boundaries by individuals and governments. With a better understanding of the connections between land management decisions and locusts, action at the level of farms or small communities may be the best option for decreasing the probability of outbreaks. For example, introduction of European livestock and agriculture to Australia over the past century may have contributed to present-day *C. terminifera* swarms (Deveson, 2012). Similarly, in Yucatán State, the main gregarisation area of *S. piceifrons* in México and Central America, deforestation due to agriculture led to a 50% increase in grassy breeding areas (Poot-Pech, 2016; see Section 10.3). This pattern for *S. piceifrons* was exacerbated by expansions of the invasive grass *P. maximum*, which creates favourable conditions for *S. piceifrons* (Poot-Pech et al., 2017). In Inner Mongolia, China, Cease et al. (2012) found that heavy livestock grazing promotes *O. asiaticus* outbreaks by lowering plant nitrogen content. Continuous high levels of livestock grazing lead to loss of topsoil and soil nitrogen (Giese et al., 2013). Reduced soil fertility decreases nitrogen content of grass leaves, which results in low-protein, high-carbohydrate leaves. In contrast to the common conception of herbivores being nitrogen limited, *O. asiaticus* prefers and grows best on low-protein, high-carbohydrate diets; high-protein diets decrease locust survivorship (Cease et al., 2012). Furthermore, these low-protein, high-carbohydrate diets, in conjunction with high population density, promote development of migratory characteristics (Cease et al., 2017), presumably promoting increased capacity for long-distance migration. While these examples illustrate mechanisms by which individual farm managers might keep locusts at bay, they may not have the resources, knowledge or incentive to maintain preventative locust management. Farm managers need to consider many potential outcomes and shocks, such as drought or market fluctuations, in addition to locust plagues. Therefore, it is important to approach sustainable locust management from a systems perspective (Fig. 2).

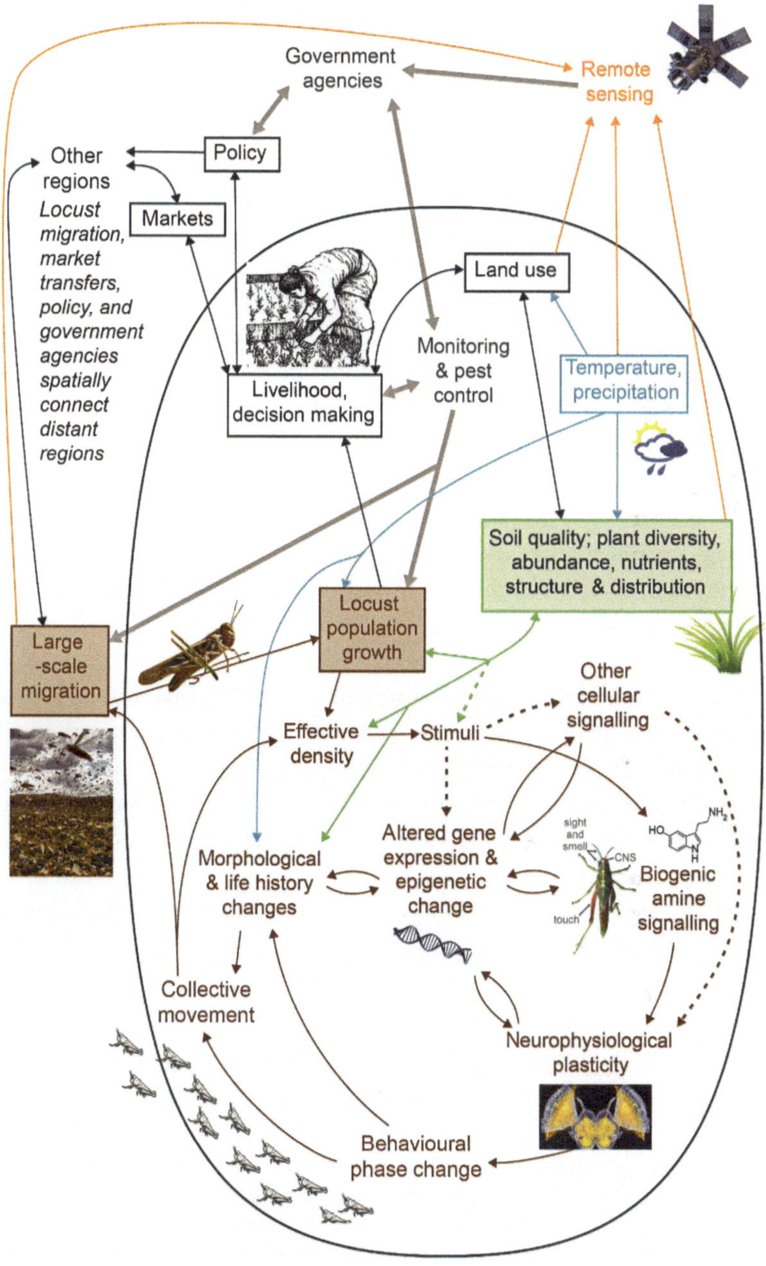

Fig. 2 Human–locust interactions. Locust (*brown lines*) population growth can lead to phase change and large-scale migration, which impacts livelihoods both locally and in distant regions. Regions are connected through markets, policy, and government agencies (*black lines*). These forces affect farmer decision making and land-use outcomes, *(Continued)*

To investigate the potential for sustainable farm-level locust management, we begin by looking at the household and farmers' decision making. Agricultural production requires the farmer to manage productive assets while jointly considering short-run (i.e. <12 months) and long-run (i.e. >12 months) planning horizons (Kay et al., 2012). Ideally, short-run decisions help the farm manager reach their long-run goals. Long-run goals may include increasing the number of animals in their grazing herd, increasing profitability, having consistent profit over time, or more generally increasing productive assets. Productive assets are agricultural items, such as livestock or land, which are managed to create revenue flows to the farm manager over time. While these assets contribute to the success of the agricultural operation, in developing countries they also provide the decision maker with a form of wealth accumulation or a 'savings account' to protect against risk, such as locust plagues. Knowing that these assets also provide a form of insurance for the farmer suggests that they will be managed in a way that emphasises long-term sustainability. The objectives and action plans for short- and long-run farm level decisions vary by farm manager. Short-run decisions are often focussed on solving a financial hardship and can overlook environmental sustainability, regulations or ecosystem effects. For example, a farmer may sell livestock or overgraze a pasture to feed the animals to a larger size to pay outstanding debt. Within this decision, they are not concerned about future output, but only what needs to be done today to survive to the next year. When farm managers make long-run decisions, they consider how the decisions they make today will affect their operation in the future. While considering these factors, farm managers can enhance the long-run productivity of their farm operation by including information on the long-term effects of stocking decisions on the quality of their land and their overall farm output. Considering the complexity of the decision-making process, it is important to approach farm management with a whole-systems perspective. For example, in China overgrazing may increase farm-level output (e.g. more sheep in the short term), but degrades soil quality and may result in

Fig. 2—Cont'd which in turn affect soil and plant parameters (*green lines*). In addition to weather (*blue lines*), these parameters influence locust population growth and the probability that locusts will undergo a series of neurophysiological, genetic, morphological, and behavioural changes that lead to collective movement and large-scale migration. Monitoring and control (*grey lines*) typically happens at the level of state or national governments or international cooperations. Remote sensing (*orange lines*) has become an increasingly important tool for research and management. *Solid lines* indicate known connections; *dotted lines* indicate suspected but unproven connections.

ecosystem externalities, such as a locust outbreak in the long term (Cease et al., 2012).

Even if farmers effectively manage locusts on their property, the remaining key problem is the familiar tragedy of the commons: if one farmer invests in preventative locust measures on their lands, they may not reap all (or even most) of the benefits of that investment. Most likely, their neighbours would benefit greatly from their actions. If multiple farm managers begin to make short-run decisions disregarding the long-term effects, ecosystem externalities such as locust outbreaks can occur more consistently. Moreover, locusts are migratory and connect not only neighbouring farms but also distant landscapes. Therefore, when thinking of locust management broadly, local farmers may be agents that can help in an overall strategy, but the management structure must fit the spatial extent of the ecological problem (Young, 2002). In the case of locusts, the scale can be continent wide.

Effective, economically efficient and sustainable management of locust plague risk requires an interdisciplinary approach in ways that parallel other integrations of natural and social sciences. On one hand, attempts to model locust management strategies that focus only on the financial cost/benefit do not reflect true economic cost. True economic value must incorporate trade-offs faced by individuals, making it incorrect to equate the economic cost of a policy with the accounting cost of an action (Bockstael et al., 2000). On the other hand, modelling efforts that omit economic cost, and instead focus solely on the natural science while ignoring the role and responses of humans, will clearly omit an integral subsystem of the social–ecological system (Ostrom, 2009). For locust social–ecological systems, these feedbacks include (but are not limited to) interactions between plant quality and locust and livestock growth rates, decisions made by farmers and ranchers on how to use land and the choices of how to control outbreaks (Ali and Mohammed, 2008; Cease et al., 2012, 2015; Lecoq, 2005). Modelling results of social–ecological systems are difficult to predict and often nonlinear, and depend on the choices made about the subsystems being modelled (Ostrom, 2009).

Bio-economic modelling, which links trade-offs resulting from complex ecosystem dynamics with the consequences of policy attempting to mitigate risk such as locust plagues, may be one field that can yield particular insight (Crocker and Tschirhart, 1992). Bio-economic modelling can be used if we can define the costs, benefits and management mechanisms that are associated with the different aspects of the natural system. General equilibrium externalities occur when there are impacts and feedbacks between various sectors and species (Carbone and Kerry Smith, 2008; Finnoff and Tschirhart, 2008). Since locusts are migratory, locust swarms may arrive and impact areas that are well managed by farmers and policymakers. Understanding the relationship between locust control and the timing and type of locust policy intervention suggests that social science could contribute to management responses by integrating human preferences for risk within locust population models.

Excluding humans from ecosystem models means that model parameters and results are accurate only within the context of existing human institutions. Humans respond to incentives, and ecosystems are subject to these changes in human behaviour. Incorporating potential risk mitigation strategies, such as preventing a plague, controlling an existing swarm, or adapting to a plague by limiting the risk to humans of adverse impacts, is important to capture the feedback loops and interactions across the ecosystem (Ehrlich and Becker, 1972; Finnoff et al., 2010). When considering the cost or benefits of any of these policies, their impacts must be compared against each other and in conjunction with each other (Kane and Shogren, 2000). When human preferences, such as negative farmer perceptions of the environmental and health costs of insecticide spraying, are included, alternatives such as insurance may be preferable (Ali and Mohammed, 2008). The benefits of each policy are going to be connected to each other—if nothing is at risk because farmers have perfectly adapted to locust outbreaks (or any adverse event), there is no benefit to preventing an outbreak from occurring. For this reason, temporal and spatial scales, in addition to the bio-physical and social contexts, need to be defined and integrated to find effective, economically efficient and sustainable solutions to locust plagues (Bockstael et al., 2000; Cease et al., 2015; Lockwood et al., 2001; Ostrom, 2009; Péloquin, 2014; Fig. 2).

12. CONCLUDING REMARKS

The last 10 years have seen dramatic progress and challenges in our understanding of locust phase change. The opportunities provided by advances in molecular biology are now being applied to locust research, and the sequencing of the *L. migratoria* genome (Wang et al., 2014) is a significant milestone. The size and complexity of the genome, together with the relatively low level of annotation and/or detectable sequence identity with those of other organisms, in some ways serves to underline the daunting task understanding the molecular basis of phase change in its entirety.

Nevertheless, techniques such as RNAi have become standard tools in experimentally manipulating putative mechanisms of phase change. The new and powerful CRISPR–Cas9 (clustered regularly interspaced short palindromic repeats) technique is a means to edit the genome and induce permanent changes in gene expression. At the time of writing only a single paper has reported this technique in locusts (Li et al., 2016), which used it to successfully delete an odorant receptor co-receptor (*Orco*) in *L. migratoria*, but it is to be expected that the use of this technique will greatly increase in the future. The demonstration of the existence of widespread DNA methylation in locusts has spurred interest in the epigenetic nature of phase change and the transmission of phase characteristics to offspring. Determining its exact nature will be a substantial research focus of the next 10 years.

Biogenic amine signalling and its role in at least the early stages of behavioural phase transition have remained a subject of interest, but the picture has become considerably more complex as more information becomes available. All the major biogenic amines have now been implicated in the process of gregarisation and/or solitarisation. In part, this probably reflects the independent evolution of density-dependent behavioural change in *S. gregaria* and *L. migratoria*, two distantly related species, which have co-opted somewhat different widespread and evolutionarily conserved mechanisms of neuronal plasticity to produce similar phenotypic outcomes. A challenge is now to unravel this complex pattern and match the roles of specific biogenic amines to particular aspects of behavioural change. This process will be assisted by recent progress in the behavioural analysis of locusts; automated tracking facilitates a greater throughput and helps generate more data more quickly, whilst new quantitative methods and statistical models will help extract more information from the data and help distinguish behavioural phase change from other kinds of behavioural response.

The interest in comparative analyses between different locust species, and between locusts and 'near locusts' that are either closely related to locusts or show some of the aspects of plasticity associated with full phase polyphenism, is to be welcomed. Such comparative analyses promise to increase our understanding of how phase polyphenism has evolved, and may help resolve some of the difficulties of how different physiological mechanisms occur in different species. As research spreads beyond the two canonical locusts, *S. gregaria* and *L. migratoria*, to other species that in their own locales may be just as important or damaging, a larger framework unifying and relating our understanding of these different species becomes essential.

Locusts have remained and will continue to be an inspiration and model species for those interested in the emergent behaviour of collective movement. The extension of experimental studies beyond the laboratory to larger scale field settings is an important development and is poised to yield important insights into the structure of locust swarms in particular, and whole field of collective motion studies in general.

There has been a renewed interest in how environmental factors affect phase change and the development of outbreaks or larger scale swarms. Most laboratory studies have focussed on simplified conditions to try and understand how locust–locust interactions initiate the process of phase change, but in natural environments, other factors greatly influence the likelihood of locusts encountering each other and may even directly influence the mechanisms of phase change. Nutritional status is emerging as an important factor in phase change itself and the propagation of locust outbreaks. Locust outbreaks and plagues, and the economic damage they inflict, exist in a wider context of human land use and management. The demonstration of the role that human activities may unwittingly play in creating conditions that facilitate gregarisation and ultimately swarm formation has been an important breakthrough.

Locust phase change research is motivated by several reasons: as a model for phenotypic plasticity generally, or neuronal and behavioural plasticity in particular, or as an exemplar of collective movement, but all of this stems from the status of locusts as major and occasionally catastrophic pest species that warrant continual monitoring and control. The inclusion of a section that analyses locusts within social–ecological systems, including the economics of locust outbreaks, is unusual in a review of this kind, but serves to underline the wider human context in which locust research occurs. The locust phase change research community is small, but the range of research interests covered within it, from molecules to management, is all encompassing as presented in this review. Major research opportunities and insights in the future will depend on crossing disciplinary boundaries and fusing different tiers of expertise to create new and deeper insights into locust phase transformation and its relationship to humans, agriculture and the environment. To this end, Arizona State University is supporting development of the Global Locust Initiative (GLI). This Initiative's purpose is to assist universities, government and non-government agencies and agricultural communities to forge new partnerships that seek to better integrate locust research and management. GLI is based on three pillars: advancing fundamental research, creating and maintaining a global network, and developing local solutions to the global challenge of locust plagues. GLI will develop and maintain a global database of interested people and organisations, along with relevant documents and data such as research methods and locust outbreak records. This network is intended to remove barriers for new people and organisations to work on locusts, and for current researchers and managers working across sectors and geographical regions to collaborate more easily. GLI will support workshops and the development of funding proposals for interdisciplinary teams advancing fundamental research, and for cross-sectoral teams fusing fundamental and applied approaches to develop new solutions to the ancient challenge of locust plagues.

ACKNOWLEDGEMENTS

We would like to thank The Entomological Society of America for hosting our two symposia at ICE 2016, the 25th International Congress of Entomology in Orlando, Florida. We would also like to gratefully thank the following funding bodies for their support:

D.A.C., J.V.d.B., R.V. and H.V. were supported by the Research Foundation of Flanders (FWO) and KU Leuven (GOA/11/02 and C14/15/050). R.V. was also supported by an FWO PhD Fellowship; D.A.C. and H.V. are supported by FWO Postdoctoral Fellowships.

A.J.C., K.B., J.C.H and B.R. were supported by the US National Science Foundation (#1313693). A.J.C. was also supported by (CHE-1313958), and B.R. was also funded by Fonds de Recherche du Québec—Société et culture (2015-NP-182013).

A.V.L. was partly funded by a Wyoming NASA Space Grant.

A.A. was partially supported by Grant No. 891-0277-13 from Israel's Ministry of Agriculture.

J.B. was funded by the Australian Research Council Future Fellowship, Discovery Projects and Linkage Projects programs.

R.D.K., T.M. and S.R.O. were supported by Research Grant BB/L02389X/1 (to S.R.O. and T.M.) from the Biotechnology and Biological Sciences Research Council, United Kingdom, and a Wellcome Trust ISSF award from the University of Leicester.

J.M.S. was supported by an MIBTP Studentship from the Biotechnology and Biological Sciences Research Council, United Kingdom.

B.F. and H.S. were supported by the US National Science Foundation (IOS-1253493).

M.A.P.-P. was supported by Yucatán Plant Health (CESVY) and the National Service of Sanitary and Agro-Alimentary Health (SENASICA), México.

G.A.S. was supported by USDA-NIFA, Texas A&M University AgriLife Research, and the Charles R. Parencia Endowment.

REFERENCES

Alessi, A.M., O'Connor, V., Aonuma, H., Newland, P.L., 2014. Dopaminergic modulation of phase reversal in desert locusts. Front. Behav. Neurosci. 8, 1–15.

Ali, H.A., Mohammed, B., 2008. Cost Benefit Analysis of Desert Locusts' Control: A Multicountry Perspective. Economic Research Forum Working Papers No. 0801.

Altizer, S., Bartel, R., Han, B.A., 2011. Animal migration and infectious disease risk. Science 331, 296–302.
Amichay, G., Ariel, G., Ayali, A., 2016. The effect of changing topography on the coordinated marching of locust nymphs. PeerJ 4, e2742.
Andiatsirevombola, J.M., Ge, J., Nyirenda, T.M., Raholijaona, 2016. Predation impact of Cattle Egret (*Bubulcus ibis*) on Migratory Locust (*Locusta migratoria capito*) and Red Locust (*Nomadacris septemfasciata*) in south and southwest regions of Madagascar. Appl. Ecol. Environ. Res. 14, 297–308.
Andretic, R., Van Swinderen, B., Greenspan, R.J., 2005. Dopaminergic modulation of arousal in *Drosophila*. Curr. Biol. 15, 1165–1175.
Angers, B., Castonguay, E., Massicotte, R., 2010. Environmentally induced phenotypes and DNA methylation: how to deal with unpredictable conditions until the next generation and after. Mol. Ecol. 19, 1283–1295.
Anstey, M.L., Rogers, S.M., Ott, S.R., Burrows, M., Simpson, S.J., 2009. Serotonin mediates behavioral gregarization underlying swarm formation in desert locusts. Science 323, 627–630.
Applebaum, S.W., Heifetz, Y., 1999. Density-dependent physiological phase in insects. Annu. Rev. Entomol. 44, 317–341.
Ariel, G., Ayali, A., 2015. Locust collective motion and its modeling. PLoS Comput. Biol. 11, e1004522.
Ariel, G., Ophir, Y., Levi, S., Ben-Jacob, E., Ayali, A., 2014. Individuals' intermittent motion is instrumental to the formation and maintenance of swarms of marching locust nymphs. PLoS One 9, e101636.
Ashall, C., Ellis, P.E., 1962. Studies on numbers and mortality in field populations of the desert locust (*Schistocerca gregaria* Forskål). Anti-Locust Bull. 38, 1–59.
Babah, M.A.O., Sword, G.A., 2004. Linking locust gregarization to local resource distribution patterns across a large spatial scale. Environ. Entomol. 33, 1577–1583.
Badisco, L., Huybrechts, J., Simonet, G., Verlinden, H., Marchal, E., Huybrechts, R., Schoofs, L., De Loof, A., Vanden Broeck, J., 2011a. Transcriptome analysis of the desert locust central nervous system: production and annotation of a *Schistocerca gregaria* EST database. PLoS One 6, e17274.
Badisco, L., Ott, S.R., Rogers, S.M., Matheson, T., Knapen, D., Vergauwen, L., Verlinden, H., Marchal, E., Sheehy, M.R.J., Burrows, M., Vanden Broeck, J., 2011b. Microarray-based transcriptomic analysis of differences between long-term gregarious and solitarious desert locusts. PLoS One 6, e28110.
Balfanz, S., Strünker, T., Frings, S., Baumann, A., 2005. A family of octopamine [corrected] receptors that specifically induce cyclic AMP production or Ca2+ release in *Drosophila melanogaster*. J. Neurochem. 93, 440–451.
Barrientos Lozano, L., 2011. Dinámica Poblacional, Biología y Ecología de la Langosta Suramericana (*Schistocerca cancellata*, Serville). Instituto Tecnológico de Ciudad Victoria, Tamaulipas, México, pp. 93–135.
Barrientos Lozano, L., Astacio, C.O., Alvarez, B.F., Poot, M.O., 1992. Manual Técnico Sobre la Langosta Voladora (*Schistocerca Piceifrons* Walker 1870) y Otros Acridoideos de Centroamérica y Sureste de México. FAO-OIRSA, El Salvador.
Bayliss, A., Roselli, G., Evans, P.D., 2013. A comparison of the signalling properties of two tyramine receptors from *Drosophila*. J. Neurochem. 125, 37–48.
Bazazi, S., Buhl, J., Hale, J.J., Anstey, M.L., Sword, G.A., Simpson, S.J., Couzin, I.D., 2008. Collective motion and cannibalism in locust migratory bands. Curr. Biol. 18, 735–739.
Bazazi, S., Romanczuk, P., Thomas, S., Schimansky-Geier, L., Hale, J.J., Miller, G.A., Sword, G.A., Simpson, S.J., Couzin, I.D., 2011. Nutritional state and collective motion: from individuals to mass migration. Proc. R. Soc. B Biol. Sci. 278, 356–363.

Bazazi, S., Bartumeus, F., Hale, J.J., Couzin, I.D., 2012. Intermittent motion in desert locusts: behavioural complexity in simple environments. PLoS Comput. Biol. 8, e1002498.
Beaulieu, J.M., Gainetdinov, R.R., 2011. The physiology, signaling, and pharmacology of dopamine receptors. Pharmacol. Rev. 63, 182–217.
Beggs, K.T., Mercer, A.R., 2009. Dopamine receptor activation by honey bee queen pheromone. Curr. Biol. 19, 1206–1209.
Beggs, K.T., Hamilton, I.S., Kurshan, P.T., Mustard, J.A., Mercer, A.R., 2005. Characterization of a D2-like dopamine receptor (AmDOP3) in honey bee, *Apis mellifera*. Insect Biochem. Mol. Biol. 35, 873–882.
Beggs, K.T., Tyndall, J.D., Mercer, A.R., 2011. Honey bee dopamine and octopamine receptors linked to intracellular calcium signaling have a close phylogenetic and pharmacological relationship. PLoS One 6, e26809.
Belayneh, Y.T., 2005. Acridid pest management in the developing world: a challenge to the rural population, a dilemma to the international community. J. Orthop. Res. 14, 187–195.
Bellés, X., 2010. Beyond *Drosophila*: RNAi in vivo and functional genomics in insects. Annu. Rev. Entomol. 55, 111–128.
Bernays, E.A., Lee, J., 1988. Food aversion learning in the polyphagous grasshopper *Schistocerca americana*. Physiol. Entomol. 13, 131–137.
Berthier, K., Chapuis, M.-P., Simpson, S.J., Ferenz, H.-J., Habib Kane, C.M., Kang, L., Lange, A., Ott, S.R., Ould Babah, M.A., Rodenburg, K.W., Rogers, S.M., Torto, B., Vanden Broeck, J., Van Loon, J.J.A., Sword, G.A., 2010. Laboratory populations as a resource for understanding the relationship between genotypes and phenotypes: a global case study in locusts. Adv. Insect Physiol. 39, 1–37.
Bewick, A.J., Vogel, K.J., Moore, A.J., Schmitz, R.J., 2016. Evolution of DNA methylation across insects. Mol. Biol. Evol. 34, 654–665.
Bischof, L.J., Enan, E.E., 2004. Cloning, expression and functional analysis of an octopamine receptor from *Periplaneta americana*. Insect Biochem. Mol. Biol. 34, 511–521.
Bischoff, J.F., Rehner, S.A., Humber, R.A., 2009. A multilocus phylogeny of the *Metarhizium anisopliae* lineage. Mycologia 101, 512–530.
Blackburn, L.M.L., 2005. Behavioural and physiological differences in the motor systems of solitarious and gregarious locusts. Unpublished PhD Thesis, University of Cambridge, UK.
Blackburn, L.M., Ott, S.R., Matheson, T., Burrows, M., Rogers, S.M., 2010. Motor neurone responses during a postural reflex in solitarious and gregarious desert locusts. J. Insect Physiol. 56, 902–910.
Blanchet, E., Lecoq, M., Sword, G.A., Berthier, K., Pages, C., Billot, C., Rivallan, R., Foucart, A., Vassal, J.-M., Risterucci, A., 2011. A comparative analysis of fine-scale genetic structure in three closely related syntopic species of the grasshopper genus *Calliptamus*. Can. J. Zool. 90, 31–41.
Blenau, W., Baumann, A., 2001. Molecular and pharmacological properties of insect biogenic amine receptors: lessons from *Drosophila melanogaster* and *Apis mellifera*. Arch. Insect Biochem. Physiol. 48, 13–38.
Blenau, W., Erber, J., 1998. Behavioural pharmacology of dopamine, serotonin and putative aminergic ligands in the mushroom bodies of the honeybee (*Apis mellifera*). Behav. Brain Res. 96, 115–124.
Blenau, W., Erber, J., Baumann, A., 1998. Characterization of a dopamine D1 receptor from *Apis mellifera*: cloning, functional expression, pharmacology, and mRNA localization in the brain. J. Neurochem. 70, 15–23.
Blenau, W., Balfanz, S., Baumann, A., 2000. Am*tyr*1: characterization of a gene from honeybee (*Apis mellifera*) brain encoding a functional tyramine receptor. J. Neurochem. 74, 900–908.

Bockstael, N.E., Freeman, A.M., Kopp, R.J., Portney, P.R., Smith, V.K., 2000. On measuring economic values for nature. Environ. Sci. Technol. 34, 1384–1389.

Boerjan, B., Sas, F., Ernst, U.R., Tobback, J., Lemière, F., Vandegehuchte, M.B., Janssen, C.R., Badisco, L., Marchal, E., Verlinden, H., Schoofs, L., De Loof, A., 2011. Locust phase polyphenism: does epigenetic precede endocrine regulation? Gen. Comp. Endocrinol. 173, 120–128.

Bouaïchi, A., Simpson, S.J., Roessingh, P., 1996. The influence of environmental microstructure on the behavioural phase state and distribution of the desert locust *Schistocerca gregaria*. Physiol. Entomol. 21, 247–256.

Brader, L., Djibo, H., Faye, F., Ghaout, S., Lazar, M., Luzietoso, P., Babah, M.O., 2006. Towards a More Effective Response to Desert Locusts and Their Impacts on Food Security, Livelihoods and Poverty: Multilateral Evaluation of the 2003–05 Desert Locust Campaign. FAO, Rome.

Bryceson, K.P., 1984. In: Discrimination of small areas of vegetation growth after localised rain in semiarid environments using Landsat. Proceedings of the 3rd Australian Remote Sensing Conference, pp. 541–550.

Bryceson, K.P., 1989. The use of Landsat MSS data to determine the distribution of locust eggbeds in the Riverina region of New South Wales, Australia. Int. J. Remote Sens. 10, 1749–1762.

Bryceson, K.P., 1990. Digitally processed satellite data as a tool in detecting potential Australian plague locust outbreak areas. J. Environ. Manage. 30, 191–207.

Bryceson, K.P., 1991. Likely locust infestation areas in western New South Wales, Australia, located by satellite. Geocarto Int. 6, 21–37.

Bryceson, K.P., Wright, D.E., 1986. An analysis of the 1984 locust plague in Australia using multitemporal Landsat multispectral data and a simulation model of locust development. Agric. Ecosyst. Environ. 16, 87–102.

Buhl, J., Rogers, S.M., 2016. Mechanisms underpinning aggregation and collective movement by insect groups. Curr. Opin. Insect Sci. 15, 125–130.

Buhl, J., Sumpter, D.J.T., Couzin, I.D., Hale, J.J., Despland, E., Miller, E.R., Simpson, S.J., 2006. From disorder to order in marching locusts. Science 312, 1402–1406.

Buhl, E., Schildberger, K., Stevenson, P.A., 2008. A muscarinic cholinergic mechanism underlies activation of the central pattern generator for locust flight. J. Exp. Biol. 211, 2346–2357.

Buhl, J., Sword, G.A., Clissold, F.J., Simpson, S.J., 2011. Group structure in locust migratory bands. Behav. Ecol. Sociobiol. 65, 265–273.

Buhl, J., Sword, G.A., Simpson, S.J., 2012. Using field data to test locust migratory band collective movement models. Interface Focus 2, 757–763.

Burgess, S.C., Baskett, M.L., Grosberg, R.K., Morgan, S.G., Strathmann, R.R., 2015. When is dispersal for dispersal? Unifying marine and terrestrial perspectives. Biol. Rev. 91, 867–882.

Burrows, M., Rogers, S.M., Ott, S.R., 2011. Epigenetic remodelling of brain, body and behaviour during phase change in locusts. Neural Syst. Circuits 1, 11.

Butler, C., Innes, J., 1936. A comparison of the rate of metabolic activity in the solitary and migratory phases of *Locusta migratoria*. Proc. R. Soc. B Biol. Sci. 119, 296–304.

Callier, S., Snapyan, M., Le Crom, S., Prou, D., Vincent, J.D., Vernier, P., 2003. Evolution and cell biology of dopamine receptors in vertebrates. Biol. Cell 95, 489–502.

Camacho, J.P.M., Ruiz-Ruano, F.J., Martín-Blázquez, R., López-León, M.D., Cabrero, J., Lorite, P., Cabral-de-Mello, D.C., Bakkali, M., 2015. A step to the gigantic genome of the desert locust: chromosome sizes and repeated DNAs. Chromosoma 124, 263–275.

Camazine, S., Deneubourg, J.-L., Franks, N.R., Sneyd, J., Theraulaz, G., Bonabeau, E., 2003. Self-Organization in Biological Systems. Princeton University Press, New Jersey.

Carbone, J.C., Kerry Smith, V., 2008. Evaluating policy interventions with general equilibrium externalities. J. Public Econ. 92, 1254–1274.
Cardinaud, B., Gilbert, J.M., Liu, F., Sugamori, K.S., Vincent, J.D., Niznik, H.B., Vernier, P., 1998. Evolution and origin of the diversity of dopamine receptors in vertebrates. Adv. Pharmacol. 42, 936–940.
Cazzamali, G., Klaerke, D.A., Grimmelikhuijzen, C.J.P., 2005. A new family of insect tyramine receptors. Biochem. Biophys. Res. Commun. 338, 1189–1196.
Cease, A.J., Hao, S., Kang, L., Elser, J.J., Harrison, J.F., 2010. Are color or high rearing density related to migratory polyphenism in the band-winged grasshopper, *Oedaleus asiaticus*? J. Insect Physiol. 56, 926–936.
Cease, A.J., Elser, J.J., Ford, C.F., Hao, S., Kang, L., Harrison, J.F., 2012. Heavy livestock grazing promotes locust outbreaks by lowering plant nitrogen content. Science 335, 467–469.
Cease, A.J., Elser, J.J., Fenichel, E.P., Hadrich, J.C., Harrison, J.F., Robinson, B.E., 2015. Living with locusts: connecting soil nitrogen, locust outbreaks, livelihoods, and livestock markets. Bioscience 65, 551–558.
Cease, A.J., Harrison, J.F., Hao, S., Niren, D.C., Zhang, G., Kang, L., Elser, J.J., 2017. Nutritional imbalance suppresses migratory phenotypes of the Mongolian locust (*Oedaleus asiaticus*). R. Soc. Open Sci. 4, 161039.
Chapman, R.F., 1954. Responses of *Locusta migratoria migratorioides* (R. & F.) to light in the laboratory. Br. J. Anim. Behav. 2, 146–152.
Chapman, J.W., Reynolds, D.R., Smith, A.D., 2003. Vertical-looking radar: a new tool for monitoring high-altitude insect migration. Bioscience 53, 503–511.
Chapuis, M.-P., Loiseau, A., Michalakis, Y., Lecoq, M., Franc, A., Estoup, A., 2009. Outbreaks, gene flow and effective population size in the migratory locust, *Locusta migratoria*: a regional scale comparative survey. Mol. Ecol. 18, 792–800.
Chapuis, M.-P., Popple, J.-A.M., Berthier, K., Simpson, S.J., Deveson, E., Spurgin, P., Steinbauer, M.J., Sword, G.A., 2011. Challenges to assessing connectivity between massive populations of the Australian plague locust. Proc. R. Soc. B Biol. Sci. 278, 3152–3160.
Chapuis, M.-P., Plantamp, C., Blondin, L., Pages, C., Vassal, J.-M., Lecoq, M., 2014. Demographic processes shaping genetic variation of the solitarious phase of the desert locust. Mol. Ecol. 23, 1749–1763.
Chapuis, M.-P., Bazelet, C.S., Blondin, L., Foucart, A., Vitalis, R., Samways, M.J., 2016. Subspecific taxonomy of the desert locust, *Schistocerca gregaria* (Orthoptera: Acrididae), based on molecular and morphological characters. Syst. Entomol. 41, 516–530.
Chapuis, M.-P., Foucart, A., Plantamp, C., Blondin, L., Leménage, N., Benoit, L., Gay, P.-E., Bazelet, C.S., 2017. Genetic and morphological variation in non-polyphenic southern African populations of the desert locust. Afr. Entomol. 25, 13–23.
Chen, H., Scherkenbeck, J., Zdobinsky, T., Antonicek, H., 2010a. Fluorescent analogues of the insect neuropeptide helicokinin I: synthesis, photophysical characterization and biological activity. Protein Pept. Lett. 17, 431–436.
Chen, S., Yang, P., Jiang, F., Wei, Y., Ma, Z., Kang, L., 2010b. De novo analysis of transcriptome dynamics in the migratory locust during the development of phase traits. PLoS One 5, e15633.
Cherlet, M., 1993. RAMSES Reconnaissance and Management System of the Environment of *Schistocerca* (Desert Locust). Programme Outline on Future Requirements in Remote Sensing and Modeling and Their Integration Using GIS to Improve Desert Locust Forecasting, Project Report GCP/INT/439/BEL. FAO, Rome.
Cherlet, M., Di Gregorio, A., 1993. Calibration and Integrated Modelling of Remote Sensing Data for Desert Locust Habitat Monitoring. RSC Series 64, Rome, FAO.

Cigliano, M.M., Braun, H., Eades, D.C., Otte, D., 2017. Orthoptera Species File. Version 5.0/5.0. [03/10/2017], http://Orthoptera.SpeciesFile.org.

Cissé, S., Ghaout, S., Mazih, A., Ebbe, B., Ould, M.A., Benahi, A.S., Piou, C., 2013. Effect of vegetation on density thresholds of adult desert locust gregarization from survey data in Mauritania. Entomol. Exp. Appl. 149, 159–165.

Cissé, S., Ghaout, S., Mazih, A., Ebbe, O.B., Abdallahi, M., Piou, C., 2015. Estimation of density threshold of gregarization of desert locust hoppers from field sampling in Mauritania. Entomol. Exp. Appl. 156, 136–148.

Cissé, S., Ghaout, S., Babah Ebbe, M.A., Kamara, S., Piou, C., 2016. Field verification of the prediction model on desert locust adult phase status from density and vegetation. J. Insect Sci. 16, 74.

Clark, L.R., 1949. Behaviour of swarm hoppers of the Australian plague locust (*Chortoicetes terminifera* Walker). CSIRO Bull. 245, 1–27.

Clobert, J., Galliard, L., Cote, J., Meylan, S., Massot, M., 2009. Informed dispersal, heterogeneity in animal dispersal syndromes and the dynamics of spatially structured populations. Ecol. Lett. 12, 197–209.

Cloudsley-Thompson, J.L., 1977. Diurnal rhythms of locomotory activity in isolated desert locusts (*Schistocerca gregaria* (Forsk)). J. Interdiscipl. Cycle Res. 8, 27–36.

Colas, J.F., Launay, J.M., Kellermann, O., Rosay, P., Maroteaux, L., 1995. *Drosophila* 5-HT2 serotonin receptor: coexpression with fushi-tarazu during segmentation. Proc. Natl. Acad. Sci. U.S.A. 92, 5441–5445.

Collett, T.S., 1978. Peering—a locust behaviour pattern for obtaining motion parallax information. J. Exp. Biol. 76, 237–241.

Collett, M., Despland, E., Simpson, S.J., Krakauer, D.C., 1998. Spatial scales of desert locust gregarization. Proc. Natl. Acad. Sci. U.S.A. 95, 13052–13055.

Contreras, S.C., Galindo, M.G., 2013. La plaga de la langosta centroamericana *schistocerca piceifrons* (Walker), en México: reconstrucción ambiental 1592–2000. In: Galindo, M.M.G., Contreras, S.C., Ibarra, Z.E. (Eds.), La Plaga de la Langosta *Schistocerca Piceifrons* Walker. Colección Sanidad Vegetal Tomo 1. Vigilancia Epidemiológica Fitosanitaria. Universidad Autónoma de San Luís Potosi, México, pp. 87–110.

Contreras, S.C., Magaña, O.C., 2013. Ficha técnica de la langosta centroamericana *schistocerca piceifrons*. In: Galindo, M.M.G., Contreras, S.C., Ibarra, Z.E. (Eds.), La Plaga de la Langosta *Schistocerca Piceifrons* Walker. Colección Sanidad Vegetal Tomo 1. Vigilancia Epidemiológica Fitosanitaria. Universidad Autónoma de San Luís Potosi, México, pp. 17–36.

Cooter, R.J., 1989. Swarm flight behavior in flies and locusts. In: Goldsworthy, G., Wheeler, C. (Eds.), Insect Flight. CRC Press, Boca Raton, pp. 165–203.

Corona, M., Libbrecht, R., Wheeler, D.E., 2016. Molecular mechanisms of phenotypic plasticity in social insects. Curr. Opin. Insect Sci. 13, 55–60.

Cressman, K., 1996. Current methods of desert locust forecasting at FAO. OEPP/EPPO Bull. 26, 577–585.

Cressman, K., 2013. Role of remote sensing in desert locust early warning. J. Appl. Remote Sens. 7, 075098.

Crocker, T.D., Tschirhart, J., 1992. Ecosystems, externalities, and economies. Environ. Resource Econ. 2, 551–567.

Cruse, H., 1985. Which parameters control the leg movement of a walking insect? Part 1. Velocity control during the stance phase. J. Exp. Biol. 116, 343–355.

Cruse, H., Schmitz, J., 1983. The control-system of the femur-tibia joint in the standing leg of a walking stick insect *Carausius morosus*. J. Exp. Biol. 102, 175–185.

Cullen, D.A., 2012. RNAi unravels the biology of the hemimetabolous and ametabolous insects. Adv. Insect Physiol. 42, 37–72.

Cullen, D.A., Sword, G.A., Dodgson, T., Simpson, S.J., 2010. Behavioural phase change in the Australian plague locust, *Chortoicetes terminifera*, is triggered by tactile stimulation of the antennae. J. Insect Physiol. 56, 937–942.

Cullen, D.A., Sword, G.A., Simpson, S.J., 2012. Optimizing multivariate behavioural syndrome models in locusts using automated video tracking. Anim. Behav. 84, 771–784.

Czirók, A., Barabasi, A.L., Vicsek, T., 1999. Collective motion of self-propelled particles: kinetic phase transition in one dimension. Phys. Rev. Lett. 82, 209.

Dean, G.J.W., 1967. Observations on the structure of hopper bands and movement of hoppers of the Red Locust (*Nomadacris septemfasciata* Serville). J. Entomol. Soc. South. Afr. 30, 1–17.

Denno, R.F., Douglass, L.W., Jacobs, D., 1985. Crowding and host plant nutrition—environmental determinants of wing-form in *Prokelisia marginata*. Ecology 66, 1588–1596.

Despland, E., 2003. Fractal index captures the role of vegetation clumping in locust swarming. Funct. Ecol. 17, 315–322.

Despland, E., Simpson, S.J., 2000. The role of food distribution and nutritional quality in behavioural phase change in the desert locust. Anim. Behav. 59, 643–652.

Despland, E., Simpson, S.J., 2005a. Food choices of solitarious and gregarious locusts reflect cryptic and aposematic antipredator strategies. Anim. Behav. 69, 471–479.

Despland, E., Simpson, S.J., 2005b. Surviving the change to warning colouration: density-dependent polyphenism suggests a route for the evolution of aposematism. Chemistry 15, 69–75.

Despland, E., Collett, M., Simpson, S.J., 2000. Small-scale processes in desert locust swarm formation: how vegetation patterns influence gregarization. Oikos 88, 652–662.

Despland, E., Rosenberg, J., Simpson, S.J., 2004. Landscape structure and locust swarming: a satellite's eye view. Ecography 27, 381–391.

Deveson, E.D., 2011. The search for a solution to Australian locust outbreaks: how developments in ecology and government responses influenced scientific research. Hist. Rec. Aust. Sci. 22, 1–31.

Deveson, E.D., 2012. Naturae amator and the grasshopper infestations of south Australia's early years. Trans. R. Soc. S. Aust. 136, 1–15.

Deveson, E.D., 2013. Satellite normalized difference vegetation index data used in managing Australian plague locusts. J. Appl. Remote Sens. 7, 075096.

Deveson, T., Hunter, D.M., 2002. The operation of a GIS-based decision support system for Australian locust management. Entomol. Sin. 9, 1–12.

Deveson, T., Woodman, J.D., 2014. Observations of *Scelio fulgidus* (Hymenoptera: Platygastridae) parasitism and development in southern NSW during the 2010 *Chortoicetes terminifera* (Orthoptera: Acrididae) locust plague. Aust. Entomol. 53, 133–137.

Deveson, E.D., Drake, V.A., Hunter, D.M., Walker, P.W., Wang, H.K., 2005. Evidence from traditional and new technologies for northward migrations of Australian plague locusts (*Chortoicetes terminifera*) (Walker) (Orthoptera: Acrididae) to western Queensland. Austral Ecol. 30, 928–943.

de Vreyer, P., Guilbert, N., Mesple-Somps, S., 2015. The impact of natural disasters on education outcomes: evidence from the 1987-89 locust plague in Mali. J. Afr. Econ. 24, 57–100.

de Wysiecki, M.L., Lange, C., 2005. La langosta *Schistocerca cancellata* Serville (orthoptera: acrididae: cyrtacanthacridinae) en argentina: biología, ecología, historia y control. In: Barrientos Lozano, L., Almaguer Sierra, P. (Eds.), Libro Manejo Integrado de la Langosta Centroamericana (*Schistocerca piceifrons piceifrons*, Walker) y Acridoideos Plaga en América Latina. Instituto Tecnológico de Ciudad Victoria, Tamaulipas, México, pp. 151–156.

de Wysiecki, M.L., Arturi, M., Torrusio, S., Cigliano, M.M., 2011. Influence of whether variables and plant communities on grasshopper density in the southern pampas, Argentina. J. Insect Sci. 11, 1–14.

Dias, B.G., Maddox, S.A., Klengel, T., Ressler, K.J., 2015. Epigenetic mechanisms underlying learning and the inheritance of learned behaviors. Trends Neurosci. 38, 96–107.

Dinku, T., Ceccato, P., Cressman, K., Connor, S.T., 2010. Evaluating detection skills of satellite rainfall estimates over desert locust recession regions. J. App. Meteorol. Climatol. 49, 1322–1332.

Dirsh, V.M., 1953. Morphometrical studies on phases of the desert locust. Anti-Locust Bull. 16, 1–34.

Dixon, A.F.G., 1998. Aphid Ecology: An Optimization Approach, 2nd ed. Chapman and Hall, London.

Drake, V.A., Wang, H., 2013. Recognition and characterization of migratory movements of Australian plague locusts, *Chortoicetes terminifera*, with an insect monitoring radar. J. Appl. Remote Sens. 7, 075095.

Drake, V.A., Gregg, P.C., Harman, I.T., Wang, H.-K., Deveson, E.D., Hunter, D.M., Rochester, W., 2001. Characterizing insect migration systems in inland Australia with novel and traditional methodologies. In: Woiwood, I.P., Reynolds, D.R., Thomas, C.D. (Eds.), Insect Movement: Mechanisms and Consequences. Proceedings of the Royal Entomological Society's 20th Symposium. CABI Publishing, London, pp. 207–234.

Dreiser, U., 1994. Mapping of desert locust habitats in Africa using landsat thematic mapper data. GeoJournal 32, 55–60.

Dukas, R., 2008a. Evolutionary biology of insect learning. Annu. Rev. Entomol. 53, 145–160.

Dukas, R., 2008b. Learning decreases heterospecific courtship and mating in fruitflies. Biol. Lett. 4, 645–647.

Dyson, L., Yates, C., Buhl, J., McKane, A., 2015. Onset of collective motion in locusts is captured by a minimal model. Phys. Rev. E 92, 052708.

Edelstein-Keshet, L., Watmough, J., Grunbaum, D., 1998. Do travelling band solutions describe cohesive swarms? An investigation for migratory locusts. J. Math. Biol. 36, 515–549.

Edney, E.B., 1937. A study of spontaneous locomotor activity in *Locusta migratoria migratorioides* (R. & F.) by the actograph method. Bull. Entomol. Res. 28, 243–278.

Ehrlich, I., Becker, G.S., 1972. Market insurance, self-insurance, and self-protection. J. Polit. Econ. 80, 623–648.

Ellis, P.E., 1951. The marching behaviour of hoppers of the African migratory locust (*Locusta migratoria migratorioides* R. & F.) in the laboratory. Anti-Locust Bull. 7, 1–46.

Ellis, P.E., 1953. Social aggregation and gregarious behaviour in hoppers of *Locusta migratoria migratorioides* (R. & F.). Behaviour 5, 225–260.

Ellis, P.E., 1959. Learning and social aggregation in locust hoppers. Anim. Behav. 7, 91–106.

Ellis, P.E., 1963a. The influence of some environmental factors on learning and aggregation in locust hoppers. Anim. Behav. 11, 142–151.

Ellis, P.E., 1963b. Changes in the social aggregation of locust hoppers with changes in rearing conditions. Anim. Behav. 11, 152–160.

Ellis, P.E., 1964a. Marching a colour in locust hoppers in relation to social factors. Behaviour 23, 177–192.

Ellis, P.E., 1964b. Changes in marching of locusts with rearing conditions. Behaviour 23, 193–202.

Ellis, P.E., Ashall, C., 1957. Field studies on diurnal behaviour, movement and aggregation in the desert locust (*Schistocerca gregaria* Forskål). Anti-Locust Bull. 25, 1–103.

Ely, S.O., Njagi, P.G.N., Bashir, M.O., El-Amin, S.E.T., Hassanali, A., 2011. Diel behavioral activity patterns in adult solitarious desert locust, *Schistocerca gregaria* (Forskål). Psyche 1–9459315.

Ernst, U.R., Van Hiel, M.B., Depuydt, G., Boerjan, B., De Loof, A., Schoofs, L., 2015. Epigenetics and locust life phase transitions. J. Exp. Biol. 218, 88–99.

Erra, U., Frola, B., Scarano, V., Couzin, I.D., 2009. In: An efficient GPU implementation for large scale individual-based simulation of collective behavior. International Workshop on High Performance Computational Systems Biology, pp. 51–58. http://dx.doi.org/10.1109/HiBi.2009.11.

Escudero, C., Yates, C.A., Buhl, J., Couzin, I.D., Erban, R., Kevrekidis, I.G., Maini, P.K., 2010. Ergodic directional switching in mobile insect groups. Phys. Rev. E 82, 011926.

Evans, P.D., Maqueira, B., 2005. Insect octopamine receptors: a new classification scheme based on studies of cloned *Drosophila* G-protein coupled receptors. Invert. Neurosci. 5, 111–118.

Evans, P.D., Bayliss, A., Reale, V., 2014. GPCR-mediated rapid, non-genomic actions of steroids: comparisons between DmDopEcR and GPER1 (GPR30). Gen. Comp. Endocrinol. 195, 157–163.

Faki, O.M.A.E., 2004. Ecological field studies on the population dynamics of the solitary Desert Locust *Schistocerca gregaria* (Forskål) (Acrididae: Orthoptera). PhD Thesis, University of Khartoum, Sudan.

Falckenhayn, C., Boerjan, B., Raddatz, G., Frohme, M., Schoofs, L., Lyko, F., 2013. Characterization of genome methylation patterns in the desert locust, *Schistocerca gregaria*. J. Exp. Biol. 216, 1423–1429.

Fan, Z., Jiang, G.-F., Liu, Y.-X., He, Q.-X., Blanchard, B., 2014. Population explosion in the Yellow-spined Bamboo Locust *Ceracris kiangsu* and inferences for the impact of human activity. PLoS One 9, e89873.

Farrow, R.A., 1990. Flight and migration in Acridoids. In: Chapman, R.F., Joern, A. (Eds.), Biology of Grasshoppers. Wiley-Interscience, New York, pp. 227–314.

Fielding, D.J., Defoliart, L.S., 2010. Embryonic developmental rates of northern grasshoppers (Orthoptera: Acrididae): implications for climate change and habitat management. Environ. Entomol. 39, 1643–1651.

Finnoff, D., Tschirhart, J., 2008. Linking dynamic economic and ecological general equilibrium models. Resour. Energy Econ. 30, 91–114.

Finnoff, D., McIntosh, C., Shogren, J.F., Sims, C., Warziniack, T., 2010. Invasive species and endogenous risk. Annu. Rev. Resour. Econ. 2, 77–100.

Fleck, M.W., Thomson, J.L., Hough, L.B., 2012. Histamine-gated ion channels in mammals? Biochem. Pharmacol. 83, 1127–1135.

Flores, G.F., 2011. Las plagas de langosta en el área maya: ambiente e historia de una calamidad en la época prehispánica. Península 6, 27–46.

Foster, W.A., Treherne, J.E., 1981. Evidence for the dilution effect in the selfish herd from fish predation on a marine insect. Nature 293, 466–467.

Franc, A., Duranton, J.-F., 2012. Red locust *Nomadacris septemfasciata* (Serville) upsurges in northern Madagascar between 1998 and 2004. Ann. Soc. Entomol. Fr. 48, 97–105.

Franc, A., Andriamaroahina, T., Legros, Y., Duranton, J.-F., 2005. Using remote sensing to improve the evaluation of the Malagasy migratory locust risk Locusta migratoria capito (Saussure, 1884). Orthoptera: Acrididae. Metaleptea, Special meeting issue, Ninth Int. Conf. Orthopterists' Soc., Canmore, Alberta, Canada, Aug 14–19, p. 41.

Franc, A., Soti, V., Tran, A., Leclair, D., Duvallet, G., Duranton, J.-F., 2008. In: Deforestation, New migration pathways and outbreaks of the red Locust *Nomadacris septemfasciata* (Orthoptera: Acrididae) in the Sofia River Basin. Conférence SDH-SAGEO, Montpellier, 23–26 June, pp. 1–17.

Fu, X.J., Hunter, D.M., Shi, W.P., 2010. Effect of *Paranosema (Nosema) locustae* (Microsporidia) on morphological phase transformation of *Locusta migratoria manilensis* (Orthoptera: Acrididae). Biocontrol Sci. Technol. 20, 683–693.

Fuchs, E., Kutsch, W., Ayali, A., 2003. Neural correlates to flight-related density dependent phase characteristics in locusts. J. Neurobiol. 57, 152–162.

Fuzeau-Braesch, S., David, J.C., 1978. Etude du taux d'octopamine chez *Locusta migratoria* (Insecte: Orthoptere) comparison entre insects gregaires, solitaires et traite au gaz carbonique. C. R. Acad. Sci. Paris D 286, 697–699.

Fuzeau-Braesch, S., Nicolas, G., 1981. Effect of carbon-dioxide on subsocial insects. Comp. Biochem. Physiol. 68, 289–297.

Gaten, E., Huston, S.J., Dowse, H.B., Matheson, T., 2012. Solitary and gregarious locusts differ in circadian rhythmicity of a visual output neuron. J. Biol. Rhythms 27, 196–205.

Geva, N., Guershon, M., Orlova, M., Ayali, A., 2010. Memoirs of a locust: density-dependent behavioral change as a model for learning and memory. Neurobiol. Learn. Mem. 93, 175–182.

Ghaout, S., 1990. Contribution à l'étude des Resources Trophiques de *Schistocerca Gregaria* (Forsk.) (Orthoptera, Acrididae) Solitaire en MAURITANIE OCCIDENTALE et Télédetection de ses Biotopes par Satellite. PhD Thesis, University of Paris XI, Orsay, France.

Giese, M., Brueck, H., Gao, Y.Z., Lin, S., Steffens, M., Kögel-Knabner, I., Glindemann, T., Susenbeth, A., Taube, F., Butterbach-Bahl, K., Zheng, X.H., Hoffmann, C., Bai, Y.F., Han, X.G., 2013. N balance and cycling of Inner Mongolia typical steppe: a comprehensive case study of grazing effects. Ecol. Monogr. 83, 195–219.

Gillett, S.D., Gonta, E., 1978. Locust as prey: factors affecting their vulnerability to predation. Anim. Behav. 26, 282–289.

Goll, M.G., Kirpekar, F., Maggert, K.A., Yoder, J.A., Hsieh, C.L., Zhang, X., Golic, K.G., Jacobsen, S.E., Bestor, T.H., 2006. Methylation of $tRNA^{Asp}$ by the DNA methyltransferase homolog Dnmt2. Science 311, 395–398.

Goosey, M.W., Candy, D.J., 1980. The D-octopamine content of the haemolymph of the locust *Schistocerca americana gregaria* and its elevation during flight. Insect Biochem. 10, 393–397.

Gordon, S.D., Jackson, J.C., Rogers, S.M., Windmill, J.F.C., 2014. Listening to the environment: hearing differences from an epigenetic effect in solitarious and gregarious locusts. Proc. R. Soc. B Biol. Sci. 281, 20141693.

Gornyy, V.I., Kritsuk, S.G., Latypov, I.S., Teplyakova, T.E., Tronin, A.A., 2011. Measuring approach of satellite locust monitoring. J. Earth Sci. Eng. 1, 126–131.

Gotham, S., Song, H., 2013. Non-swarming grasshoppers exhibit density-dependent phenotypic plasticity reminiscent of swarming locusts. J. Insect Physiol. 59, 1151–1159.

Gray, L.J., Sword, G.A., Anstey, M.L., Clissold, F.J., Simpson, S.J., 2009. Behavioural phase polyphenism in the Australian plague locust (*Chortoicetes terminifera*). Biol. Lett. 5, 306–309.

Greenwood, M., Chapman, R.F., 1984. Differences in numbers of sensilla on the antennae of solitarious and gregarious *Locusta migratoria* L. (Orthoptera: Acrididae). Int. J. Insect Morphol. Embryol. 13, 295–301.

Grohmann, L., Blenau, W., Erber, J., Ebert, P.R., Strünker, T., Baumann, A., 2003. Molecular and functional characterization of an octopamine receptor from honeybee (*Apis mellifera*) brain. J. Neurochem. 86, 725–735.

Guo, W., Wang, X., Ma, Z., Xue, L., Han, J., Yu, D., Kang, L., 2011. *CSP* and *takeout* genes modulate the switch between attraction and repulsion during behavioral phase change in the migratory locust. PLoS Genet. 7, e1001291.

Guo, X., Ma, Z., Kang, L., 2013. Serotonin enhances solitariness in phase transition of the migratory locust. Front. Behav. Neurosci. 7, 129.
Guo, X., Ma, Z., Kang, L., 2015. Two dopamine receptors play different roles in phase change of the migratory locust. Front. Behav. Neurosci. 9, 80.
Guttal, V., Romanczuk, P., Simpson, S.J., Sword, G.A., Couzin, I.D., 2012. Cannibalism can drive the evolution of behavioural phase polyphenism in locusts. Ecol. Lett. 15, 1158–1166.
Hamilton, W.D., 1971. Geometry for the selfish herd. J. Theor. Biol. 31, 295–311.
Han, K.A., Millar, N.S., Grotewiel, M.S., Davis, R.L., 1996. DAMB, a novel dopamine receptor expressed specifically in *Drosophila* mushroom bodies. Neuron 16, 1127–1135.
Hansen, M.J., Buhl, J., Bazazi, S., Simpson, S.J., Sword, G.A., 2011. Cannibalism in the lifeboat—collective movement in Australian plague locusts. Behav. Ecol. Sociobiol. 65, 1715–1720.
Harrell Jr., F.E., 2001. Regression Modeling Strategies: With Applications to Linear Models, Logistic Regression, and Survival Analysis. Springer, New York.
Harvey, A.W., 1983. *Schistocerca piceifrons* (Walker) (Orthoptera: Acrididae), the swarming locust of tropical America: a review. Bull. Entomol. Res. 73, 171–184.
Hassenstein, B., Hustert, R., 1999. Hiding responses of locusts to approaching objects. J. Exp. Biol. 202, 1701–1710.
Hauser, F., Cazzamali, G., Williamson, M., Blenau, W., Grimmelikhuijzen, C.J., 2006. A review of neurohormone GPCRs present in the fruitfly *Drosophila melanogaster* and the honey bee *Apis mellifera*. Prog. Neurobiol. 80, 1–19.
Hauser, F., Cazzamali, G., Williamson, M., Park, Y., Li, B., Tanaka, Y., Predel, R., Neupert, S., Schachtner, J., Verleyen, P., Grimmelikhuijzen, C.J., 2008. A genome-wide inventory of neurohormone GPCRs in the red flour beetle *Tribolium castaneum*. Front. Neuroendocrinol. 29, 142–165.
Hearn, M.G., Ren, Y., McBride, E.W., Reveillaud, I., Beinborn, M., Kopin, A.S., 2002. A *Drosophila* dopamine 2-like receptor: molecular characterization and identification of multiple alternatively spliced variants. Proc. Natl. Acad. Sci. U.S.A. 99, 14554–14559.
Heifetz, Y., 1997. The control of phase transition in the desert locust *Schistocerca Gregaria* (Forskål). PhD Thesis, Hebrew University of Jerusalem, Israel.
Heifetz, Y., Applebaum, S.W., 1995. Density-dependent physiological phase in a non-migratory grasshopper *Aiolopus thalassinus*. Entomol. Exp. Appl. 77, 251–262.
Hernández-Zul, M.I., Quijano-Carranza, J.A., Yañez-López, R., Ocampo-Velazquez, R.V., Torrés-Pacheco, I., Guevara-Gonzalez, R.G., Castro Ramírez, A.E., 2013. Dynamic simulation model of central American locust *Schistocerca piceifrons* (Othoptera: Acrididae). Fla. Entomol. 96, 1274–1283.
Hielkema, J.U., 1981. Desert locust habitat monitoring with satellite remote sensing: a new technology for an old problem. ITC J. 4, 387–417.
Hielkema, J.U., Roffey, J., Tucker, C.J., 1986. Assessment of ecological conditions associated with the 1980/81 desert locust plague upsurge in west Africa using environmental satellite data. Int. J. Remote Sens. 7, 1609–1622.
Hill, C.A., Doyle, T., Nuss, A.B., Ejendal, K.F.K., Meyer, J.M., Watts, V.J., 2016. Comparative pharmacological characterization of D1-like dopamine receptors from *Anopheles gambiae*, *Aedes aegypti* and *Culex quinquefasciatus* suggests pleiotropic signaling in mosquito vector lineages. Parasit. Vectors 9, 192.
Hosmer, D.W., Hosmer, T., Le Cessie, S., Lemeshow, S., 1997. A comparison of goodness-of-fit tests for the logistic regression model. Stat. Med. 16, 965–980.
Hoste, B., Simpson, S.J., Tanaka, S., De Loof, A., Breuer, M., 2002. A comparison of phase-related shifts in behavior and morphometrics of an albino strain, deficient in [His7]-corazonin, and a normally colored *Locusta migratoria* strain. J. Insect Physiol. 48, 791–801.

Houghton, J.T., Ding, Y., Griggs, D.J., Noguer, N., van der Linden, P.J., Xiaosu, D., Maskell, K., Johnson, C.A. (Eds.), 2001. Climate Change 2001: The Scientific Basis. Cambridge University Press, Cambridge. 881 pp.

Huang, J., Ohta, H., Inoue, N., Takao, H., Kita, T., Ozoe, F., Ozoe, Y., 2009. Molecular cloning and pharmacological characterization of a *Bombyx mori* tyramine receptor selectively coupled to intracellular calcium mobilization? Insect Biochem. Mol. Biol. 39, 842–849.

Huang, J., Hamasaki, T., Ozoe, Y., 2010. Pharmacological characterization of a *Bombyx mori* alpha-adrenergic-like octopamine receptor stably expressed in a mammalian cell line. Arch. Insect Biochem. Physiol. 73, 74–86.

Humphries, M.A., Mustard, J.A., Hunter, S.J., Mercer, A., Ward, V., Ebert, P.R., 2003. Invertebrate D2 type dopamine receptor exhibits age-based plasticity of expression in the mushroom bodies of the honeybee brain. J. Neurobiol. 55, 315–330.

Hunter, D.M., 2004. Advances in the control of locusts (Orthoptera: Acrididae) in eastern Australia: from crop protection to preventative control. Aust. J. Entomol. 43, 293–303.

Hunter, D., Cosenzo, E., 1990. The origin of plagues and recent outbreaks of the South American locust, *Schistocerca cancellata* (Orthoptera: Acrididae) in Argentina. Bull. Entomol. Res. 80, 295–300.

Hunter, D., McCulloch, L., Wright, D., 1981. Lipid accumulation and migratory flight in the Australian plague locust, *Chortoicetes terminifera* (Walker) (Orthoptera: Acrididae). Bull. Entomol. Res. 71, 543–546.

Hunter, D.M., McCulloch, L., Spurgin, P.A., 2008. Aerial detection of nymphal bands of the Australian plague locust (*Chortoicetes terminifera* (Walker)(Orthoptera: Acrididae)). Crop Prot. 27, 118–123.

IPPC, 2007. Climate change 2007. In: Pachauri, R.K., Reisinger, A. (Eds.), Synthesis Report. Contribution of Working Groups I, II and III to the Fourth Assessment Report of the Intergovernmental Panel on Climate Change. IPCC, Geneva, Switzerland. 104 pp.

Isles, A.R., 2015. Neural and behavioral epigenetics; what it is, and what is hype. Genes Brain Behav. 14, 64–72.

Jech, L., 2015. In: Grasshoppers in Arizona in 2015. Proceedings of Annual Meeting of the National Grasshopper Management Board, Denver, CO.

Jeltsch, A., Ehrenhofer-Murray, A., Jurkowski, T.P., Lyko, F., Reuter, G., Ankri, S., Nellen, W., Schaefer, M., Helm, M., 2016. Mechanism and biological role of Dnmt2 in nucleic acid methylation. RNA Biol. 1–16. http://dx.doi.org/10.1080/15476286.2016.1191737.

Ji, R., Xie, B.Y., Li, D.M., Li, Z., Zhang, X., 2004. Use of MODIS data to monitor the oriental migratory locust plague. Agric. Ecosyst. Environ. 104, 615–620.

Jiang, F., Yang, M., Guo, W., Wang, X., Kang, L., 2012. Large-scale transcriptome analysis of retroelements in the migratory locust, *Locusta migratoria*. PLoS One 7, e40532.

Johnson, C., 1969. Migration and Dispersal of Insects by Flight. Methuen, London.

Johnson, O., Becnel, J., Nichols, C.D., 2009. Serotonin 5-HT(2) and 5-HT(1A)-like receptors differentially modulate aggressive behaviors in *Drosophila melanogaster*. Neuroscience 158, 1292–1300.

Kane, S., Shogren, J.F., 2000. Linking adaptation and mitigation in climate change policy. Clim. Change 45, 75–102.

Kang, L., Chen, X., Zhou, Y., Liu, B., Zheng, W., Li, R., Wang, J., Yu, J., 2004. The analysis of large-scale gene expression correlated to the phase changes of the migratory locust. Proc. Natl. Acad. Sci. U.S.A. 101, 17611–17615.

Kay, R.D., Edwards, W.M., Duffy, P.A., 2012. Farm Management, seventh ed. McGraw-Hill Education, New York.

Key, K., 1950. A critique on the phase theory of locusts. Q. Rev. Biol. 25, 363–407.

Kibasa, R., 2006. Use of SPOT 5 Image to Identify Presence of Locusts. International Institute for Geoinformation Science and Earth Observation, Enschede, Netherlands.
Kita, T., Irie, T., Nomura, K., Ozoe, F., Ozoe, Y., 2017. Pharmacological characterization of histamine-gated chloride channels from the housefly *Musca domestica*. Neurotoxicology 60, 245–253.
Koh, L.P., Wich, S.A., 2012. Dawn of drone ecology: low-cost autonomous vehicles for conservation. Trop. Conserv. Sci. 5, 121–132.
Kuitert, L.C., Connin, R.V., 1952. Biology of the American grasshopper in the southeastern United States. Fla. Entomol. 35, 22–33.
Latchininsky, A.V., 2013. Locusts and remote sensing: a review. J. Appl. Remote Sens. 7, 075099.
Latchininsky, A.V., 2017a. Climate change and locusts: what to expect? Proc. Russ. State Hydrometeorol. Univ. 46, 134–143.
Latchininsky, A.V., 2017b. Locusts. In: Reference Module in Life Sciences. Elsevier.
Latchininsky, A.V., Launois-Luong, M.H., 1992. Le Criquet marocain, *Dociostaurus maroccanus* (Thunberg, 1815), dans la partie orientale de son aire de distribution. In: Etude Monographique Relative à l'ex-URSS et aux Pays Proches, International Center of Agronomic Research for Development CIRAD-GERDAT-PRIFAS, Montpellier, France/All-Russian Research Institute for Plant Protection VIZR, St. Petersburg, Russia.
Latchininsky, A.V., Sivanpillai, R., 2010. Locust habitat monitoring and risk assessment using remote sensing and GIS technologies. In: Ciancio, A., Mukerji, K.G. (Eds.), Integrated Management of Arthropod Pests and Insect Borne Diseases. Springer, Dordrecht–Heidelberg–London–New York, pp. 163–188.
Latchininsky, A.V., Sivanpillai, R., Driese, K.L., Wilps, H., 2007. Can early season Landsat image improve locust habitat monitoring in the Amudarya River Delta, Uzbekistan? J. Orthop. Res. 16, 167–173.
Latchininsky, A.V., Kokanova, E.O., Gapparov, F.A., Childebaev, M.K., Temreshev, I.I., 2015. Locusts and climate change. Bull. Al-Farabi Kazakhstan State Univ. Ecol. Ser. 2/2 (44), 641–648. in Russian with English summary.
Latchininsky, A.V., Piou, C., Franc, A., Soti, V., 2016. Applications of remote sensing to locust management. In: Baghdadi, N., Zribi, M. (Eds.), Land Surface Remote Sensing, Environment and Risks. ISTE Press/Elsevier, London/Oxford, pp. 263–293.
Lechner, M., Marz, M., Ihling, C., Sinz, A., Stadler, P.F., Krauss, V., 2013. The correlation of genome size and DNA methylation rate in metazoans. Theor. Biosci. 132, 47–60.
Lecoq, M., 2005. Desert locust management: from ecology to anthropology. J. Orthop. Res. 14, 179–186.
Lecoq, M., Balanca, G., 1998. Field trials of fipronil for control of *Rhammatocerus schistocercoides* (Rehn, 1906) hopper bands in Brazil. Crop Prot. 17, 105–110.
Lecoq, M., Foucart, A., Balanca, G., 1999. Behaviour of *Rhammatocerus schistocercoides* (Rehn, 1096) hopper bands in Mato Grosso, Brazil (Orthoptera: Acrididae: Gomphocerinae). Ann. Soc. Entomol. Fr. 35, 217–228.
Leitch, B., Judge, S., Pitman, R.M., 2003. Octopaminergic modulation of synaptic transmission between an identified sensory afferent and flight motoneuron in the locust. J. Comp. Neurol. 462, 55–70.
Lenoir, J., Gégout, J.C., Marquet, P.A., de Ruffray, P., Brisse, H., 2008. A significant upward shift in plant species optimum elevation during the 20th century. Science 320, 1768–1771.
Li, Y., Zhang, J., Chen, D., Yang, P., Jiang, F., Wang, X., Kang, L., 2016. CRISPR/Cas9 in locusts: successful establishment of an olfactory deficiency line by targeting the mutagenesis of an odorant receptor co-receptor (*Orco*). Insect Biochem. Mol. Biol. 79, 27–35.

Libersat, F., Pflueger, H.-J., 2004. Monoamines and the orchestration of behavior. Bioscience 54, 17–25.
Lihoreau, M., Raine, N.E., Reynolds, A.M., Stelzer, R.J., Lim, K.S., Smith, A.D., Osborne, J.L., Chittka, L., 2012. Radar tracking and motion-sensitive cameras on flowers reveal the development of pollinator multi-destination routes over large spatial scales. PLoS Biol. 10, e1001392.
Liu, X., 1975. Old Book of Tang. Zhonghua Book Company, Beijing, China.
Lockwood, J.A., Showler, A.T., Latchininsky, A.V., 2001. Can we make locust and grasshopper management sustainable? J. Orthop. Res. 10, 315–329.
Lorenz, M.W., 2009. Migration and trans-Atlantic flight of locusts. Quat. Int. 196, 4–12.
Louveaux, A., Ghaout, S., Gillon, Y., 1990. In: Fonctionnement de l'aire de Reproduction Hivernale du Criquet Pèlerin en Mauritanie. ANPP–Deuxième Conférence Internationale sur les Ravageurs de l'Agriculture. ANPP, Versailles, France.
Lovejoy, N.R., Mullen, S.P., Sword, G.A., Chapman, R.F., Harrison, R.G., 2006. Ancient trans-Atlantic flight explains locust biogeography: molecular phylogenetics of *Schistocerca*. Proc. R. Soc. B Biol. Sci. 273, 767–774.
Löw, F., Navratil, P., Kotte, K., Schöler, H.F., Bubenzer, O., 2013. Remote sensing based analysis of landscape change in the desiccated seabed of the Aral Sea—a potential tool for assessing the hazard degree of dust and salt storms. Environ. Monit. Assess. 185, 8303–8319.
Löw, F., Waldner, F., Latchininsky, A.V., Biradar, C., Bolkart, M., Colditz, R., 2016. Timely monitoring of Asian migratory locust habitats in the Amudarya delta, Uzbekistan using time series of satellite remote sensing vegetation index. J. Environ. Manage. 183, 562–575.
Lyko, F., Foret, S., Kucharski, R., Wolf, S., Falckenhayn, C., Maleszka, R., 2010. The honey bee epigenomes: differential methylation of brain DNA in queens and workers. PLoS Biol. 8, e1000506.
Ma, S., 1958. The population dynamics of the oriental migratory locust (*Locusta migratoria manilensis* Meyen) in China. Acta Entomol. Sin. 8, 1–40(In Chinese).
Ma, J., Han, X., Hasibagan, Wang, C., Zhang, Y., Tang, J., Xie, Z., Deveson, T., 2005. Monitoring east Asian migratory locust plagues using remote sensing data and field investigations. Int. J. Remote Sens. 26, 629–634.
Ma, Z., Yu, J., Kang, L., 2006. LocustDB: a relational database for the transcriptome and biology of the migratory locust (*Locusta migratoria*). BMC Genomics 7, 11.
Ma, Z., Guo, W., Guo, X., Wang, X., Kang, L., 2011. Modulation of behavioral phase changes of the migratory locust by the catecholamine metabolic pathway. Proc. Natl. Acad. Sci. U.S.A. 108, 3882–3887.
Ma, C., Yang, P., Jiang, F., Chapuis, M.P., Shali, Y., Sword, G.A., Kang, L., 2012. Mitochondrial genomes reveal the global phylogeography and dispersal routes of the migratory locust. Mol. Ecol. 21, 4344–4358.
Ma, Z., Guo, X., Lei, H., Li, T., Hao, S., Kang, L., 2015. Octopamine and tyramine respectively regulate attractive and repulsive behavior in locust phase changes. Sci. Rep. 5, 8036.
Maeno, K., Tanaka, S., 2011. Phase-specific responses to different qualities of food in the desert locust, *Schistocerca gregaria*: developmental, morphological and reproductive characteristics. J. Insect Physiol. 57, 514–520.
Maeno, K.O., Piou, C., Ely, S.O., Mohamed, S.O., Jaavar, M.E.H., Babah, M.A.O., Nakamura, S., 2012. Field observations of the sheltering behavior of the solitarious phase of the desert locust, *Schistocerca gregaria*, with particular reference to antipredator strategies. JARQ—Jpn. Agric. Res. Q. 46, 339–345.
Maeno, K.O., Piou, C., Babah, M.A.O., Nakamura, S., 2013. Eggs and hatchlings variations in desert locusts: phase related characteristics and starvation tolerance. Front. Physiol. 4, 345.

Maeno, K.O., Ely, S.O., Nakamura, S., Abdellaoui, K., Cissé, S., El Hacen Jaavar, M., Mohamed, S.A.O., Atheimine, M., Babah, M.A.O., 2016. Daily microhabitat shifting of solitarious-phase desert locust adults: implications for meaningful population monitoring. SpringerPlus 5, 107.

Magor, J.I., Lecoq, M., Hunter, D.M., 2008. Preventive control and desert locust plagues. Crop Prot. 27, 1527–1533.

Mallon, E.B., Amarasinghe, H.E., Ott, S.R., 2016. Acute and chronic gregarisation are associated with distinct DNA methylation fingerprints in desert locusts. Sci. Rep. 6, 35608.

Maqueira, B., Chatwin, H.M., Evans, P.D., 2005. Identification and characterization of a novel family of *Drosophila* beta-adrenergic-like octopamine G-protein coupled receptors. J. Neurochem. 94, 547–560.

Marvin, D.C., Koh, L.P., Lynam, A.J., Wich, S., Davies, A.B., Krishnamurthy, R., Stokes, E., Starkey, R., Asner, G.P., 2016. Integrating technologies for scalable ecology and conservation. Glob. Ecol. Conserv. 7, 262–275.

Matheson, T., Rogers, S.M., Krapp, H.G., 2004. Plasticity in the visual system is correlated with a change in lifestyle of solitarious and gregarious locusts. J. Neurophysiol. 91, 1–12.

McCulloch, L., Hunter, D.M., 1983. Identification and monitoring of Australian plague locust habitats from Landsat. Remote Sens. Environ. 13, 95–102.

Mery, F., Kawecki, T.J., 2004. An operating cost of learning in *Drosophila melanogaster*. Anim. Behav. 68, 589–598.

Mery, F., Kawecki, T.J., 2005. A cost of long-term memory in *Drosophila*. Science 308, 1148.

Michel, R., 1970. Étude expérimentale des variations de la tendance au vol chez le criquet pèlerin *Schistocerca gregaria* (Forsk), élevé isolément pendant plusieurs générations. Ins. Soc. 17, 21–38.

Missale, C., Nash, S.R., Robinson, S.W., Jaber, M., Caron, M.G., 1998. Dopamine receptors: from structure to function. Physiol. Rev. 78, 189–225.

Müller, C.B., Williams, I.S., Hardie, J., 2001. The role of nutrition, crowding and interspecific interactions in the development of winged aphids. Ecol. Entomol. 26, 330–340.

Musolin, D.L., Saulich, A.H., 2012. Reactions of insects to recent climate change: from physiology and behavior to distribution area shift. Entomol. Rev. 91, 3–35(in Russian with English summary).

Mustard, J.A., Blenau, W., Hamilton, I.S., Ward, V.K., Ebert, P.R., Mercer, A.R., 2003. Analysis of two D1-like dopamine receptors from the honey bee *Apis mellifera* reveals agonist-independent activity. Mol. Brain Res. 113, 67–77.

Mustard, J.A., Beggs, K.T., Mercer, A.R., 2005. Molecular biology of the invertebrate dopamine receptors. Arch. Insect Biochem. Physiol. 59, 103–117.

Nathanson, J.A., Hunnicutt, E.J., 1981. N-demethylchlordimeform: a partial agonist of octopamine-sensitive adenylate cyclase. Mol. Pharmacol. 20, 68–75.

Navratil, P., Wilps, H., 2013. Object-based locust habitat mapping using high resolution multispectral satellite data in the southern Aral Sea basin. J. Appl. Remote Sens. 7, 075097.

Nespolo, R.F., Roff, D.A., Fairbairn, D.J., 2008. Energetic trade-off between maintenance costs and flight capacity in the sand cricket (*Gryllus firmus*). Funct. Ecol. 22, 624–631.

Ngumbi, E., Jordan, M., Fadamiro, H., 2012. Comparison of associative learning of host-related plant volatiles in two parasitoids with different degrees of host specificity, *Cotesia marginiventris* and *Microplitis croceipes*. Chemoecology 22, 207–215.

Nichols, D.E., Nichols, C.D., 2008. Serotonin receptors. Chem. Rev. 108, 1614–1641.

Nishide, Y., Tanaka, S., 2016. Desert locust, *Schistocerca gregaria*, eggs hatch in synchrony in a mass but not when separated. Behav. Ecol. Sociobiol. 70, 1507–1515.

Nishide, Y., Tanaka, S., Saeki, S., 2015a. Adaptive difference in daily timing of hatch in two locust species, *Schistocerca gregaria* and *Locusta migratoria*: the effects of thermocycles and phase polyphenism. J. Insect Physiol. 72, 79–87.

Nishide, Y., Tanaka, S., Saeki, S., 2015b. Egg hatching of two locusts, *Schistocerca gregaria* and *Locusta migratoria*, in response to light and temperature cycles. J. Insect Physiol. 76, 24–29.

Noldus, L.P.J.J., Spink, A.J., Tegelenbosch, R.A.J., 2001. Etho vision: a versatile video tracking system for automation of behavioral experiments. Behav. Res. Methods Instrum. Comput. 33, 398–414.

Nolte, D., 1967. Phase transformation and chiasma formation in locusts. Chromosoma 21, 123–139.

Nufio, C.R., McGuire, C.R., Bowers, M.D., Guralnick, R.P., 2010. Grasshopper community response to climatic change: variation along an elevational gradient. PLoS One 5, e12977.

Ochieng', S.A., Hallberg, E., Hansson, B.S., 1998. Fine structure and distribution of antennal sensilla of the desert locust, *Schistocerca gregaria* (Orthoptera: Acrididae). Cell Tissue Res. 291, 525–536.

Ohta, H., Utsumi, T., Ozoe, Y., 2003. B96Bom encodes a *Bombyx mori* tyramine receptor negatively coupled to adenylate cyclase. Insect Mol. Biol. 12, 217–223.

Ohta, H., Tsuchihara, K., Mitsumasu, K., Yaginuma, T., Ozoe, Y., Asaoka, K., 2009. Comparative pharmacology of two D1-like dopamine receptors cloned from the silkworm *Bombyx mori*. Insect Biochem. Mol. Biol. 39, 342–347.

Ohtani, A., Arai, Y., Ozoe, F., Ohta, H., Narusuye, K., Huang, J., Enomoto, K., Kataoka, H., Hirota, A., Ozoe, Y., 2006. Molecular cloning and heterologous expression of an alpha-adrenergic-like octopamine receptor from the silkworm *Bombyx mori*. Insect Mol. Biol. 15, 763–772.

Olfert, O., Weiss, R.M., Kriticos, D., 2011. Application of general circulation models to assess the potential impact of climate change on potential distribution and relative abundance of *Melanoplus sanguinipes* (Fabricius) (Orthoptera: Acrididae) in North America. Psyche, Article ID 980372.

Oppenheim, S.J., Baker, R.H., Simon, S., DeSalle, R., 2015. We can't all be supermodels: the value of comparative transcriptomics to the study of non-model insects: comparative transcriptomics of non-model insects. Insect Mol. Biol. 24, 139–154.

Ostrom, E., 2009. A general framework for analyzing sustainability of social-ecological systems. Science 325, 419–422.

Ott, S.R., Rogers, S.M., 2010. Gregarious desert locusts have substantially larger brains with altered proportions compared with the solitarious phase. Proc. Biol. Sci. 277, 3087–3096.

Ott, S.R., Verlinden, H., Rogers, S.M., Brighton, C.H., Quah, P.S., Vleugels, R.K., Verdonck, R., Vanden Broeck, J., 2012. Critical role for protein kinase A in the acquisition of gregarious behavior in the desert locust. Proc. Natl. Acad. Sci. U.S.A. 109, 381–387.

Padgham, D.E., 1981. Hatching rhythms in the desert locust, *Schistocerca gregaria*. Physiol. Entomol. 6, 191–198.

Pedgley, D.E., 1974. In: Frieden, S.C., Mercanti, E.P., Becker, M.A. (Eds.), ERTS surveys a 500 km^2 locust breeding site in Saudi Arabia. Third Earth Resources Technology Satellite—Symposium (December 1973), NASA, Maryland. vol. 1, pp. 233–246.

Pekel, J.F., Ceccato, P., Vancutsem, C., Cressman, K., Vanbogaert, E., Defourny, P., 2011. Development and application of multi-temporal colorimetric transformation to monitor vegetation in the desert locust habitat. IEEE J. Sel. Top. Appl. Earth Observ. Remote Sens. 4, 318–326.

Pélissié, B., Piou, C., Jourdan-Pineau, H., Pagès, C., Blondin, L., Chapuis, M.-P., 2016. Extra molting and selection on nymphal growth in the desert locust. PLoS One 11, e0155736.

Péloquin, C., 2014. Unruly nature and technological authority: governing locust swarms in the Sahel. PhD Thesis, University of Arizona, Tempe, AZ.
Pener, M.P., 1983. Endocrine aspects of phase polymorphism in locusts. In: Downer, R.G.H., Laufer, H. (Eds.), Invertebrate Endocrinology. In: Endocrinology of Insects, vol. 1. Alan R. Liss Inc., New York, pp. 379–394
Pener, M.P., 1991. Locust phase polymorphism and its endocrine relations. Adv. Insect Physiol. 23, 1–79.
Pener, M.P., Simpson, S.J., 2009. Locust phase polyphenism: an update. Adv. Insect Physiol. 36, 1–286.
Pener, M.P., Yerushalmi, Y., 1998. The physiology of locust phase polymorphism: an update. J. Insect Physiol. 44, 365–377.
Pener, M.P., Ayali, A., Golenser, E., 1997. Adipokinetic hormone and flight fuel related characteristics of density-dependent locust phase polymorphism: a review. Comp. Biochem. Physiol. B 117, 513–524.
Penick, C.A., Brent, C.S., Dolezal, K., Liebig, J., 2014. Neurohormonal changes associated with ritualized combat and the formation of a reproductive hierarchy in the ant *Harpegnathos saltator.* J. Exp. Biol. 217, 1496–1503.
Peroutka, S.J., Howell, T.A., 1994. The molecular evolution of G protein-coupled receptors: focus on 5-hydroxytryptamine receptors. Neuropharmacology 33, 319–324.
Pfadt, R.E., 1988. Field Guide to the Common Western Grasshoppers. University of Wyoming Agricultural Experiment Station Bulletin 912, 145 pp.
Pflüger, J.-H., Stevenson, P.A., 2005. Evolutionary aspects of octopaminergic systems with emphasis on arthropods. Arthropod Struct. Dev. 34, 379–396.
Piou, C., Lebourgeois, V., Benahi, A.S., Bonnal, V., Jaavar, M.E.H., Lecoq, M., Vassal, J.M., 2013. Coupling historical prospection data and a remotely-sensed vegetation index for the preventative control of desert locusts. Basic Appl. Ecol. 14, 593–604.
Poels, J., Suner, M.M., Needham, M., Torfs, H., De Rijck, J., De Loof, A., Dunbar, S.J., Vanden Broeck, J., 2001. Functional expression of a locust tyramine receptor in murine erythroleukaemia cells. Insect Mol. Biol. 10, 541–548.
Poot-Pech, M.A., 2016. La langosta voladora *Schistocerca piceifrons* (Orthoptera: Acrididae): hacia un manejo sustentable. In: El Patrimonio, su Importancia y Conservación. Conociendo el Patrimonio, Cuerpo Académico Patrimonio y Desarrollo Sustentable, Editorial TECCIS, San Francisco de Campeche. , pp. 58–66.
Poot-Pech, M.A., 2017. Meeting on the locust situation in South America and the OIRSA region. Metaleptea 37, 2.
Poot-Pech, M.A., Ruiz-Sánchez, E., Ballina-Gómez, H.S., Gamboa-Angulo, M.M., Reyes-Ramírez, A., 2016. Olfactory response and host plant feeding of the central American locust *Schistocerca piceifrons piceifrons* Walker to common plants in a gregarious zone. Neotrop. Entomol. 45, 382–388.
Poot-Pech, M.A., Ruíz, S.E., Ballina, G.H., 2017. Indicator plants in solitary phase and migration behavior of *Schistocerca piceifrons* in Yucatán, México. Metaleptea 37, 21.
Popov, G.B., 1958. Ecological studies on oviposition by swarms of the desert locust (*Schistocerca gregaria* Forskål) in Eastern Africa. Anti-Locust Bull. 31, 1–70.
Popov, G.B., Duranton, J.-F., Gigault, J., 1991. Etude écologique des biotopes du criquet pèlerin *Schistocerca gregaria* (Forskal, 1775) en Afrique Nord-Occidentale: mise en évidence et description des unités territoriales écologiquement homogènes CIRAD-PRIFAS, Montpellier.
Popova, E.N., 2014. The increase in locust threat following climate warming in Russia: calculating estimations. Agro XXI, 7–9.
Popova, E.N., Semenov, S.M., Popov, I.O., 2016. Assessment of possible expansion of the climatic range of Italian locust (*Calliptamus italicus* L.) in Russia in the 21st century at simulated climate changes. Russ. Meteorol. Hydrol. 41, 213–217.

Propastin, P., 2012. Multisensor monitoring system for assessment of locust hazard risk in the Lake Balkhash drainage basin. Environ. Manag. 50, 1234–1246.
Propastin, P., 2013. Satellite-based monitoring system for assessment of vegetation vulnerability to locust hazard in the river Ili delta (Lake Balkhash, Kazakhstan). J. Appl. Remote Sens. 7, 075094.
Qi, Y.X., Xia, R.Y., Wu, Y.S., Stanley, D., Huang, J., Ye, G.Y., 2014. Larvae of the small white butterfly, *Pieris rapae*, express a novel serotonin receptor. J. Neurochem. 131, 767–777.
Raszick, T.J., Song, H., 2016. The ecotype paradigm: testing the concept in an ecologically divergent grasshopper. Insect Syst. Evol. 47, 363–387.
Reale, V., Hannan, F., Hall, L.M., Evans, P.D., 1997. Agonist-specific coupling of a cloned *Drosophila melanogaster* D1-like dopamine receptor to multiple second messenger pathways by synthetic agonists. J. Neurosci. 17, 6545–6553.
Reim, T., Balfanz, S., Baumann, A., Blenau, W., Thamm, M., Scheiner, R., 2017. AmTAR2: functional characterization of a honeybee tyramine receptor stimulating adenylyl cyclase activity. Insect Biochem. Mol. Biol. 80, 91–100.
Renier, C., Waldner, F., Jacques, D., Babah Ebbe, M., Cressman, K., Defourny, P., 2015. A dynamic vegetation senescence indicator for near-real-time desert locust habitat monitoring with MODIS. Remote Sens. (Basel) 7, 7545–7570.
Reynolds, D.R., 1988. Twenty years of radar entomology. Antenna 12, 44–49.
Riley, J.R., 1989. Remote sensing in entomology. Annu. Rev. Entomol. 34, 247–271.
Riley, J.R., Reynolds, D.R., 1997. Vertical-looking radar as a means to improve forecasting and control of desert locusts. In: Krall, S., Peveling, R., Ba Diallo, D. (Eds.), New Strategies in Locust Control. Birkhauser Verlag, Basel, pp. 47–54.
Riley, J.R., Greggers, U., Smith, A.D., Reynolds, D.R., Menzel, R., 2005. The flight paths of honeybees recruited by the waggle dance. Nature 435, 205–207.
Rillich, J., Stevenson, P.A., 2014. A fighter's comeback: dopamine is necessary for recovery of aggression after social defeat in crickets. Horm. Behav. 66, 696–704.
Robb, S., Cheek, T.R., Hannan, F.L., Hall, L.M., Midgley, J.M., Evans, P.D., 1994. Agonist-specific coupling of a cloned *Drosophila* octopamine/tyramine receptor to multiple second messenger systems. EMBO J. 13, 1325–1330.
Robinet, C., Roques, A., 2010. Direct impacts of recent climate warming on insect populations. Integr. Zool. 5, 132–142.
Robinson, K.L., Tohidi-Esfahani, D., Lo, N., Simpson, S.J., Sword, G.A., 2011. Evidence for widespread genomic methylation in the migratory locust, *Locusta migratoria* (Orthoptera: Acrididae). PLoS One 6, e28167.
Robinson, K.L., Tohidi-Esfahani, D., Ponton, F., Simpson, S.J., Sword, G.A., Lo, N., 2016. Alternative migratory locust phenotypes are associated with differences in the expression of genes encoding the methylation machinery. Insect Mol. Biol. 25, 105–115.
Roessingh, P., Simpson, S.J., 1994. The time-course of behavioural phase change in nymphs of the desert locust, *Schistocerca gregaria*. Physiol. Entomol. 19, 191–197.
Roessingh, P., Simpson, S.J., James, S., 1993. Analysis of phase-related changes in behavior of desert locust nymphs. Proc. R. Soc. B Biol. Sci. 252, 43–49.
Roessingh, P., Bouaïchi, A., Simpson, S.J., 1998. Effects of sensory stimuli on the behavioural phase state of the desert locust, *Schistocerca gregaria*. J. Insect Physiol. 44, 883–893.
Roffey, J., 1963. Observation on night flight in the desert locust *Schistocerca gregaria* (Forskål). Anti-Locust Bull. 39, 1–32.
Roffey, J., 1969. Report on radar studies on the desert locust, *Schistocerca Gregaria* (Forskål) in the Niger Republic, (September–October 1968). Occasional Report no. 17, Anti-Locust Research Centre, London.
Roffey, J., Popov, G., 1968. Environmental and behavioural processes in a desert locust outbreak. Nature 219, 446–450.

Rogers, S.M., Ott, S.R., 2015. Differential activation of serotonergic neurons during short- and long-term gregarization of desert locusts. Proc. R. Soc. B Biol. Sci. 282, 20142062.
Rogers, S.M., Simpson, S.J., 2014. Thanatosis. Curr. Biol. 24, R1031–R1033.
Rogers, S.M., Matheson, T., Despland, E., Dodgson, T., Burrows, M., Simpson, S.J., 2003. Mechanosensory-induced behavioural gregarization in the desert locust *Schistocerca gregaria*. J. Exp. Biol. 206, 3991–4002.
Rogers, S.M., Matheson, T., Sasaki, K., Kendrick, K., Simpson, S.J., Burrows, M., 2004. Substantial changes in central nervous system neurotransmitters and neuro-modulators accompany phase change in the locust. J. Exp. Biol. 207, 3603–3617.
Rogers, S.M., Krapp, H.G., Burrows, M., Matheson, T., 2007. Compensatory plasticity at an identified synapse tunes a visuomotor pathway. J. Neurosci. 27, 4621–4633.
Rogers, S.M., Harston, G.W.J., Kilburn-Toppin, F., Matheson, T., Burrows, M., Gabbiani, F., Krapp, H.G., 2010. Spatiotemporal receptive field properties of a looming-sensitive neuron in solitarious and gregarious phases of the desert locust. J. Neurophysiol. 103, 779–792.
Rogers, S.M., Cullen, D.A., Anstey, M.L., Burrows, M., Despland, E., Dodgson, T., Matheson, T., Ott, S.R., Stettin, K., Sword, G.A., Simpson, S.J., 2014. Rapid behavioural gregarization in the desert locust, *Schistocerca gregaria* entails synchronous changes in both activity and attraction to conspecifics. J. Insect Physiol. 65, 9–26.
Romanczuk, P., Couzin, I.D., Schimansky-Geier, L., 2009. Collective motion due to individual escape and pursuit response. Phys. Rev. Lett. 102, 010602.
Röser, C., Jordan, N., Balfanz, S., Baumann, A., Walz, B., Baumann, O., Blenau, W., 2012. Molecular and pharmacological characterization of serotonin 5-HT2α and 5-HT7 receptors in the salivary glands of the blowfly *Calliphora vicina*. PLoS One 7, e49459.
Rotte, C., Krach, C., Balfanz, S., Baumann, A., Walz, B., Blenau, W., 2009. Molecular characterization and localization of the first tyramine receptor of the American cockroach (*Periplaneta americana*). Neuroscience 162, 1120–1133.
Saudou, F., Amlaiky, N., Plassat, J.L., Borelli, E., Hen, R., 1990. Cloning and characterization of a *Drosophila* tyramine receptor. EMBO J. 9, 3611–3617.
Saudou, F., Boschert, U., Amlaiky, N., Plassat, J.L., Hen, R., 1992. A family of *Drosophila* serotonin receptors with distinct intracellular signalling properties and expression patterns. EMBO J. 11 (1), 7–17.
Schaefer, G.W., 1969. Radar studies of locust, moth and butterfly migration in the Sahara. Proc. R. Entomol. Soc. Lond. C34 (33), 39–40.
Schaefer, G.W., 1976. In: Rainey, R.C. (Ed.), Radar observations of insect flight. Insect Flight. Symposia of the Royal Entomological Society. In: vol. 7. Blackwell, Oxford, UK, pp. 157–197.
Schlenstedt, J., Balfanz, S., Baumann, A., Blenau, W., 2006. Am5-HT7: molecular and pharmacological characterization of the first serotonin receptor of the honeybee (*Apis mellifera*). J. Neurochem. 98, 1985–1998.
Shand, J.D., 2016. The effects of density-dependent polyphenism on circadian biology of the desert locust *Schistocerca gregaria*. Unpublished. PhD Thesis, University of Leicester, UK.
Shand, J.D., Matheson, T., Rosato, E., 2015. Circadian rhythmicity in behavioural and neuronal sensitivity in locusts. Proceedings of the 35th Göttingen Neurobiology Conference. T14-1B.
Simões, P.M.V., Ott, S.R., Niven, J.E., 2011. Associative olfactory learning in the desert locust, *Schistocerca gregaria*. J. Exp. Biol. 214, 2495–2503.
Simões, P.M.V., Ott, S.R., Niven, J.E., 2012. A long-latency aversive learning mechanism enables locusts to avoid odours associated with the consequences of ingesting toxic food. J. Exp. Biol. 215, 1711–1719.
Simões, P.M., Niven, J.E., Ott, S.R., 2013. Phenotypic transformation affects associative learning in the desert locust. Curr. Biol. 23, 2407–2412.

Simões, P.M.V., Ott, S.R., Niven, J.E., 2016. Environmental adaptation, phenotypic plasticity, and associative learning in insects: the desert locust as a case study. Integr. Comp. Biol. 56, 914–924.
Simpson, S.J., 1981. An oscillation underlying feeding and a number of other behaviours in fifth instar *Locusta migratoria* nymphs. Physiol. Entomol. 6, 315–324.
Simpson, S.J., 1982. Patterns in feeding: a behavioural analysis using *Locusta migratoria* nymphs. Physiol. Entomol. 7, 325–336.
Simpson, S.J., Sword, G.A., 2009. Phase polyphenism in locusts: mechanisms, population consequences, adaptive significance and evolution. In: Whitman, D.W., Ananthakrishnan, T.N. (Eds.), Phenotypic Plasticity of Insects Mechanisms and Consequences. Science Publishers, Enfield, NH, pp. 140–190.
Simpson, S.J., McCaffery, A.R., Hägele, B.F., 1999. A behavioural analysis of phase change in the desert locust. Biol. Rev. 74, 461–480.
Simpson, S.J., Despland, E., Hägele, B.F., Dodgson, T., 2001. Gregarious behavior in desert locusts is evoked by touching their back legs. Proc. Natl. Acad. Sci. U.S.A. 98, 3895–3897.
Sivanpillai, R., Latchininsky, A.V., 2007. Mapping locust habitats in Amudarya River Delta, Uzbekistan with multi-temporal MODIS imagery. Environ. Manag. 39, 876–886.
Sivanpillai, R., Latchininsky, A.V., 2008. Can late summer Landsat data be used for mapping Asian migratory locust, *Locusta migratoria migratoria*, oviposition sites in the Amudarya River delta, Uzbekistan? Entomol. Exp. Appl. 128, 346–353.
Sivanpillai, R., Latchininsky, A.V., Driese, K.L., Kambulin, V.E., 2006. Mapping locust habitats in River Ili Delta, Kazakhstan, using Landsat imagery. Agric. Ecosyst. Environ. 117, 128–134.
Smith, J., 2012. Application of Remote Sensing to Moroccan locust (*Dociostaurus maroccanus* Thunberg, 1815) Habitat in Southern Uzbekistan. Unpublished M.S. Thesis, University of Wyoming, Laramie, WY.
Smith, A.D., Riley, J.R., Gregory, R.D., 1993. A method for routine monitoring of the aerial migration of insects by using a vertical-looking radar. Philos. Trans. R. Soc. B 340, 393–404.
Song, H., 2005. Phylogenetic perspectives on the evolution of locust phase polyphenism. J. Orthop. Res. 14, 235–245.
Song, H., 2011. Density-dependent phase polyphenism in nonmodel locusts: a minireview. Psyche, Article ID 741769.
Song, H., Wenzel, J.W., 2008. Phylogeny of bird-grasshopper subfamily *Cyrtacanthacridinae* (Orthoptera: Acrididae) and the evolution of locust phase polyphenism. Cladistics 24, 515–542.
Song, H., Moulton, M.J., Hiatt, K.D., Whiting, M.F., 2013. Uncovering historical signature of mitochondrial DNA hidden in the nuclear genome: the biogeography of *Schistocerca* revisited. Cladistics 29, 643–662.
Song, H., Foquet, B., Mariño-Pérez, R., Woller, D.A., 2017. Phylogeny of locusts and grasshoppers reveals complex evolution of density-dependent phenotypic plasticity. Sci. Rep. 7, 6606.
Spielman, S.J., Kumar, K., Wilke, C.O., 2015. Comprehensive, structurally-informed alignment and phylogeny of vertebrate biogenic amine receptors. PeerJ 3, e773.
Srivastava, D.P., Yu, E.J., Kennedy, K., Chatwin, H., Reale, V., Hamon, M., Smith, T., Evans, P.D., 2005. Rapid, nongenomic responses to ecdysteroids and catecholamines mediated by a novel Drosophila G-protein-coupled receptor. J. Neurosci. 25 (26), 6145–6155.
Stamo, P.D., Kovalenkov, V.G., Kuznetsova, O.V., Nikitenko, Y.V., 2013. Moroccan locust again in Stavropol. Plant Protect. Quarantine 2, 14–20(in Russian).

Standage, D.S., Berens, A.J., Glastad, K.M., Severin, A.J., Brendel, V.P., Toth, A.L., 2016. Genome, transcriptome and methylome sequencing of a primitively eusocial wasp reveal a greatly reduced DNA methylation system in a social insect. Mol. Ecol. 25, 1769–1784.

Steedman, A. (Ed.), 1990. Locust Handbook, third ed. Natural Resources Institute, Chatham.

Stettin, K., 2015. Mechanisms driving the acquisition and maintenance of gregarious behaviour in the desert locust (*Schistocerca gregaria*). Unpublished PhD Thesis, University of Cambridge, UK.

Stettin, K., Ott, S.R., 2012. In: Phase change in the desert locust: a novel computerised behavioural assay to study the effects of anaesthesia on gregarisation. Frontiers in Behavioral Neuroscience. Conference Abstract: 10th International Congress of Neuroethology. http://dx.doi.org/10.3389/conf.fnbeh.2012.27.00278.

Stige, L.C., Chan, K.S., Zhang, Z., Frank, D., Stenseth, N.C., 2007. Thousand-year-long Chinese time series reveals climate forcing of decadal locust dynamics. Proc. Natl. Acad. Sci. U.S.A. 104, 16188–16193.

Stower, W.J., 1963. Photographic techniques for the analysis of locust "hopper" behaviour. Anim. Behav. 11, 199–205.

Sugahara, R., Saeki, S., Jouraku, A., Shiotsuki, T., Tanaka, S., 2015. Knockdown of the corazonin gene reveals its critical role in the control of gregarious characteristics in the desert locust. J. Insect Physiol. 79, 80–87.

Sugahara, R., Tanaka, S., Jouraku, A., Shiotsuki, T., 2016. Functional characterization of the corazonin-encoding gene in phase polyphenism of the migratory locust, *Locusta migratoria* (Orthoptera: Acrididae). Appl. Entomol. Zool. 51, 225–232.

Sugamori, K.S., Demchyshyn, L.L., McConkey, F., Forte, M.A., Niznik, H.B., 1995. A primordial dopamine D1-like adenylyl cyclase-linked receptor from *Drosophila melanogaster* displaying poor affinity for benzazepines. FEBS Lett. 362, 131–138.

Sumpter, D.J., 2010. Collective Animal Behaviour. Princeton University Press, Princeton.

Sun, W., Dong, H., Gao, Y.-B., Su, Q.-F., Qian, H.-T., Bai, H.-Y., Zhang, Z.-T., Cong, B., 2015. Genetic variation and geographic differentiation among populations of the nonmigratory agricultural pest *Oedaleus infernalis* (Orthoptera: Acridoidea) in China. J. Insect Sci. 15, 150.

Sword, G.A., 1999. Density-dependent warning coloration. Nature 397, 217.

Sword, G.A., 2002. A role of phenotypic plasticity in the evolution of aposematism. Proc. R. Soc. B Biol. Sci. 269, 1639–1644.

Sword, G.A., 2003. To be or not to be a locust? A comparative analysis of behavioral phase change in nymphs of *Schistocerca americana* and *S. gregaria*. J. Insect Physiol. 49, 709–717.

Sword, G.A., Simpson, S.J., 2008. Locusts. Curr. Biol. 18, 364–366.

Sword, G.A., Simpson, S.J., El Hadi, O.M., Wilps, H., 2000. Density-dependent aposematism in the desert locust. Proc. R. Soc. B Biol. Sci. 267, 63–68.

Sword, G.A., Lorch, P.D., Gwynne, D.T., 2005. Insect behaviour: migratory bands give crickets protection. Nature 433, 703.

Sword, G.A., Lecoq, M., Simpson, S.J., 2010. Phase polyphenism and preventative locust management. J. Insect Physiol. 56, 949–957.

Symmons, P.M., 1992. Strategies to combat desert locust. Crop Prot. 11, 206–212.

Szyf, M., McGowan, P., Meaney, M.J., 2008. The social environment and the epigenome. Environ. Mol. Mutagen. 49, 46–60.

Tahir, N., Brooker, G., 2015. Toward the development of millimeter wave harmonic sensors for tracking small insects. IEEE Sens. J. 15, 5669–5676.

Tanaka, S., Nishide, Y., 2013. Behavioral phase shift in nymphs of the desert locust, *Schistocerca gregaria*: special attention to attraction/avoidance behaviors and the role of serotonin. J. Insect Physiol. 59, 101–112.

Thamm, M., Balfanz, S., Scheiner, R., Blenau, W., 2010. Characterization of the 5-HT1A receptor of the honeybee (*Apis mellifera*) and involvement of serotonin in phototactic behavior. Cell. Mol. Life Sci. 67, 2467–2479.

Thamm, M., Rolke, D., Jordan, N., Balfanz, S., Schiffer, C., Baumann, A., Blenau, W., 2013. Function and distribution of 5-HT2 receptors in the honey bee (*Apis mellifera*). PLoS One 8, e82407.

Tian, H.D., Ji, R., Xie, B.Y., Li, X.H., Li, D.M., 2008. Using multi-temporal Landsat ETM + data to monitor the plague of oriental migratory locust. Int. J. Remote Sens. 29, 1685–1692.

Tian, H., Stige, L.C., Cazelles, B., Kausrud, K.L., Svarverud, R., Stenseth, N.C., Zhang, Z., 2011. Reconstruction of a 1,910-y-long locust series reveals consistent associations with climate fluctuations in China. Proc. Natl. Acad. Sci. U.S.A. 108, 14521–14526.

Tierney, A.J., 2001. Structure and function of invertebrate 5-HT receptors: a review. Comp. Biochem. Physiol. A Mol. Integr. Physiol. 128, 791–804.

Tobback, J., Boerjan, B., Vandersmissen, H.P., Huybrechts, R., 2011. The circadian clock genes affect reproductive capacity in the desert locust *Schistocerca gregaria*. Insect Biochem. Mol. Biol. 41, 313–321.

Tobback, J., Boerjan, B., Vandersmissen, H.P., Huybrechts, R., 2012. Male reproduction is affected by RNA interference of period and timeless in the desert locust *Schistocerca gregaria*. Insect Biochem. Mol. Biol. 42, 109–115.

Tokuda, M., Tanaka, S., Zhu, D.H., 2010. Multiple origins of *Locusta migratoria* (Orthoptera: Acrididae) in the Japanese Archipelago and the presence of two major clades in the world: evidence from a molecular approach. Biol. J. Linn. Soc. 99, 570–581.

Tomioka, K., Matsumoto, A., 2010. A comparative view of insect circadian clock systems. Cell. Mol. Life Sci. 67, 1397–1406.

Tomioka, K., Wakatsuki, T., Shimono, K., Chiba, Y., 1991. Circadian control of hatching in the cricket, *Gryllus bimaculatus*. J. Insect Physiol. 37, 365–371.

Tong, X.-W., Chen, B., Huang, L.-H., Feng, Q.-L., Kang, L., 2015. Proteomic analysis reveals that COP9 signalosome complex subunit 7A (CSN7A) is essential for the phase transition of migratory locust. Sci. Rep. 5, 12542.

Torney, C.J., Berdahl, A., Couzin, I.D., 2011. Signalling and the evolution of cooperative foraging in dynamic environments. PLoS Comput. Biol. 7, e1002194.

Tratalos, J.A., Cheke, R.A., 2006. Can NDVI GAC imagery be used to monitor desert locust breeding areas? J. Arid Environ. 64, 342–356.

Tronin, A.A., Gornyy, V.I., Kiselev, A.V., Kritsuk, S.G., Latypov, I.S., 2014. Forecasting of locust mass breeding by using satellite data. Mod. Probl. Earth Remote Sens. Space 11, 137–150.

Troppmann, B., Balfanz, S., Baumann, A., Blenau, W., 2010. Inverse agonist and neutral antagonist actions of synthetic compounds at an insect 5-HT1 receptor. Br. J. Pharmacol. 159, 1450–1462.

Troppmann, B., Balfanz, S., Krach, C., Baumann, A., Blenau, W., 2014. Characterization of an invertebrate-type dopamine receptor of the American cockroach, *Periplaneta americana*. Int. J. Mol. Sci. 15, 629–653.

Tucker, C.J., Hielkema, J.U., Roffey, J., 1985. The potential of satellite remote sensing of ecological conditions for survey and forecasting desert-locust activity. Int. J. Remote Sens. 6, 127–138.

Tyrer, N.M., Pozza, M.F., Humbel, U., Peters, B.H., Bacon, J.P., 1988. The tritocerebral commissure dwarf (TCD)—a major GABA-immunoreactive descending interneurone in the locust. J. Comp. Physiol. A 164, 141–150.

Uvarov, B., 1966. Grasshoppers and Locusts. vol. 1. Cambridge University Press, Cambridge.
Uvarov, B., 1977. Grasshoppers and Locusts. vol. 2. Centre for Overseas Pest Research, London.
Vanden Broeck, J., Vulsteke, V., Huybrechts, R., De Loof, A., 1995. Characterization of a cloned locust tyramine receptor cDNA by functional expression in permanently transformed *Drosophila* S2 cells. J. Neurochem. 64, 2387–2395.
Van der Werf, W., Woldewahid, G., Van Huis, A., Butrous, M., Kora, K.Y., 2005. Plant communities can predict the distribution of solitarious desert locust *Schistocerca gregaria*. J. Appl. Ecol. 42, 989–997.
Veran, S., Simpson, S.J., Sword, G.A., Deveson, E., Piry, S., Hines, J.E., Berthier, K., 2015. Modeling spatiotemporal dynamics of outbreaking species: influence of environment and migration in a locust. Ecology 96, 737–748.
Vergoz, V., McQuillan, H.J., Geddes, L.H., Pullar, K., Nicholson, B.J., Paulin, M.G., Mercer, A.R., 2009. Peripheral modulation of worker bee responses to queen mandibular pheromone. Proc. Natl. Acad. Sci. U.S.A. 106, 20930–20935.
Verlinden, H., Badisco, L., Marchal, E., Van Wielendaele, P., Vanden Broeck, J., 2009. Endocrinology of reproduction and phase transition in locusts. Gen. Comp. Endocrinol. 162, 79–92.
Verlinden, H., Vleugels, R., Marchal, E., Badisco, L., Pflüger, H., Blenau, W., Broeck, V., 2010a. The role of octopamine in locusts and other arthropods. J. Insect Physiol. 56, 854–867.
Verlinden, H., Vleugels, R., Marchal, E., Badisco, L., Pflüger, H., Blenau, W., Broeck, V., 2010b. The cloning, phylogenetic relationship and distribution pattern of two new putative GPCR-type octopamine receptors in the desert locust (*Schistocerca gregaria*). J. Insect Physiol. 56, 868–875.
Verlinden, H., Vleugels, R., Zels, S., Dillen, S., Lenaerts, C., Crabbé, K., Spit, J., Vanden Broeck, J., 2014. Receptors for neuronal of endocrine signalling molecules as potential targets for the control of insect pests. Adv. Insect Physiol. 46, 167–303.
Verlinden, H., Vleugels, R., Verdonck, R., Urlacher, E., Vanden Broeck, J., Mercer, A., 2015. Pharmacological and signalling properties of a D2-like dopamine receptor (Dop3) in *Tribolium castaneum*. Insect Biochem. Mol. Biol. 56, 9–20.
Vicsek, T., Czirók, A., Ben-Jacob, E., Cohen, I., Shochet, O., 1995. Novel type of phase transition in a system of self-driven particles. Phys. Rev. Lett. 75, 1226–1229.
Vinatier, F., Tixier, P., Duyck, P.-F., Lescuarret, F., 2011. Factors and mechanisms explaining spatial heterogeneity: a review of methods for insect populations. Methods Ecol. Evol. 2, 11–22.
Vleugels, R., Lenaerts, C., Baumann, A., Vanden Broeck, J., Verinden, H., 2013. Pharmacological characterization of a 5-HT1-type serotonin receptor in the red flour beetle, *Tribolium castaneum*. PLoS One 8, e65052.
Vleugels, R., Lenaerts, C., Vanden, B., Verlinden, H., 2014. Signalling properties and pharmacology of a 5-HT7-type serotonin receptor from *Tribolium castaneum*. Insect Mol. Biol. 23, 230–243.
Vleugels, R., Verlinden, H., Vanden Broeck, J., 2015. Serotonin, serotonin receptors and their actions in insects. Neurotransmitter 2, e314.
Voss, F., Dreiser, U., 1994. Mapping of desert locust and other migratory pests habitats using remote sensing techniques. In: Krall, S., Wilps, H. (Eds.), New Trends in Locust Control. In: vol. 245. GTZ Schriftenreiche, Eschborn Control, pp. 23–29.
Voss, F., Dreiser, U., 1997. Mapping of desert locust habitats using remote sensing techniques. In: Krall, S., Peveling, R., Ba Diallo, D. (Eds.), New Strategies in Locust Control. Birkhauser Verlag, Basel, pp. 37–45.

Waldner, F., Babah Ebbe, M.A., Cressman, K., 2015. Geo-information operational monitoring of the desert locust habitat with earth observation: an assessment. ISPRS Int. J. Geo-Inform. 4, 2379–2400.

Waloff, Z., Pedgley, D.E., 1986. Comparative biogeography and biology of the south American locust, *Schistocerca cancellata* (Serville), and the south African desert locust, *S. gregaria flaviventris* (Burmeister) (Orthoptera: Acrididae): a review. Bull. Entomol. Res. 76, 1–20.

Wang, X., Kang, L., 2014. Molecular mechanisms of phase change in locusts. Annu. Rev. Entomol. 59, 225–244.

Wang, H.S., Wang, X.H., Zhou, C.S., Huang, L.H., Zhang, S.F., Guo, W., Kang, L., 2007. cDNA cloning of heat shock proteins and their expression in the two phases of the migratory locust. Insect Mol. Biol. 16, 207–219.

Wang, X., Fang, X., Yang, P., Jiang, X., Jiang, F., Zhao, D., Li, B., Cui, F., Wei, J., Ma, C., Wang, Y., He, J., Luo, Y., Wang, Z., Guo, X., Guo, W., Wang, X., Zhang, Y., Yang, M., Hao, S., Chen, B., Ma, Z., Yu, D., Xiong, Z., Zhu, Y., Fan, D., Han, L., Wang, B., Chen, Y., Wang, J., Yang, L., Zhao, W., Feng, Y., Chen, G., Lian, J., Li, Q., Huang, Z., Yao, X., Lv, N., Zhang, G., Li, Y., Wang, J., Wang, J., Zhu, B., Kang, L., 2014. The locust genome provides insight into swarm formation and long-distance flight. Nat. Commun. 5, 2957.

Wei, Y., Chen, S., Yang, P., Ma, Z., Kang, L., 2009. Characterization and comparative profiling of the small RNA transcriptomes in two phases of locust. Genome Biol. 10, R6.

Weiss, J.E.R., 2016. Do locusts seek greener pastures? An evaluation of MODIS vegetation indices to predict presence, abundance and impact of the Australian plague locust in south-eastern Australia. Unpublished PhD Thesis, University of Melbourne, Australia.

Westerman, M., Barton, N.H., Hewitt, G.M., 1987. Differences in DNA content between two chromosomal races of the grasshopper *Podisma pedestris*. Heredity 58, 221–228.

Witz, P., Amlaiky, N., Plassat, J.L., Maroteaux, L., Borelli, E., Hen, R., 1990. Cloning and characterization of a *Drosophila* serotonin receptor that activates adenylate cyclase. Proc. Natl. Acad. Sci. U.S.A. 87, 8940–8944.

Woldewahid, G., Van Der Werf, W., Van Huis, A., Steins, A., 2004. Spatial distribution of populations solitarious adult desert locust (*Schistocerca gregaria* Forsk.) on coastal plain of Sudan. Agric. For. Entomol. 6, 181–191.

Woldewahid, G., Van der Werf, W., Sykora, K., Abate, T., Mostafa, B., Van Huis, A., 2007. Description of plant communities on the Red Sea coastal plain of Sudan. J. Arid Environ. 68, 113–131.

Wu, S.F., Yao, Y., Huang, J., Ye, G., 2012a. Characterization of a beta-adrenergic-like octopamine receptor from the rice stem borer (*Chilo suppressalis*). J. Exp. Biol. 215, 2646–2652.

Wu, T., Hao, S., Sun, O.J., Kang, L., 2012b. Specificity responses of grasshoppers in temperate grasslands to diel asymmetric warming. PLoS One 7, e41764.

Wu, S.F., Huang, J., Ye, G.Y., 2013. Molecular cloning and pharmacological characterisation of a tyramine receptor from the rice stem borer, *Chilo suppressalis* (Walker). Pest Manag. Sci. 69, 126–134.

Wu, S.F., Xu, G., Xia, R.Y., Huang, J., Ye, G.Y., 2014. Two splicing variants of a novel family of octopamine receptors with different signaling properties. J. Neurochem. 129, 37–47.

Wu, S.F., Xu, G., Ye, G.Y., 2015. Characterization of a tyramine receptor type 2 from hemocytes of rice stem borer, *Chilo suppressalis*. J. Insect Physiol. 75, 39–46.

Wynant, N., Verlinden, H., Breugelmans, B., Simonet, G., Vanden Broeck, J., 2012. Tissue-dependence and sensitivity of the systemic RNA interference response in the desert locust, *Schistocerca gregaria*. Insect Biochem. Mol. Biol. 42, 911–917.

Wynant, N., Santos, D., Subramanyam, S.H., Verlinden, H., Vanden Broeck, J., 2015. Drosha, Dicer-1 and Argonaute-1 in the desert locust: phylogenetic analyses, transcript profiling and regulation during phase transition and feeding. J. Insect Physiol. 75, 20–29.

Xian-Lei, Q., Xian-Hui, W., Hong-Fu, X., Kang, L., 2007. Influence of soil moisture on egg cold hardiness in the migratory locust *Locusta migratoria* (Orthoptera: Acrididae). Physiol. Entomol. 32, 219–224.

Xu, L., Li, L., Yang, P., Ma, Z., 2017. Calmodulin as a downstream gene of octopamine-OAR a1 signalling mediates olfactory attraction in gregarious locusts. Insect Mol. Biol. 26, 1–12.

Yan, H., Bonasio, R., Simola, D.F., Liebig, J., Berger, S.L., Reinberg, D., 2015. DNA methylation in social insects: how epigenetics can control behavior and longevity. Annu. Rev. Entomol. 60, 435–452.

Yates, C.A., Erban, R., Escudero, C., Couzin, I.D., Buhl, J., Kevrekidis, I.G., Sumpter, D.J., 2009. Inherent noise can facilitate coherence in collective swarm motion. Proc. Natl. Acad. Sci. U.S.A. 106, 5464–5469.

Young, O.R., 2002. The Institutional Dimensions of Environmental Change: Fit, Interplay and Scale. MIT Press, Cambridge, MA.

Yu, G., Shen, H., Liu, J., 2009. Impacts of climate change on historical locust outbreaks in China. J. Geophys. Res. 114, D18104.

Zha, Y., Gao, J., Ni, S., Shen, N., 2005. Temporal filtering of successive MODIS data in monitoring a locust outbreak. Int. J. Remote Sens. 26, 5665–5674.

Zhang, Z., Cazelles, B., Tian, H., Stige, L.C., Bräuning, A., Stenseth, N.C., 2009. Periodic temperature-associated drought/flood drives locust plagues in China. Proc. R. Soc. B Biol. Sci. 276, 823–831.

Zhang, Z., Peng, Z.-Y., Yi, K., Cheng, Y., Xia, Y., 2012. Identification of representative genes of the central nervous system of the locust, *Locusta migratoria manilensis* by deep sequencing. J. Insect Sci. 12, 1–15.

CHAPTER EIGHT

The Future of Environmental Epigenetics: Insights Using the Clonal Water Flea Model

Marcin W. Wojewodzic*,‡, Margaret J. Beaton†
*University of Birmingham, School of Biosciences, Birmingham, United Kingdom
†Mount Allison University, Sackville, NB, Canada
‡Centre for Epigenetics, Development and Evolution, University of Oslo, Oslo, Norway

Contents

Abstract

Here, we review the fast-changing state-of-the-art and future perspectives for environmental epigenetics where the model organism for environment, *Daphnia*, is in focus. First, we describe why we believe the genus *Daphnia* possesses such remarkable characteristics for usage in epigenetic studies and especially in environmental epigenetics. Further, we consider current epigenetic research that employ these animals, followed by a discourse on the methods used and hint at other methods available. As part of this exercise, we identify the gaps in current knowledge and offer some questions that require attention. Finally, we offer our perspective on how, we think, this particular model organism could be exploited in the future in the field of environmental epigenetics.

Advances in Insect Physiology, Volume 53
ISSN 0065-2806
http://dx.doi.org/10.1016/bs.aiip.2017.05.002

ABBREVIATIONS

5-mC 5-methylcytosine
5-hmC 5-hydroxymethylcytosine
ChIP-seq chromatin immunoprecipitation followed by next-generation sequencing
DNMT1 (cytosine-5)-methyltransferase 1
EE environmental epigenetics
epiGBS reference-free reduced representation bisulphite sequencing
EpiRADseq restriction-site associated DNA sequencing representation bisulphite sequencing
E(z) histone-lysine *N*-methyltransferase
H3K4me2 histone 3 dimethylated in the lysine 4
H3K27me histone 3 methylated in the lysine 27
lncRNA, long ncRNAs long noncoding RNAs are defined as nonprotein coding transcripts longer than 200 nucleotides
miRNA microRNA is a small noncoding RNA molecule (containing about 22 nucleotides) found also in *Daphnia*, that functions in RNA silencing and posttranscriptional regulation of gene expression
ncRNAs noncoding ribonucleic acids causing RNA silencing and posttranscriptional regulation of gene expression
orthoDB the hierarchical catalogue of orthologs
piRNA piwi-interacting RNA (piRNA) is the largest class of small noncoding RNA molecules expressed in animal cells
RRBS reduced representation bisulphate sequencing
SAM *S*-adenosylmethionine
siRNA short-interfering RNA molecule is the most commonly used RNA interference tool for inducing short-term silencing of protein-coding genes
SNPs single-nucleotide polymorphisms
TET ten–eleven translocation enzyme
UniProtKB UniProt: the universal protein knowledgebase
WGBS whole-genome bisulphite sequencing

1. INTRODUCTION

Since 1785, when Danish naturalist Otto F. Müller described the genus *Daphnia*—the water flea, the research community has gathered an enormous body of knowledge pertaining to the ecology and evolution of these keystone freshwater species. Sometimes called 'water cows', they play a central role in trophic cascades and nutrient flow dynamics of lakes, and their intra- and interspecific interactions have led to new concepts in modern ecology and evolution. *Daphnia* are routinely used as the focal species in ecotoxicology studies, often acting as a surrogate for mammalian models when testing is desired according to the 3Rs idea: replace, reduce and refine.

Moreover, the National Institute of Health (NIH) recently recognized the species as one of the top model organisms to be used for human health studies, thus opening many, so far unexplored doors. *Daphnia* has become the key target for a handful of groups worldwide serving as the model organism for research in toxicological genomics, metagenomics, ecology and evolution (Colbourne et al., 2011; Shaw et al., 2008). Last but not least, this freshwater planktonic microcrustacean has also been proposed as a model organism for studying developmental (re)programming via epigenetics, as clonal lines contain genetically identical but phenotypically divergent individuals (Harris et al., 2012). It is definitively the combined phenomena of parthenogenesis and phenotypic plasticity that has attracted the attention for epigenetics experiments in this model organism.

Our working definition of epigenetics for the purpose of this chapter will follow the classical Waddlngton (1942) definition, whereby it encompasses heritable changes in gene expression without altering underlying DNA sequences and includes, e.g., DNA methylation (including 5-mC and 5-hmC), histone modifications (i.e. H3K4me and H3K27me) and changes in the positioning of nucleosomes. The symphony of these marks ultimately regulates the 3D organization (compaction) of the chromatin, and thus accessibility of proteins (including transcription factors and polymerase) to DNA. Sometimes, the definition of epigenetics has been expanded to include noncoding RNAs (ncRNAs) such as: miRNA, siRNA, piRNA and long noncoding RNA (lncRNA) and any modification of mRNA (i.e. methylation of adenosines) or other methylations that involve protein methylation of arginine, excluding histones. For simplicity and due to a lack of knowledge on our favourite pet, we do not discuss any of these modifications in this chapter. We also prefer the use of the term *environmental epigenetics* (EE), the influence of the environment on epigenetic mechanisms and marks that translate to fitness-related phenomena and covers evolutionary aspects of variance in epialleles.

It is widely regarded that DNA methylation is susceptible to external environmental cues and that these changes may result in phenotypic alterations, often dysregulated by age, stress, diet or exposure to different toxins (Burris and Baccarelli, 2014; Pacchierotti and Spano, 2015; Ruiz-Hernandez et al., 2015). Although the epigenetic state can be modified by these mentioned factors, it is in development and differentiation that epigenetics truly plays crucial roles by placing epigenetic marks in the correct location, at the right time. Any perturbation in critical developmental windows will have immense consequences on fitness. For *Daphnia*, the scientific community has yet to satisfactorily describe the position of epigenetic marks during development (time-course experiments or even time-specific snapshots are lacking). Indeed, we are missing answers to basic questions, such as

I. How are epigenetic marks maintained throughout all life stages. Is there any accumulation of epimutations during development and aging and if so is that a random process?

II. Which histone marks are the most important during development and how do they interact with methylation marks?

III. Are epigenetic marks during sexual reproduction reset?

IV. Are epigenetic epimutations/epialleles accumulated during asexual reproduction and passed on to subsequent generations?

V. Finally, can the differences in methylation patterns in the genome of resting eggs potentially contribute to rapid evolution in the face of changed environments, allowing natural selection to act on these random epimutations/metastable epialleles?

What we do know today is that (i) the major histone methylation marks are predicted from antibody studies (Harris et al., 2012); (ii) based on genomic data, methylation accumulates during aging; and (iii) *Daphnia magna* sperm cells carry methylation marks (Gomez et al., 2016).

2. *DAPHNIA* AS A MODEL ORGANISM

It is perhaps surprising to some that the use of *Daphnia* as an effective animal model for epigenetic research is in its infancy. Gene–environment interactions are integral to multiple facets of its biology, and a wide range of its life history, genomic and ecological traits have combined to create what should be a valuable research asset. First, *Daphnia* has a short generation time (similar to *Drosophila*) with a life cycle (for most species) that includes both an asexual and a sexual component (Fig. 1), and the transition between the two can be manipulated easily in the laboratory. For much of its life, asexual reproduction dominates and is accomplished via abortive meiosis (Hiruta et al., 2010), resulting in clutches of eggs that develop immediately. As ambient conditions become less favourable, environmental signals generate a cascade of reproductive events that lead to the production of morphologically distinct males, each possessing an identical DNA sequence to their mother. Furthermore, females appear to switch reproductive pathways from production of asexual diploid egg clutches to the formation of one or two haploid eggs via typical meiosis, which when fertilized, generate

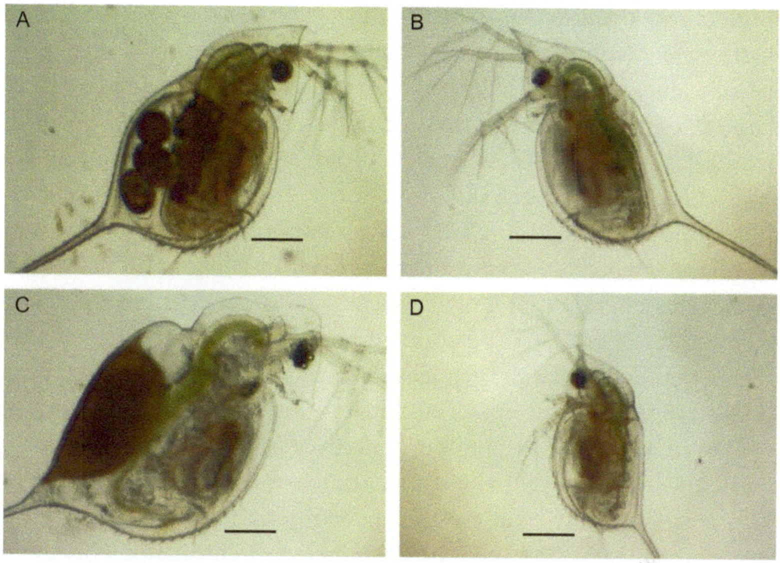

Fig. 1 Various life stages of *Daphnia lumholtzi*, a laboratory strain originally collected from Bull Shoals (Arkansas, USA). (A) Parthenogenetic female, (B) juvenile female, (C) ephippial female (carrying resting eggs inside darkly tinted case), (D) male. Scale bars represent 250 μm.

diapausing eggs. As a consequence of this polyphenism, it should be possible to compare epigenetic marks in genetically identical males and females as well as their offspring.

Second, environmental links to the genome are also responsible for multiple polyphenisms as part of observed inducible defences. Neckteeth, modest to extravagant helmets and crests, and elongated tailspines and fornices are only a partial list of the morphological reconfigurations that are created without DNA sequence changes. Females that are alerted to the presence of their specific predator type (some are aquatic insects and some are fish) via unidentified kairomones initiate the development of their particular defensive structure. Neckteeth, which are typically observed in certain species' youngest instars begin development during embryogenesis. The robust helmets produced by *Daphnia longicephala*, on the other hand, have not been observed prior to the fourth instar (Weiss et al., 2015b). Furthermore, Agrawal et al. (1999) has shown that the maternal environment influences the extent of the defence for two or more generations. Others have found that even among species that do not form obvious defensive structures, size at maturity and brood size can be affected by the maternal environment

(Mikulski and Pijanowska, 2010; Walsh et al., 2014). The precise genetic pathways involved in creating defensive polyphenisms have not yet been fully elucidated, but links between neuronal signals and the induction of some defensive structures have been uncovered (Barry, 2002; Weiss et al., 2015a,b). For example, Weiss et al. (2015a), using dopaminergic and cholinergic stimulants, induced the formation of helmets in *D. longicephala*. In addition, they identified localized concentrations of dopamine in specific cephalic cells, which appear to match previously identified polyploid cells (Beaton and Hebert, 1997). Polyploid cells naturally occur throughout most somatic tissues of *Daphnia* (Beaton and Hebert, 1989). Beaton and Hebert (1997) hypothesized that the cephalic epidermal polyploid cells (and perhaps even those in the thoracic epidermis) function as control centres to influence cell division rates in surrounding diploid cells and in so doing, control growth. One question yet to be addressed, that is of interest to a wider readership including those focused on cancer research, is whether epigenetic marks (both in term of methylation pattern and chromatin organization) differ between diploid and polyploid cells and how they translate to gene expression.

In addition to growth, reproductive success and overall health, nutritional status is likely linked to both male production and helmet formation. Kusari et al. (2017) observed a relationship between vitamin B_{12} availability or methotrexate (MTX) exposure and levels of DNA methylation. Since MTX inhibits a pathway required for DNA methylation, the reduced fitness that was observed in *Daphnia* exposed to this chemical was presumed to be linked to global methylation loss. Evidence of nutritional methylation mechanisms in a nonmammalian species reinforces the efficacy of using invertebrate species, particularly *Daphnia*, to examine molecular mechanisms related to environmental and transgenerational epigenetic inheritance. These findings also suggest that one-carbon metabolism may be limited by vitamins in the natural environment; we propose that future experiments are performed to test this hypothesis.

Third, *Daphnia* has been employed as a model system in toxicological studies for more than 20 years. As an environmentally sensitive aquatic invertebrate (effectively the canary in the mine), this organism (typically it has been *D. magna*) has been the celebrated focus for much of our elementary and fundamental information on a host of inorganic and organic contaminants. Kim et al.'s (2015) recent review identified the significant contribution that the Vandegehuchte lab has played over the last decade in establishing connections between ecotoxicology and epigenetics (see, for example, Vandegehuchte et al., 2009a, 2010; Vandegehuchte and Janssen, 2011). Yet, issues of accessibility to a comprehensively annotated genome have impeded the overall progress in the epigenetic field (questions of the quality of the reference genome that reached the public in 2010 have recently been raised as explained later). As independent projects take off, we expect faster progress in the near future.

From a genomics perspective, chromosome number across the genus is conserved, with diploid counts of 20 or 24 (Beaton and Hebert, 1994), yet despite these moderate numbers, the genome itself is rather modest and only about two times larger than *Drosophila* (Beaton and Peters, 2013). The genetic sequence and annotations (reference genomes) of two lineages of *Daphnia pulex* and one of *D. magna* have been completed (Colbourne et al., 2011; Orsini et al., 2016; Ye et al., 2017), although annotations of a great many of the genes are still lacking and the number of genes itself remains disputed. The difference in gene number seems to be caused by substantial copy number variations found between clones of the same species, as well as different strategies (conservativeness criteria) for gene predictions used in gene assemblies. Epigenetic studies depend on good annotations and well-described relative positions of epigenetic features of the genes that they influence, so we recommend using the last genome assembly for *D. pulex* (Ye et al., 2017). In addition, many molecular tools have been employed to address an array of questions. As will be made clear later and in subsequent sections of this chapter, surveys of specific genomic regions in order to identify changes in epigenetic marks has already begun, including examinations of DNA methylation (Vandegehuchte et al., 2009a,b, 2010; Vandegehuchte and Janssen, 2011) and histone modifications (Robichaud et al., 2012).

Chromatin states are representative of gene activity (euchromatin) and gene repression (heterochromatin) (Allis and Jenuwein, 2016). An interesting finding coming from the first, and perhaps only, histone study was that the localization of euchromatin and heterochromatin in the nucleus of embryonic cells does not follow a conventional pattern. In mammals, heterochromatin is typically concentrated around the margins of the nucleus. But in *Daphnia*, it is the stained euchromatin, based on the marker H3 dimethylated at lysine 4 (H3K4me2), that appeared to be the material clustered around the nuclear periphery (Robichaud et al., 2012). In mammals, H3K4me2 defines transcription factor binding regions (Wang et al., 2014). Why this atypical chromatin organization should occur or the consequence of this difference has never been investigated. We propose that this altered configuration might be of great consequence to (epi)mutation rates and a potential source of increased copy number variation in *Daphnia*. Any connection to the widespread presence of endopolyploidy in these animals (Beaton and Hebert, 1989) to both of these phenomena should be addressed in future studies.

Comparative methylomics recently revealed an association between one histone modification and methylation level; gene-body H3K36me3 in *Drosophila melanogaster* seems to be predictive of DNA methylation and CpG landscapes in other invertebrates (Nanty et al., 2011). Another comparative study done across social insects has shown that DNA methylation patterns are mirrored by histone modifications (Hunt et al., 2013). We await confirmation as to whether these predictions hold for *Daphnia*. Using flow cytometry, Jalal et al. (2014) has provided some indirect evidence based on generational changes in DNA compaction in response to food quality and temperature. Although not investigated by the authors, such compaction driven by environmental conditions could be mediated by global histone modifications, but ChIP-seq (chromatin immunoprecipitation followed by next-generation sequencing) and RNA-seq (mRNA next-generation sequencing) experiments are necessary to confirm this prediction regarding such striking DNA organizational changes. Expression of the following histones appeared to be differentially regulated under hypoxic conditions: H2B, H3 and H4 and histone deacetylase 4 (HDAC4) (Lai et al., 2016). HDAC4 regulates HIF1 protein acetylation and stability and leads to increased haemoglobin expression in *Daphnia* limited by oxygen (Table 1). This study suggested that generational

Table 1 Differentially Regulated Histone Clusters in *D. pulex* Under Hypoxic Conditions

Gene ID	Log2 Fold Change (Normoxia/Hypoxia)	*P*-value	Gene Description
DAPPUDRAFT_43976	−3.74	<0.05	Histone H2B
DAPPUDRAFT_442965	−1.31		Histone deacetylase HDAC4 protein-like
DAPPUDRAFT_43920	+2.54	<0.01	Histone H4
DAPPUDRAFT_304645	+2.00	<0.01	Histone H3

Modified from Lai, K.P., Li, J.W., Chan, C.Y.S., Chan, T.F., Yuen, K.W.Y., Chiu, J.M.Y., 2016. Transcriptomic alterations in *Daphnia magna* embryos from mothers exposed to hypoxia. *Aquat. Toxicol.* 177, 454–463.

effects of hypoxia in *Daphnia* might be mediated through epigenetic regulations of massive histone modifications and since the ten–eleven translocation enzyme (TET, Dapma7bEVm018501t1, DAPPUDRAFT_200491) was not altered, hypoxia probably did not act via DNA hypermethylation to reduce TET activity (and so the ratio between hydroxymethylation and methylation), as has been recently shown for hypoxic cancers (Thienpont et al., 2016). Yet, these studies only point to the importance of histones in the regulation of haemoglobin expression without elaborating on the exact mechanisms, such as modifications to N- and C-terminal tails (i.e. acetylation, methylation, ubiquitination and phosphorylation). These hypotheses can be studied typically using ChIP-seq experiments.

3. EPIGENETIC MACHINERY IN *DAPHNIA*: EVIDENCE AND PREDICTIONS

It is clear that *Daphnia* has at least as much complicated epigenetic machinery as *Drosophila*, as is predicted from their genomic sequences. For instance, cytosine-5-methyltransferases (DNMT1, DNMT2 and DNMT3), which place de novo methylation marks or maintain marks during cell division, are present in *Daphnia* (Fig. 2) and potentially might be regulated by different stressors (Orsini et al., 2016) and during development. Using these published results, we gathered and displayed DNMT1 gene expression values for *D. magna* for the 12 stressors as genomic tracks on the *Daphnia* genome browser (Fig. 3). There is substantial variation seen in the relative expression across the stressors, suggesting deregulation of this important methylation enzyme.

The presence of ten–eleven translocation enzymes (TET1–2, Dap ma7bEVm018501t1, DAPPUDRAFT_200491) also suggests that active demethylation in *Daphnia* might exist and prediction of such enzymatic activity can be done based on Cys-rich and DSBH (double-stranded beta-helix) regions in *Daphnia* orthologs. Strepetkaitė et al. (2016) revealed that 5-hydroxymethylation in cytosines occurs. There is evidence that a gene with high homology to the sel-1 suppressor of the lin-12 gene correlates negatively to global methylation in animals exposed to the heavy metal zinc. This gene is known to alter the Notch signalling pathway and, therefore, Vandegehuchte et al. (2009b) proposed that this mechanism may be

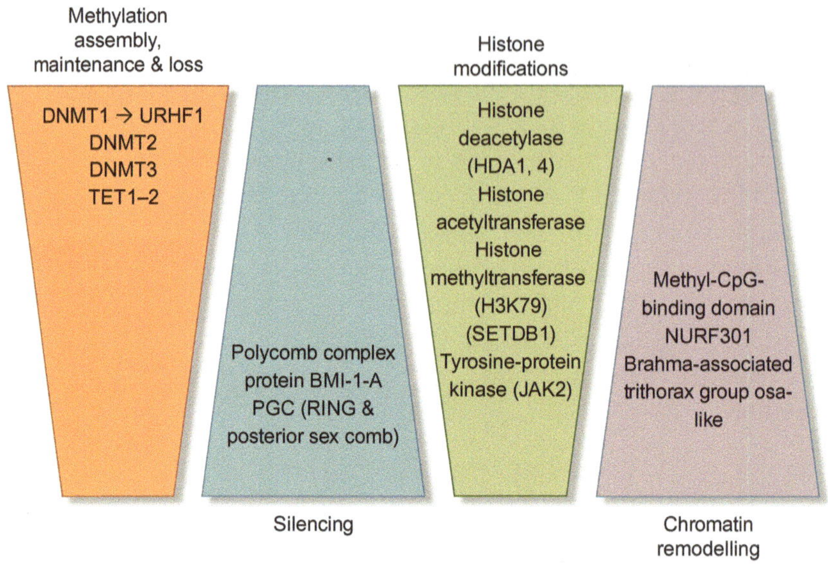

Fig. 2 Predicted epigenetic machinery present in the *Daphnia* genome based on presumed functions: methylation, gene silencing, histone modifications and chromatin remodelling. Presence of factors is based on sequence homology (UNIPROT, FLEABASE and orthoDB) or from the literature (Beh et al., 2012; Kim et al., 2015; Vandegehuchte et al., 2009a).

important for the hypermethylation of genes in *Daphnia*. The same authors observed the downregulation of a gene homologue to a Ser/Thr protein phosphatase PP1 with a protein from polycomb repressive complex (Fig. 2) thus, for the first time, showing the importance of histones from a transcriptomic perspective. The necessary changes in its expression profiles are frequently associated with variation in chromatin structure and this could be done via H3K4 or H3K27 methylations (open and closed chromatin marks). As reviewed previously, the conformation of chromatin is profoundly influenced by the posttranslational modification of the histone proteins, the incorporation of histone variants, the activity of nucleosome remodelling factors and the association of nonhistone chromatin proteins (Imhof, 2006).

Also, histone-modifying enzymes play a major causal role in setting up a particular chromatin structure and they are present in *Daphnia* as indicated by the UniProt knowledge database (2017). These enzymes are often categorized in three groups: writers, readers and erasers. They lay down epigenetic marks

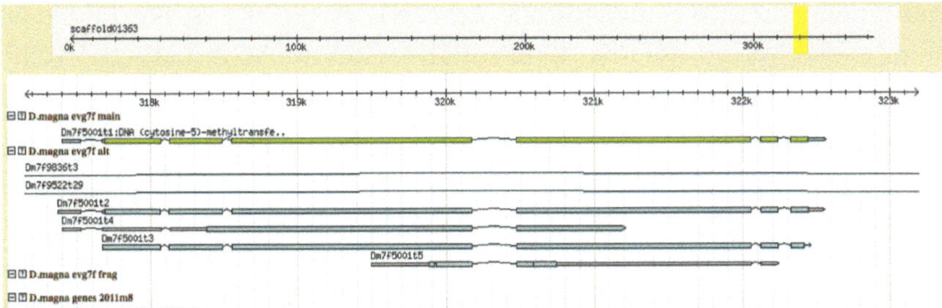

Fig. 3 Cytosine-5-methyltransferase 1 (DNMT1) expression in *Daphnia magna* under different stressor conditions. The RNA-seq expression values from 12 different stressors are shown in different colours (*green* low, *red* high). We gathered the values that were published in Orsini et al. (2016) and displayed them as genomic tracks using the *Daphnia* genome browser. Notice that the expression values of this epigenetically important enzyme differs according to the stressors.

on amino acid residues on histone tails. Among *writers* we found histone acetyltransferases (HATs, UniProtKB—A0A164VDA8{A0A164VDA8_9CRUS}), histone methylytransferases (HMTs, UniProtKB—E9GAA53 {E9GA53_DAPPU}), protein methyltransferases (PRMTs, UniProtKB—E9GJW0{E9GjW0_DAPPU} and kinases. The epigenetic *readers* bind to these epigenetic marks and are often bromodomains (ZFAM10, Seetharam et al., 2010), chromodomains (UniProtKB—A0A0P6IQJ9{A0A0P6IQJ9_9CRUS}) and Tudor domains (UniProtKB—A0A0P6D0I6{A0A0P6D0I6_9CRUS}. Epigenetic *erasers* catalyse the removal of epigenetic marks and among these we find histone deacetylases (HDACs, like HDAC4), lysine demethylases (KDMs, UniProtKB—A0A165ACJ9 (A0A165ACJ9_9CRUS)) and phosphatases.

A very important group of proteins that shape the epigenome via histone marks is the Polycomb–Trithorax complex (PcG–trxG, Fig. 2). In *Drosophila*, the Trithorax-group (trxG) and Polycomb-group (PcG) proteins act antagonistically and interact with chromosomal elements. TrxG proteins maintain the active state of gene expression, while the PcG proteins counteract this activation with a repressive function that is stable over many cell generations and can only be overcome by germline differentiation processes. In mammals, polycomb gene complexes, or PcG silencing, consists of at least three kinds of multiprotein complexes: polycomb repressive complex 1 (PRC1), PRC2 and PhoRC (Pho-repressive complex) that work together. PcG proteins are evolutionarily conserved and exist in at least two separate protein complexes; the PcG repressive complex 1 (PRC1) and the PcG repressive complex 2–4 (PRC2/3/4). In *Daphnia* PRC1 is present, but we did not locate the other complexes. We performed a search for the both Polycomb and Thrithorax components in *Daphnia* using OrthoDB database (Zdobnov et al., 2016) and UniProt. The UniProt search for 'Daphnia' and the term 'Trithorax' resulted in 2 hits (Trithorax group osa-like protein, an AT-rich interactive domain-containing protein 1A (Trithorax group osa-like protein)) and with the term 'Polycomb' resulted 266 hits for enhancer of polycomb-like protein (including, DAPPUDRAFT_214661, or polycomb protein ASX). These are, however, unreviewed, have very low annotation scores, and the proteins are inferred from sequence homology, so further confirmation of their presence and characterization is required as their presence can have a huge implication as to how marks are being transferred between cell divisions as well as in germ cells. Among others proteins that belong to the PcG–trxG complex, which has been shown to be very important in term of epigenetic memory, in orthoDB we found histone-lysine *N*-methyltransferase, E(z) (homologue of Ezh1), enhancer of zeta and sexcombs. DAPPUDRAFT_304723 seems to be an ortholog of the esc gene. In *Drosophila* E(z)/esc complex, the catalytic subunit of this complex methylates 'Lys-9' and 'Lys-27' of histone H3, leading to transcriptional repression of the affected target genes in *Drosophila* (UniProt). We also see variation in expression across 12 different stressors for E(z), suggesting deregulation of this important epigenetic-related enzyme (Fig. 4).

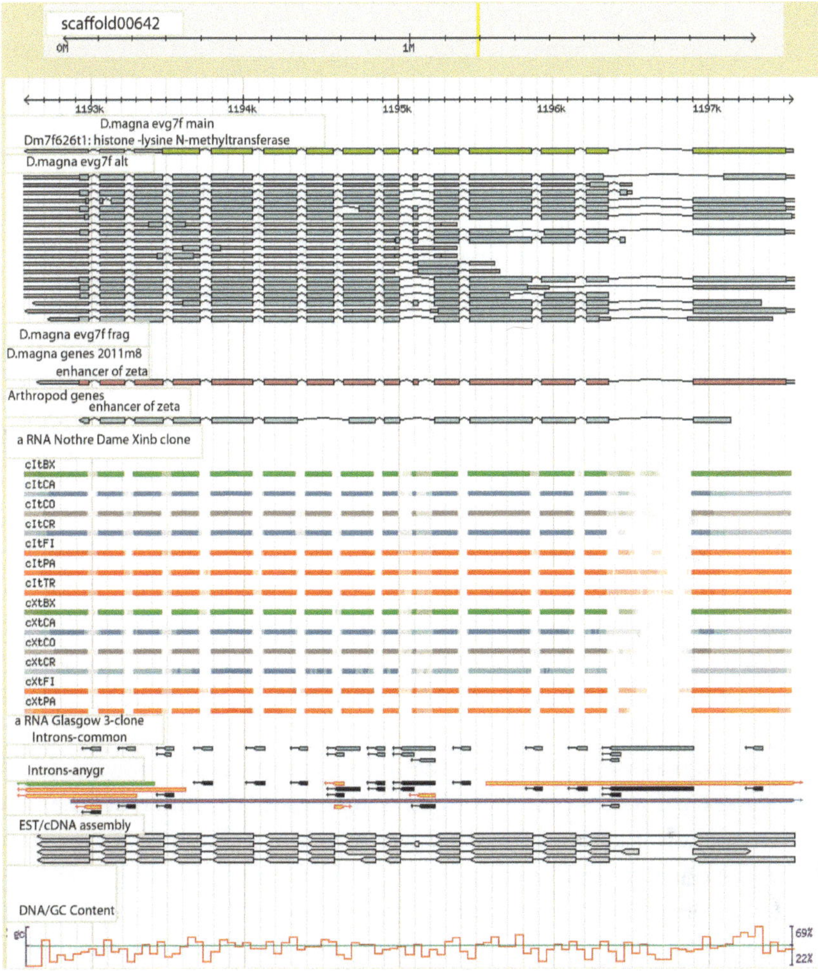

Fig. 4 E(z) expression in *Daphnia magna* as a response to different stressors. The RNA-seq expression values from 12 different stressors are shown in different colours (*green* low, *red* high). We gathered the values that were published in Orsini et al. (2016) and displayed them as genomic tracks using the *Daphnia* genome browser. Notice that the expression values of this epigenetically important enzyme differs according to the stressors.

4. INSIGHTS FROM EPIGENETIC STUDIES

DNA methylation is an epigenetic modification often involved with gene regulation, and it has been extensively studied in mammals (Li and Zhang, 2014) and only recently in invertebrates (Roberts and Gavery,

2012), also reviewed in this book. There is an order of magnitude difference in methylation levels between vertebrate and invertebrates and the evolutionary reasons for this difference remain unknown. Much of our understanding of invertebrate epigenetics comes from methylation studies on a handful of species (a nematode, a mollusc, insects (bees, ants, *Drosophila*) and an echinoderm) revealing disparate levels within the meagre dataset (Roberts and Gavery, 2012).

DNA methyltransferases (DNMTs) use methyl groups that are supplied by one-carbon metabolism (Anderson et al., 2012), which is comprised of the folate and methionine metabolism pathway (Ducker and Rabinowitz, 2017). It is this pathway that generates the methyl donor, *S*-adenosylmethionine (SAM) (Mentch and Locasale, 2016), and its presence is nutrient- and vitamin-dependent (Ducker and Rabinowitz, 2017). Mammalian studies have shown that poor diets can alter methylation patterns (Dolinoy et al., 2007; Dominguez-Salas et al., 2014; Lim and Song, 2012; Sandovici et al., 2011; Soubry et al., 2015; Wolff et al., 1998). Parallel invertebrate experiments are lacking, although Keating (1985) did establish that reproductive performance in *Daphnia* is linked to availability of vitamin B_{12}. Extending this early work, Kusari et al. (2017) used a low resolution comet assay to show that B_{12} supplements nearly doubled global methylation levels.

Most regulatory regions in vertebrates span promoter regions (CpG islands and CpG shores), but for invertebrates it seems that this is not the case. In the past few years, invertebrate genome-wide differential methylation studies have identified the primary methylation targets as the gene bodies in a CpG context, with promoter, intergenic regions and non-CpG islands only weakly affected (Roberts and Gavery, 2012). Published studies suggest that methylation may function in alternative splicing, enhancing expression of the genes, or is somehow related to the gene family size (Roberts and Gavery, 2012). In *Daphnia*, it has been predicted that the genome-wide methylation status may change depending on factors like: presence of predators, prey, or toxicants, as well as age, dietary status or sex (Asselman et al., 2015). We are beginning to unravel the effects of toxicants and obviously, it is the ecotoxicological research community that are currently a driving force for much of the epigenetics progress in this genus.

Daphnia was recognized as an important species for epigenetics research a few years back (Harris et al., 2012; Vandegehuchte et al., 2009a), but despite this, research using this organism has only recently gained momentum and is far behind the rapidly expanding literature for other nonmammalian model species like *Arabidopsis*, *Drosophila* or *Apis*. Only the last few years has seen advances in characterizing the *Daphnia* epigenome, although no research has connected DNA methylation status to ecological function directly (Asselman et al., 2016, 2017). *Daphnia* methylation studies started with surveys of global methylation content, which varies depending on the stimuli used, species or even strain (Asselman et al., 2015), and generally suggests a low global methylation percentage. This has been achieved using HPLC studies with observed methylation levels between 0.49% and 0.52% under control conditions, and fivefold difference across genotypes (0.21%–1.02%) exposed to predation cues (Asselman et al., 2015). A very important discovery from this work is that authors observed a significant interaction between a genotype and environment effects, suggesting that small differences in methylation level might have a high impact on the outcome.

Most recently, an analysis in two species of *Daphnia* revealed both highly methylated and unmethylated gene bodies, with the methylation level appearing to be negatively associated with gene family size (Asselman et al., 2016). The most comprehensive epigenetic study appeared only recently (Asselman et al., 2017) and focused on genome-wide methylation patterns in *D. magna* exposed to the toxic cyanobacterium, *Microcystis aeruginosa*. Here, differential gene body methylation (specifically exonic methylation) occurred in *Daphnia* exposed to the algae (Asselman et al., 2017). Methylation of gene bodies has been described for other invertebrates previously (as also discussed elsewhere in this book). Previously, it was observed that significant changes in global methylation occurs when animals are exposed to a cyanobacterium (the same one used by Asselman et al., 2017) compared to a green algae. For all genes and both treatments, promoter methylation decreased as the distance to the first exon decreased (Fig. 5). Furthermore, Asselman et al. (2017) observed an increase in methylation along the length of the first exon, but decreases along subsequent exons for all gene models. Finally, introns showed lower methylation levels and seemed to serve as break points. Also, Strepetkaitė et al. (2016) have shown that 5-hydroxymethylcytosine (5-hmC) in exons is twice as frequent as in introns and their functional analysis revealed that 5-hmC abundance is associated with genes that are involved in the adenylate cyclase-activating G-protein-coupled receptor signalling pathway, moulting cycles, morphogenesis and cell fate determination. Genes that lack 5-hmC tend to be involved in the regulation of the transforming growth factor beta receptor signalling pathway and in many mRNA-related processes.

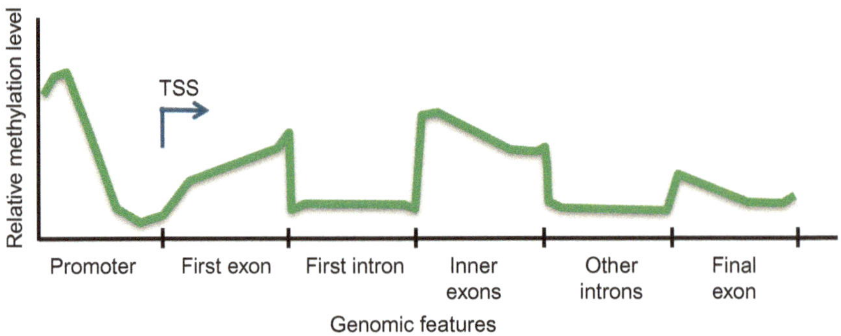

Fig. 5 Relative methylation profile across genomic features for *Daphnia*. Promoter regions were defined as putative promoters regions, 1500 basepairs upstream of the first exon. *Redrawn and modified from Asselman, J., De Coninck, D.I., Beert, E., Janssen, C.R., Orsini, L., Pfrender, M.E., Decaestecker, E., De Schamphelaere, K.A., 2017. Bisulfite sequencing with* Daphnia *highlights a role for epigenetics in regulating stress response to microcystis through preferential differential methylation of serine and threonine amino acids.* Environ. Sci. Technol. *51, 924–931.*

Changes in exon–intron structure during vertebrate evolution is known to affect the splicing pattern of exons. However, to confirm the presence of this relationship in *Daphnia*, we need to verify that the correlation of methylation with transcriptomic data is strong and changes when the environment changes, if adaptive.

Although Asselman's epigenomic study has a several weaknesses, it offers a good introduction for epigenetic discoveries in *Daphnia*. First, their study suffers from low coverage that is obviously very important when the ratio between C:T is used to define methylation level per CG site. Here, a coverage of 5× was the cutoff, whereas often for human studies at least 10× coverage is required to properly calculate the state of methylation for heterogonous samples (different tissue and animals), which is not the case in this study. This is a serious weakness for the Asselman et al. (2017) study and future experiments should avoid this limitation (for only one difference in five sequencing reads at the position of a methylation can result in an error of 20%). With the frequent single-nucleotide polymorphisms (SNPs) that are being discovered in the *Daphnia* genome, this method does not allow us to distinguish SNPs from true methylated cytosines.

The serious problem in *Daphnia* studies, however, remains the poorly annotated genome estimated as only ~150 Mb genome (Ye et al., 2017) depending on the species. As a result, Asselman et al. (2017) seem to have doubts about the genome model, and therefore, the full interpretation of these results must be guarded. Obviously, development of curated genomes and their annotations is happening (recently it has been discovered that as much as 7000 genes were false positives in *D. pulex*) so we might expect this to affect any results (Ye et al., 2017). However, the work of Asselman et al. (2017) is a nice contribution to our understanding of the epigenome of *D. magna*. The authors studied differential methylation in 21,870 gene models; for 12,201 genes and 13,509 putative promotor regions and no methylated reads were identified in any of their three biological replicates for both treatments. Differential methylation was observed between *Microcystis*-stressed and control treatments in 876 genes and 247 putative promoter regions.

5. TECHNOLOGY-DRIVEN RESEARCH: METHODS AND BIOINFORMATICS

Modern technology has advanced both genetic and epigenetic studies in crustaceans. With the promise that comes with the use of clonal organisms for studies, the annotation of the *Daphnia* genomic regions should have reached a more advanced state and should be in much better shape, than it is today. Despite this, our research community has begun to link epigenetic changes with animal phenotypes, being aware of possible biases and changes expected with future reanalysis. Our definition of epigenetics contains three major categories of the epigenetic mechanisms: miRNA, methylation marks and histone marks. Much of the progress over the last few years has mainly focused on descriptive miRNA and methylation.

There exists a variety of methods available to access methylation status of genomic regions. The gold standard for the methylation analysis is whole-genome bisulphite sequencing (WGBS); extracted DNA is exposed to a bisulphite treatment, which converts nonmethylated C sites to U, resulting in a T read at sequencing, while methylated C sites remain unconverted. Examination of a reference genome allows one to assess the status of methylation for particular locations. Asselman et al. (2017) employed this method, and as the genome of *Daphnia* is quite small, likely this may become the preferred method in future.

With difficulties related to reference genomes and economical issues associated with the comparison of genome-wide methylation patterns across several samples, an alternate methodology is epiGBS, a reference-free reduced representation bisulphite sequencing (van Gurp et al., 2016). This technique is cost-effective, as it focuses only on the regions that are cut by

MspI (or alternatively other restriction enzymes). In this case, de novo reconstruction, mapping and variant calling brings the possibility of detecting SNPs and methylation variation. Employing this method with *D. magna*, CHG methylation was absent and the frequency of CG methylation was very low (van Gurp et al., 2016). *Daphnia*-related proteins deposited at NCBI were used here as a reference.

Yet another method is EpiRADseq, a reducing representation bisulphite sequencing, which uses methylation-sensitive restriction enzymes (such as HpaII). This method identifies changes in the genomic methylation state of cytosines by sampling loci that are not methylated within a sample. Using *D. ambigua*, Schield et al. (2016) found shifts in the methylation state of loci that were sampled across two generations in response to exposure to fish predator cues. Thus, epiGBS and EpiRADseq offer alternative tools to the existing methods and provide the means to complete comprehensive studies on differential DNA methylation levels using many samples in a cost-effective way.

Dobrovic (2016) previously gave a comprehensive overview of methods for analysis of DNA methylation. Since DNA methylation information is lost once the DNA is amplified during PCR, the information must be preserved in a different way for sequencing. 5-methylcytosine, but not cytosine, is resistant to deamination during sodium bisulphite treatment. As such, instances of methylation can be determined by comparison to the original sequence. This can be made on a global scale (using WGBS—whole bisulphate genome bisulphate sequencing or RRBS—reduced representation bisulphate sequencing) or locally in specific loci using PCR-based methods for the detection of DNA methylation via a so-called methylation-specific PCR (MSP) which comes with a different flavour for this method. MSP is a rapid method to screen for methylation in regions that are highly methylated. However, it is prone to false positives and therefore respective controls always need to be implemented. Quantitative MSP can also be performed using real-time PCR for methylation analysis with bisulphite-treated DNA and monitored by a TaqMan assay with reporting probes containing 5′ fluorescence reporters and 3′ quenchers that bind inside the amplified region (for instance MethylLight probes). DNA melting analysis is another, more labour-intensive method for methylation estimation that uses the general property that methylated DNA has a higher melting temperature compared to unmethylated DNA. Melting methodologies have become more widely adopted since the introduction of high-resolution melting-capable machines. Methylation-sensitive high-resolution melting is now widely used but the downside of this method is that it does not gives resolution down to individual C sites.

A very attractive, alternative for epigenetic analysis uses Nanopore sequencing where huge progress in base calling has been made recently allowing detection of the epigenetic state of Cs passing through nanopores (Schatz, 2017). The challenging part of this method, however, will be the relatively large amount of DNA required for this type of sample preparation for further sequencing, unless new ways of preparing the library can be implemented. This type of sequencing would help immensely during fieldwork.

We predict that the methylation-sensitive assays will take off in the next 5 years, as the genome annotation improves, allowing the design of probes targeting sites in genome with methylation differences. These assays will be used in future epigenetic testing with potentially beneficial environmental outcomes. For instance, male formation disturbance is an attractive candidate for epigenetic tests of new drugs with potential endocrine disruption requiring extremely high sensitivity and specificity.

However, both WGBS and RRBS in different modes (see epiGBS) will be instrumental as they are gold standards for epigenetic studies. We also predict that epiGBS, which samples DNA regions enriched in methylation and recognized by methylation-sensitive restriction enzymes while at the same time preserving information about SNPs, will be a powerful tool for use in population studies. Here, the limitation again will be in the required amount of material and the heterogeneity of the material, rather than in sequencing costs. Yet, substantial investments from agencies will be needed before the research community can exploit these techniques to collect the baseline data, allowing the switch to locus specific approaches.

6. *QUO VADIS?*

Recently, many labs, using different models, focused on elucidating the role of epigenetic processes in ecology and evolution following the ideas of evolutionary biologists Jablonka and Lamb (1999). Jablonka and Lamb predicted, a long time ago, that adaptive changes on a different level than genetics might be passed to subsequent generations, speeding up the rate of evolution. They pointed out that, in certain environments, the Darwinian predictions can be inadequate since mutations and selection may not be sufficient to explain rapid evolution. It took a while, however, for the field to understand that her prediction had little to do with the traditional

Lamarckian postulate, as well as for technology to catch up with their bright idea. In contrast to plants, we are just beginning to collect evidence of epigenetic mechanisms in metazoans and to link it to intragenerational and transgenerational effects. Indeed, it is becoming increasingly evident that changes in epigenetic profiles could allow for rapid, heritable adaptations to environmental cues without the need for the more slowly evolving changes in DNA sequences (Darwinian mutation accumulation and selection concept) and epigenetic diversity in the absence of genetic diversity enables invasive species to successfully adapt to new environments (Deakin et al., 2014) for instance via the concept of 'epimutations'.

Confirming the intragenerational and transgenerational inheritance of DNA methylation is more plausible in *Daphnia* due to the possibility for asexual reproduction and because in sexual reproduction gametes are derived from almost fully matured tissue. In asexual lines of *Daphnia*, we predict that epigenetics might be a driving force for local adaptation and clonal invasiveness. Transmission of DNA methylation from parent(s) to offspring may also be more adaptive in asexual animals than in those which reproduce sexually, since gametes and resting eggs are derived from tissue that has been subjected to the environmental stress that occurred during most of the life history of the parental generation.

Ecologists and evolutionary biologists are likely to be most interested in intragenerational and transgenerational epigenetic changes that shape fitness-related phenotypes. Often, research in ecological systems has difficulty identifying the control (Verhoeven et al., 2016), thus a survey of natural epigenetic variation across wild populations is highly desirable for quantitative studies in epigenetics. In the case of *Daphnia*, we have an exceptional DNA archive, deposited in lake sediments as dormant eggs. Akin to the achievement by Frisch et al. (2014), we now need to combine knowledge from both fields: resurrection biology and epigenetics and link it to changes in selection pressure (temperature shift, trophic change, predation, etc.). The dormant egg archive will be a significant resource to further insights into evolution, with current work in progress that uses whole genome amplification of dormant eggs dating back several centuries from a lake in Greenland (Frisch and Wojewodzic, in preparation). Especially exiting will be Frisch's contribution to the epigenetic signatures from whole genome amplification of dormant eggs in epigenetic mode (WGBS) of asexual clonal association.

We predict that environmentally induced phenotypic changes, observed in *Daphnia* spp., will be highly associated with underlying epigenetic alterations to the genome. To test this hypothesis, perturbation experiments, targeting epigenetic machinery directly: DNMTs enzymes and/or SAM (specific inhibitors, known antagonist of one-carbon metabolism) or indirectly (heavy metals affecting enzymes, diet manipulation, etc.) will be essential. Yet, these experiments preferentially should be accompanied by RNA-seq studies as well as physiological data such as clutch size or fitness.

This initiative has begun through the European Union sponsored dGHOST project (project no. 629892) in Birmingham, UK. One way of studying male formation and sex determination is to use two cancer drugs that block the activity of DNMTs and alter sex ratio in exposed animals (Wojewodzic and Colbourne, in preparation). Also this project has resulted in a protocol for ChIP-seq that will speed up chromatin studies in *Daphnia* (Wojewodzic and Colbourne, in preparation). Further public outreach of this project can be found here: http://cordis.europa.eu/result/rcn/187658_en.html.

Hopefully, the *Daphnia* community is nearing the end of the descriptive analysis of epigenetics of this ecoresponsive model organism, giving us better a understanding of similarities and differences in epigenetic signatures between it and other organisms. In the near future, we hope to witness a number of experiments using a clustered, regularly interspaced, short palindromic repeat (CRISPR)–Cas system-based approach for targeting DNA demethylation (Xu et al., 2016). These experiments will become pivotal in demonstrating the regulation of gene expression in ecoresponsive processes. Once we know the potential candidates (loci) that cause a given phenotypic change, we can target these, conditionally, to observe their effects on gene expression. We expect that this will come from toxicological applications. The CRISPR technique has already been successfully used in other contexts for *Daphnia* in Osaka and Tokyo, with low efficiency but high specificity (Nakanishi et al., 2014), and new more efficient technologies will soon follow.

For the field of EE to progress, we need to actively use a systems biology approach and make the connection between RNA-seq experiments, phenotypes and epigenome itself. This can be achieved by studying gene pathways, network analysis and finally by using the CRISPR–Cas system in targeted analysis. In the near future, we will first design in silico experiments that will pinpoint the genomic regions that are crucial (the hub methylation features) to elucidate the interplay between epigenome, transcriptome and phenotype networks. Based on these predictions, targeted experiments will be performed. These will be the era of the 21st century experiments where ‘conditional mutants’ in ecoresponsive settings will be the focus. Such connections should be made for both abiotic and biotic disturbances and their interactions to give a better understanding of how these networks operate in nature and how evolution has shaped them.

The future of EE is bright. Changes in the epigenome seem to be the most sensitive marks to be discovered, when perturbations in the environment occur. Therefore the field, most probably, will bring new guidelines for risk assessment in exotoxicology so we can protect us and our environment from impact of unwanted substances around us. The field potentially may give us the most sensitive warning system that exists with both high specificity and sensitivity.

ACKNOWLEDGEMENTS

M.W. is grateful to the European Union from the Intra-European Fellowship (IEF) for Career Development, Independent Research Fellowship under Marie Skłodowska Curie (FP7-PEOPLE-2013-IEF) for the project, Ghosts in parthenogenetic daughters—epigenetic effects on clonal organisms can reveal the degree of phenotypic plasticity due to biotic cues (dGHOST, project no. 629892) and also to 'TOPP finansiering' from the Norwegian Research Council to M.W.

REFERENCES

Agrawal, A.A., Laforsch, C., Tollrian, R., 1999. Transgenerational induction of defences in animals and plants. Nature 401, 60–63.

Allis, C.D., Jenuwein, T., 2016. The molecular hallmarks of epigenetic control. Nat. Rev. Genet. 17, 487–500.

Anderson, O.S., Sant, K.E., Dolinoy, D.C., 2012. Nutrition and epigenetics: an interplay of dietary methyl donors, one-carbon metabolism and DNA methylation. J. Nutr. Biochem. 23, 853–859.

Asselman, J., De Coninck, D.I., Vandegehuchte, M.B., Jansen, M., Decaestecker, E., De Meester, L., De Schamphelaere, K.A., 2015. Global cytosine methylation in *Daphnia magna* depends on genotype, environment, and their interaction. Environ. Toxicol. Chem. 34, 1056–1061.

Asselman, J., De Coninck, D.I., Pfrender, M.E., De Schamphelaere, K.A., 2016. Gene body methylation patterns in *Daphnia* are associated with gene family size. Genome Biol. Evol. 8, 1185–1196.

Asselman, J., De Coninck, D.I., Beert, E., Janssen, C.R., Orsini, L., Pfrender, M.E., Decaestecker, E., De Schamphelaere, K.A., 2017. Bisulfite sequencing with *Daphnia* highlights a role for epigenetics in regulating stress response to microcystis through preferential differential methylation of serine and threonine amino acids. Environ. Sci. Technol. 51, 924–931.

Barry, M.J., 2002. Progress toward understanding the neurophysiological basis of predator-induced morphology in *Daphnia pulex*. Physiol. Biochem. Zool. 75, 179–186.

Beaton, M.J., Hebert, P.D., 1989. Miniature genomes and endopolyploidy in cladoceran crustaceans. Genome 32, 1048–1053.

Beaton, M.J., Hebert, P.D., 1994. Variation in chromosome numbers of *Daphnia* (Crustacea, Cladocera). Hereditas 120, 275–279.

Beaton, M.J., Hebert, P.D., 1997. The cellular basis of divergent head morphologies in *Daphnia*. Limnol. Oceanogr. 42, 346–356.

Beaton, M.J., Peters, C.M., 2013. A cytological perspective. In: Smirnov, N.S. (Ed.), The Physiology of Cladocera. Academic Press, Oxford.

Beh, L.Y., Colwell, L.J., Francis, N.J., 2012. A core subunit of Polycomb repressive complex 1 is broadly conserved in function but not primary sequence. Proc. Natl. Acad. Sci. U.S.A. 109, E1063–E1071.

Burris, H.H., Baccarelli, A.A., 2014. Environmental epigenetics: from novelty to scientific discipline. J. Appl. Toxicol. 34, 113–116.

Colbourne, J.K., Pfrender, M.E., Gilbert, D., Thomas, W.K., Tucker, A., Oakley, T.H., Tokishita, S., Aerts, A., Arnold, G.J., Basu, M.K., Bauer, D.J., Caceres, C.E., Carmel, L., Casola, C., Choi, J.H., Detter, J.C., Dong, Q., Dusheyko, S., Eads, B.D., Frohlich, T., Geiler-Samerotte, K.A., Gerlach, D., Hatcher, P., Jogdeo, S., Krijgsveld, J., Kriventseva, E.V., Kultz, D., Laforsch, C., Lindquist, E., Lopez, J., Manak, J.R., Muller, J., Pangilinan, J., Patwardhan, R.P., Pitluck, S., Pritham, E.J., Rechtsteiner, A., Rho, M., Rogozin, I.B., Sakarya, O., Salamov, A., Schaack, S., Shapiro, H., Shiga, Y., Skalitzky, C., Smith, Z., Souvorov, A., Sung, W., Tang, Z., Tsuchiya, D., Tu, H., Vos, H., Wang, M., Wolf, Y.I., Yamagata, H., Yamada, T., Ye, Y., Shaw, J.R., Andrews, J., Crease, T.J., Tang, H., Lucas, S.M., Robertson, H.M., Bork, P., Koonin, E.V., Zdobnov, E.M., Grigoriev, I.V., Lynch, M., Boore, J.L., 2011. The ecoresponsive genome of *Daphnia pulex*. Science 331, 555–561.

Deakin, J.E., Domaschenz, R., Lim, P.S., Ezaz, T., Rao, S., 2014. Comparative epigenomics: an emerging field with breakthrough potential to understand evolution of epigenetic regulation. AIMS Genet. 1, 34–54.

Dobrovic, A., 2016. Analysis of DNA methylation in clinical samples: methods and applications. In: Lakhani, S.R., Fox, S.B. (Eds.), Molecular Pathology in Cancer Research. Springer, New York, pp. 261–277.

Dolinoy, D.C., Huang, D., Jirtle, R.L., 2007. Maternal nutrient supplementation counteracts bisphenol A-induced DNA hypomethylation in early development. Proc. Natl. Acad. Sci. U.S.A. 104, 13056–13061.

Dominguez-Salas, P., Moore, S.E., Baker, M.S., Bergen, A.W., Cox, S.E., Dyer, R.A., Fulford, A.J., Guan, Y., Laritsky, E., Silver, M.J., Swan, G.E., Zeisel, S.H., Innis, S.M., Waterland, R.A., Prentice, A.M., Hennig, B.J., 2014. Maternal nutrition at conception modulates DNA methylation of human metastable epialleles. Nat. Commun. 5, 3746.

Ducker, G.S., Rabinowitz, J.D., 2017. One-carbon metabolism in health and disease. Cell Metab. 25, 27–42.

Frisch, D., Wojewodzic, M.W., Epigenetic memory of environmental stress imprinted in 400-year-old dormant asexual *Daphnia* lineages isolated from Arctic Lake Sediment (West Greenland), in preparation.

Frisch, D., Morton, P.K., Chowdhury, P.R., Culver, B.W., Colbourne, J.K., Weider, L.J., Jeyasingh, P.D., 2014. A millennial scale chronicle of evolutionary responses to cultural eutrophication in *Daphnia*. Ecol. Lett. 17, 360–368.

Gomez, R., Van Damme, K., Gosálvez, J., Morán, E.S., Colbourne, J.K., 2016. Male meiosis in Crustacea: synapsis, recombination, epigenetics and fertility in *Daphnia magna*. Chromosoma 125, 769–787.

Harris, K.D., Bartlett, N.J., Lloyd, V.K., 2012. *Daphnia* as an emerging epigenetic model organism. Genet. Res. Int. 2012, 147892.

Hiruta, C., Nishida, C., Tochinai, S., 2010. Abortive meiosis in the oogenesis of parthenogenetic *Daphnia pulex*. Chromosome Res. 18, 833–840.

Hunt, B., Glastad, K.M., Yi, S.V., Goodisman, M.A.D., 2013. Pattering and regulatory associations of DNA methylation are mirrored by histone modifications in insects. Genome Biol. Evol. 5, 591–598.

Imhof, A., 2006. Epigenetic regulators and histone modification. Brief. Funct. Genomic. Proteomic. 5, 222–227.

Jablonka, E., Lamb, M.J., 1999. Epigenetic Inheritance and Evolution. The Lamarckian Dimension. Oxford University Press, New York.

Jalal, M., Shala, N.K., Wojewodzic, M.W., Andersen, T., Hessen, D.O., 2014. Multigenerational genomic responses to dietary phosphorus and temperature in *Daphnia*. Genome 57, 439–448.
Keating, K.I., 1985. The influence of vitamin-B12 deficiency on the reproduction of *Daphnia pulex* Leydig (Cladocera). J. Crustac. Biol. 5, 130–136.
Kim, H.J., Koedrith, P., Seo, Y.R., 2015. Ecotoxicogenomic approaches for understanding molecular mechanisms of environmental chemical toxicity using aquatic invertebrate, *Daphnia* model organism. Int. J. Mol. Sci 16, 12261–12287. http://dx.doi.org/10.3390/ijms160612261.
Kusari, F., O'Doherty, A., Hodges, N., Wojewodzic, M., 2017. Impact of one-carbon metabolism manipulation on *Daphnia magna* fitness and genomic methylation. Ecol. Lett.(submitted 21 Feb).
Lai, K.P., Li, J.W., Chan, C.Y.S., Chan, T.F., Yuen, K.W.Y., Chiu, J.M.Y., 2016. Transcriptomic alterations in *Daphnia magna* embryos from mothers exposed to hypoxia. Aquat. Toxicol. 177, 454–463.
Li, E., Zhang, Y., 2014. DNA methylation in mammals. Cold Spring Harb. Perspect. Biol. 6, a019133.
Lim, U., Song, M.A., 2012. Dietary and lifestyle factors of DNA methylation. Methods Mol. Biol. 863, 359–376.
Mentch, S.J., Locasale, J.W., 2016. One-carbon metabolism and epigenetics: understanding the specificity. Ann. N. Y. Acad. Sci. 1363, 91–98.
Mikulski, A., Pijanowska, J., 2010. When and how can *Daphnia* prepare their offspring for the threat of predation? Hydrobiologia 643, 21–26.
Nakanishi, T., Kato, Y., Matsuura, T., Watanabe, H., 2014. CRISPR/Cas-mediated targeted mutagenesis in *Daphnia magna*. PLoS One, 9, e98363.
Nanty, L., Carbajosa, G., Heap, G.A., Ratnieks, F., van Heel, D.A., Down, T.A., Rakyan, V.K., 2011. Comparative methylomics reveals gene-body H3K36me3 in *Drosophila* predicts DNA methylation and CpG landscapes in other invertebrates. Genome Res. 21, 1841–1850.
Orsini, L., Gilbert, D., Podicheti, R., Jansen, M., Brown, J.B., Solari, O.S., Spaier, K.I., Colbourne, J.K., Rush, D., Decaestecker, E., Asselman, J., De Schamphelaere, K.A.C., Ebert, D., Haag, C.R., Kvist, J., Laforsch, C., Petrusek, A., Beckerman, A.P., Little, T.J., Chaturvedi, A., Pfrender, M.E., De Meester, L., Frilander, M.J., 2016. *Daphnia magna* transcriptome by RNA-Seq across 12 environmental stressors. Sci. Data 3, 170006.
Pacchierotti, F., Spano, M., 2015. Environmental impact on dna methylation in the germline: state of the art and gaps of knowledge. Biomed. Res. Int. 2015, 123484.
Roberts, S.B., Gavery, M.R., 2012. Is there a relationship between DNA methylation and phenotypic plasticity in invertebrates? Front. Physiol. 2, 116.
Robichaud, N.F., Sassine, J., Beaton, M.J., Lloyd, V.K., 2012. The epigenetic repertoire of Daphnia magna includes modified histones. Genet. Res. Int. 2012, 174860.
Ruiz-Hernandez, A., Kuo, C.C., Rentero-Garrido, P., Tang, W.Y., Redon, J., Ordovas, J.M., Navas-Acien, A., Tellez-Plaza, M., 2015. Environmental chemicals and DNA methylation in adults: a systematic review of the epidemiologic evidence. Clin. Epigenetics 7, 55.
Sandovici, I., Smith, N.H., Nitert, M.D., Ackers-Johnson, M., Uribe-Lewis, S., Ito, Y., Jones, R.H., Marquez, V.E., Cairns, W., Tadayyon, M., O'Neill, L.P., Murrell, A., Ling, C., Constancia, M., Ozanne, S.E., 2011. Maternal diet and aging alter the epigenetic control of a promoter-enhancer interaction at the Hnf4a gene in rat pancreatic islets. Proc. Natl. Acad. Sci. U. S. A. 108, 5449–5454.
Schatz, M.C., 2017. Nanopore sequencing meets epigenetics. Nat. Methods 14, 347–348.

Schield, D.R., Walsh, M.R., Card, D.C., Andrew, A.L., Adams, R.H., Castoe, T.A., 2016. EpiRADseq: scalable analysis of genomewide patterns of methylation using next generation sequencing. Meth. Ecol. Evol. 7, 60–69.

Seetharam, A., Bai, Y., Stuart, G.W., 2010. A survey of well conserved families of C2H2 zinc-finger genes in *Daphnia*. BMC Genomics 11, 276.

Shaw, J.R., Pfrender, M.E., Eads, B.D., Klaper, R., Callaghan, A., Sibly, R.M., Colson, I., Jansen, B., Gilbert, D., Colbourne, J.K., 2008. *Daphnia* as an emerging model for toxicological genomics. Adv. Exp. Biol.-Comp. Toxicogen. 2, 165–219.

Soubry, A., Murphy, S.K., Wang, F., Huang, Z., Vidal, A.C., Fuemmeler, B.F., Kurtzberg, J., Murtha, A., Jirtle, R.L., Schildkraut, J.M., Hoyo, C., 2015. Newborns of obese parents have altered DNA methylation patterns at imprinted genes. Int. J. Obes. (Lond) 39, 650–657. http://dx.doi.org/10.1038/ijo.2013.193.

Strepetkaitė, D., Alzbutas, G., Astromskas, E., Lagunavičius, A., Sabaliauskaitė, R., Arbačiauskas, K., Lazutka, J., 2016. Analysis of DNA methylation and hydroxymethylation in the genome of crustacean *Daphnia pulex*. Genes (Basel) 7, 1.

Thienpont, B., Steinbacher, J., Zhao, H., D'Anna, F., Kuchnio, A., Ploumakis, A., Ghesquière, B., Van Dyck, L., Boeckx, B., Schoonjans, L., Hermans, E., 2016. Tumour hypoxia causes DNA hypermethylation by reducing TET activity. Nature 537, 63–68.

UniProtKB, 2016. UniProt: the universal protein knowledgebase. Nucleic Acids Res. 45, 158–169.

van Gurp, T.P., Wagemaker, N.C.A.M., Wouters, B., Vergeer, P., Ouborg, J.N.J., Verhoeven, K.J.F., 2016. epiGBS: reference-free reduced representation bisulfite sequencing. Nat. Methods 13, 322–324.

Vandegehuchte, M.B., Janssen, C.R., 2011. Epigenetics and its implications for ecotoxicology. Ecotoxicology 20, 607–624.

Vandegehuchte, M.B., Kyndt, T., Vanholme, B., Haegeman, A., Gheysen, G., Janssen, C.R., 2009a. Occurrence of DNA methylation in *Daphnia magna* and influence of multigeneration Cd exposure. Environ. Int. 35, 700–706.

Vandegehuchte, M.B., Lemiere, F., Janssen, C.R., 2009b. Quantitative DNA-methylation in *Daphnia magna* and effects of multigeneration Zn exposure. Comp. Biochem. Physiol. C Toxicol. Pharmacol. 150 (2009b), 343–348.

Vandegehuchte, M.B., Lemière, F., Vanhaecke, L., Berghe, W.V., Janssen, C.R., 2010. Direct and transgenerational impact on *Daphnia magna* of chemicals with a known effect on DNA methylation. Comp. Biochem. Physiol. C Toxicol. Pharmacol. 151, 278–285.

Verhoeven, K.J.F., Vonholdt, B.M., Sork, V.L., 2016. Epigenetics in ecology and evolution: what we know and what we need to know. Mol. Ecol. 25, 1631–1638.

Waddlngton, C.H., 1942. The pupal contraction as an epigenetic crisis in *Drosophila*. Proc. Zool. Soc. Lond. 111, 181–188.

Walsh, M.R., Whittington, D., Funkhouser, C., 2014. Thermal transgenerational plasticity in natural populations of *Daphnia*. Integr. Comp. Biol. 54, 822–829.

Wang, Y., Li, X.L., Hu, W., 2014. H3K4me2 reliably defines transcription factor binding regions in different cells. Genomics 103, 222–228.

Weiss, L.C., Leese, F., Laforsch, C., Tollrian, R., 2015a. Dopamine is a key regulator in the signalling pathway underlying predator-induced defences in *Daphnia*. Proc. R. Soc. B 282, 20151440.

Weiss, L.C., Leimann, J., Tollrian, R., 2015b. Predator-induced defences in *Daphnia longicephala*: location of kairomone receptors and timeline of sensitive phases to trait formation. J. Exp. Biol. 218, 2918–2926.

Wojewodzic, M.W., Colbourne, J.K., Chromatin immunoprecipitation and chromatin immunoprecipitation with massive parallel sequencing (ChiP-Seq) on *Daphnia* tissues reveals further effects of P limitation, in preparation.

Wolff, G.L., Kodell, R.L., Moore, S.R., Cooney, C.A., 1998. Maternal epigenetics and methyl supplements affect agouti gene expression in Avy/a mice. FASEB J. 12, 949–957.
Xu, X., Tao, Y., Gao, X., Zhang, L., Li, X., Zou, W., Ruan, K., Wang, F., Xu, G., Hu, R., 2016. A CRISPR-based approach for targeted DNA demethylation. *Cell Discov.* 2, 16009. http://dx.doi.org/10.1038/celldisc.2016.9.
Ye, Z., Xu, S., Spitze, K., Asselman, J., Jiang, X., Ackerman, M.S., Lopex, J., Harker, B., Raborn, R.T., Ramsdell, J., Pfrender, M.E., Lynch, M., 2017. A new reference genome assembly for the microcrustacean *Daphnia pulex*. G3 (Bethesda) 7, 1405–1416.
Zdobnov, E.M., Tegenfeldt, F., Kuznetsov, D., Waterhouse, R.M., Simão, F.A., Ioannidis, P., Seppey, M., Loetscher, A., Kriventseva, E.V., 2016. OrthoDB v9. 1: cataloging evolutionary and functional annotations for animal, fungal, plant, archaeal, bacterial and viral orthologs. Nucleic Acids Res. gkw1119.

FURTHER READING

Karasov, W.H., Martínez del Rio, C., 2007. Physiological Ecology: How Animals Process Energy, Nutrients, and Toxins. Princeton University Press, Princeton.
Sterner, R.W., Elser, J.J., 2002. Ecological Stoichiometry: The Biology of Elements From Molecules to the Biosphere. Princeton University Press, Princeton.

CHAPTER NINE

Epigenetics: A Hidden Target of Insecticides

Ann-Marie Oppold*,†, Ruth Müller‡,§

*Molecular Ecology Group, Senckenberg Biodiversity and Climate Research Centre, Frankfurt am Main, Hessen, Germany

†Institute for Ecology, Evolution and Diversity, Goethe-University, Frankfurt am Main, Germany

‡GM Mosquitoes Confined Release Facilities, PoloGGB, Terni, Italy

§Environmental Toxicology and Medical Entomology, Institute of Occupational, Social and Environmental Medicine, Goethe-University, Frankfurt am Main, Germany

Contents

Abstract

Epigenetic mechanisms are a responsive element at the genotype-to-phenotype interface, since they are involved in the regulation of developmental processes, mediation of homeostasis, induction of responses to the external environment, and even in transgenerational inheritance of gene expression patterns. The scenario of insecticidal exposure sets a strong pressure on insect species, and without any means of resistance mechanisms they will not survive—what is the purpose of insecticide applications. For reasons of public health and industrial agriculture, the use of insecticides in vector and pest control programmes is inevitable. However, there is no insecticide class left without a documented resistance response in insect species. Evidence suggests that among the known resistance mechanisms, epigenetic mechanisms may also play an important role in their development. Expression control via DNA methylation patterns and, by implication, the control of DNA methylation patterns might create a sensitive and reactive mode of action to enhance early on adaptation. Our understanding of how exactly insecticide resistance mechanisms involve the epigenetic system is still scarce and in-depth investigations are needed. Moreover, the epigenetic system contains promising targets that could be used in vector genetics and biochemistry to design novel insecticidal chemicals for a more sustainable vector and pest management in the future.

Advances in Insect Physiology, Volume 53
ISSN 0065-2806
http://dx.doi.org/10.1016/bs.aiip.2017.04.002

1. INSECTICIDES

The modern industrial agriculture and vector control programmes for public health purposes have a great need for chemical insecticides to control manifold insect pests and disease carriers (Davis, 2017). Insecticides are intensively used to combat insect pests, such as fruit and corn borers (e.g. tomato fruit borer *Helicoverpa armigera*, eggplant fruit borer *Leucinodes orbonalis*, and the European corn borer *Ostrinia nubilalis*) or the diamondback moth *Plutella xylostella*, to avoid economic damage. For instance, only the costs for insecticides to control the invasive spotted wing fruit fly *Drosophila suzukii* in sweet cherry production in Switzerland amount to approximately US$ 313/ha (Mazzi et al., 2017). Furthermore, insecticides provide a key management tool against anopheline and aedine mosquito disease carriers for malaria and dengue, but also against tsetse flies, fleas, and others (WHO, 2012). Disease agents transmitted by animal vectors (mainly insects) cause altogether more than 1 million human deaths per year (WHO, 2016).

Though the principal purpose of insecticides is always to kill insects, they represent a very heterogeneous group of chemicals. One may distinguish between naturally occurring insecticides, so-called botanicals such as nicotine, pyrethrum, toxins of the bacterium *Bacillus thuringiensis israelensis* (B.t.i.), and various plant-based essential oils; and chemical (often synthetic) insecticides such as chlorinated hydrocarbons (e.g. DDT), organophosphates (e.g. parathion), carbamates, pyrethroids (e.g. deltamethrin), and neonicotinoids (e.g. imidacloprid). One of the earliest synthesised insecticides and probably the most famous is DDT (dichlorodiphenyltrichloroethane, synthesis in 1874). DDT was heavily applied to combat typhus during World War II. From 1945 onwards, DDT was extensively used by farmers and from 1955, by public health managers in order to control malaria. Indeed, thanks to the widespread use of DDT, malaria disease could be successfully eliminated in several regions of the world (at least for few decades, reviewed in Hemingway and Ranson, 2000; Woodwell et al., 1971).

Carson drew public attention to the detrimental environmental and health effects of DDT, and especially to its transfer within food chains (Carson, 2002, first published in 1962). Follow-up studies proved the high persistence of DDT in the environment (reviewed in Woodwell et al., 1971). The heated public debate about pesticides and the consequential foundation of environmental agencies in several countries finally led to the ban of initially 12 persistent organic pollutants (the so-called dirty dozen, including many insecticides of the first generation, i.e. DDT, aldrin, chlordane, mirex) under UN's Stockholm convention, which is in force since 2004. The next generations of insecticides were intentionally designed to have a lower persistence in the environment, a maximum lethal effect on target species at very low doses but a minimum effect on environmental and human health. However, the design of environmentally harmless synthetic insecticides is still far from perfect, as substantiated by the case of the newest class of insecticides, the neonicotinoids. Neonicotinoids are nowadays under suspicion to cause significant mortality in honeybees and bumblebees, when applied in the field (Goulson, 2013). They may thereby disrupt plant pollination and pose a risk to the sustainability of apiculture and finally agriculture (EFSA, 2012). Currently, neonicotinoids are reviewed on ban from the European Union (EU, 2017).

Due to environmental and health risks arising from the overuse of insecticides (e.g. Palmquist et al., 2012) and increasing public awareness, alternative strategies need to be pursued to stop the ever-growing chemical pesticide and fertiliser usage in agriculture. The Chinese Ministry of Agriculture announced 'the zero-growth plans to 2020 for the application of chemical pesticides and fertilizers' on 18 March 2015. Integrated pest management programmes that combine a set of measures, including preventive practices, monitoring of pest populations, mechanic and biological control, and responsible insecticide treatment, gain increasing acceptance. Some authors predict a bright future for biopesticides such as botanical insecticides (e.g. Glare et al., 2016). In public health, the One Health concept focusing on both, the health of humans and ecosystems, is currently spreading and one may expect a rise of local, integrative, and interdisciplinary vector prevention and control strategies, which may result in a reduced usage of insecticides. Such strategies will certainly include an integrated insecticide resistance management, with focus on planned and reduced usage of insecticide products by private and public consumers (Roush and Tabashnik, 1990; WHO, 2012).

2. INSECTICIDE RESISTANCE

The rapid evolution of insecticide resistance starkly counteracts insecticidal crop protection and public health programmes. Insecticide resistance describes a state of decreased susceptibility of an insect population towards an insecticide which was previously shown as effective control agent for the species. The development of insecticide resistance can become reinforced by cross-selecting mechanisms after exposure to allelochemicals, agricultural fungicides and herbicides, as well as heterogeneous insecticide applications in urban areas (Nkya et al., 2013). As one of the first observed cases, *Aedes tritaeniorhynchus* and *Ae. sollicitans* became resistant against DDT, documented in 1947, after only 1 year of DDT spraying (Brown, 1986). Finally, there is no insecticide class left without a documented resistance response in insect species (e.g. Hemingway and Ranson, 2000; Nauen and Denholm, 2005). The widespread resistance to pyrethroids is of special concern, since this insecticide class has a unique portfolio regarding environmental and human health safety, with a high effectiveness in parallel. This portfolio led the WHO to recommend exclusively pyrethroids for insecticidal bed net treatment, which is one of the most effective tools in malaria control programmes (WHO, 2012). Pest control is further challenged by the development of insecticidal multiresistance, for instance, observed in the diamondback moth in Asia (Srinivasan et al., 2011) and the malaria vector *Anopheles gambiae* s.l. in Burkina Faso, Africa (Namountougou et al., 2012). Unfortunately, the number of species becoming resistant to promising botanical insecticides (B.t.i. toxins) is also going to increase (Tabashnik, 1994; Sheikh et al., 2017).

Insects become resistant to insecticides by different resistance mechanisms, i.e., target-site resistance, metabolic resistance, penetration resistance, and behavioural resistance (reviewed in Ffrench-Constant, 2013; Hemingway and Ranson, 2000; WHO, 2012). *Target-site resistance* is characterised by nonsynonymous single or multiple point mutations in structural genes, resulting in an amino acid substitution that leads to changed secondary structure of the protein, finally preventing the insecticide molecules to bind or interact with their target site. Knockdown resistance mutations cause a reduction in neuronal sensitivity by preventing organophosphates and carbamates to target acetylcholinesterases, pyrethroids and DDT to target voltage-gated sodium channels, or dieldrin to target GABA receptors (heteromultimeric gated chloride ion channel, e.g. Assogba et al., 2014; Hirata et al., 2014; Kasai et al., 2011). The *metabolic resistance* describes the ability of an insect to enhance detoxification capacity, as overexpression of enzymes, amplification of genes coding for enzymes, or increased affinity of detoxification enzymes to the insecticide molecules.

Detoxification enzymes involved in metabolic insecticide resistance are members of three protein superfamilies: the multifunction oxidases (family of cytochrome P450 monooxygenases), glutathione-*S*-transferases, and esterases. Metabolic resistance mechanisms are based on gene amplification, transcriptional regulation, and alternative splicing (e.g. Després et al., 2014; Ihara et al., 2017; Rider et al., 1998). The *penetration resistance*, also called cuticle resistance, is defined by a slower penetration of insecticides into mosquito cuticle and is probably strongly related to metabolic resistance. Differentially regulated gene transcription of structural constituents of the cuticle has been observed, for instance, in the yellow-fever mosquito *Aedes aegypti* after exposure to pyrethroid, neonicotinoid, carbamate insecticides, and B.t. i. toxin (Després et al., 2014). Though the link to differential gene regulation remains speculative, a thicker cuticle has been measured, for instance, in permethrin-resistant vs -sensitive *Anopheles funestus* strains (Wood et al., 2010). The *behavioural resistance* results from recognition of insecticides by the insect (shown for organochlorines, organophosphates, carbamates, and pyrethroids) and its following avoidance behaviour. Insects do not feed on insecticide-polluted food sources, do not oviposit in insecticide-polluted water, fly away, and/or search unpolluted shelters, e.g., underneath leaves (reviewed in Haynes, 1988).

Rider et al. (1998), Rahman et al. (2010), and Oppold et al. (2015) suspect an *additional mode of resistance functioning on the epigenetic level.* The few available research on how epigenetics is involved in insecticide response, and how insecticides could speed up the development of insecticide resistance, will be presented in the next sections.

3. EPIGENETICS INVOLVED IN INSECTICIDE RESPONSE

First evidence for the role of epigenetic mechanisms, i.e., DNA methylation, in insecticide resistance came from the peach-potato aphid *Myzus persicae* (Field et al., 1989). This globally distributed and economically relevant crop pest has a remarkable ability to evolve different mechanisms to avoid or overcome toxic effects of pesticides, and resistance was documented for a wide range of insecticides (organophosphates, carbamates, dimethyl carbamates, pyrethroids, cyclodienes, nicotinoids, and neonicotinoids; reviewed in Bass et al., 2014). Until now there are seven different mechanisms of insecticide resistance investigated in *M. persicae*, one of which involves regulation via DNA methylation as an epigenetic feature of metabolic resistance (Bass et al., 2014). Amplification of one of two possible carboxylesterase genes, *E4* and *FE4*, allows for the overproduction of the respective carboxylesterases that sequester and hydrolyse insecticidal esters and confer a broad spectrum of resistance to organophosphate and carbamate insecticides (Devonshire and Moores, 1982). The level of amplification hereby correlates with the resistance phenotype, with a serial fourfold increase in gene copy number, and thus, resistance is not based on overexpression of amplified genes (Field et al., 1999). Either the *E4* or *FE4* gene underwent a single amplification event and became dispersed thereafter by insecticide selection and migration (Field and Devonshire, 1998). The amplification of *E4* is linked to a chromosomal translocation, the autosomal 1,3 translocation event, and builds a tandem array of head-to-tail amplicons, containing a CpG island in the promotor sequence (Blackman et al., 1995; Devonshire et al., 1998; Field and Devonshire, 1998). Demethylation of 5-methylcytosine in this sequence results in *E4* gene silencing and is therefore a decisive transcriptional control element (Field et al., 1989; Hick et al., 1996). Through this epigenetic mechanism, fitness costs due to enzyme overproduction can be reduced in the absence of insecticide selection (Bass et al., 2014) and the resistance mechanism is nevertheless inherited to future generations (Field et al., 1989). The *E4* carboxylesterase overproduction under epigenetic control is the only inducible and reversible resistance mechanism described for this species and confers rapid response to spontaneous short-term insecticide treatments in the field.

DNA methylation was long assumed to play a minor or even no role in insect biology, when *Drosophila melanogaster*, the best investigated insect model organism, was found to have almost no detectable DNA methylation (Lyko et al., 2000). The example from *M. persicae*, however, proves the importance of transcriptional control via DNA methylation as a mechanistic basis of insecticide resistance. Moreover, Ffrench-Constant (2013) proposed that altered DNA methylation and epigenetics might even play a larger role than simply switching on or off amplified esterase genes in aphids.

A study with the Asian tiger mosquito *Ae. albopictus* could show that alterations on the level of global DNA methylation, as a genome-wide nontarget effect, correlated with decreasing insecticide sensitivity towards the neonicotinoid imidacloprid on the phenotypic level (Oppold et al., 2015). Artificial modification of DNA methylation levels happened in a parental generation, whereas insecticide sensitivity of the mosquitoes was investigated in following offspring generations. It seems plausible that altered DNA methylation was inherited to offspring generations, inducing an early stage of resistance phenotype. Even though the precise molecular mechanism is pending deeper investigations, Oppold et al. (2015) proposed the cellular balance of homocysteine synthesis as mechanistic model (Fig. 1): on the one hand homocysteine is necessary for the synthesis of glutathione that will

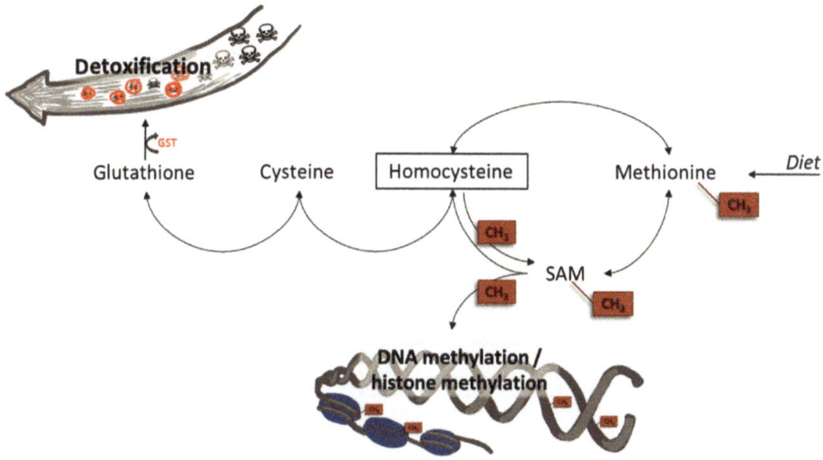

Fig. 1 Homocysteine balance between synthesis of *S*-adenosyl methionine (SAM) for methylation of DNA and histones, and synthesis of glutathione for catalysis of detoxification processes involving the glutathione-*S*-transferase (GST). CH_3 shows circulation of methyl groups. Hypothesis based on *Lee, D.H., Jacobs, D.R., Porta, M., 2009. Hypothesis: a unifying mechanism for nutrition and chemicals as lifelong modulators of DNA hypomethylation. Environ. Health Perspect. 117, 1799–1802;* modified figure *Oppold, A.-M., Kreß, A., Vanden Bussche, J., Diogo, J.B., Kuch, U., Oehlmann, J., Vandengehuchte, M.B., Müller, R., 2015. Epigenetic alterations and decreasing insecticide sensitivity of the Asian tiger mosquito* Aedes albopictus. *Ecotoxicol. Environ. Saf. 122, 45–53.*

be consumed by the glutathione-*S*-transferase during detoxification processes; on the other hand, homocysteine donates free methyl groups to *S*-adenosyl methionine (SAM), an important methyl group carrier to DNA and histones (Lee et al., 2009). In case of exposure to insecticides, detoxification processes will be favoured, exploiting homocysteine for the synthesis of glutathione. Resulting undersupply of methyl groups might hence lead to alterations of DNA methylation patterns or histone modifications (Vandegehuchte and Janssen, 2011).

The homocysteine/SAM balance model provides a mechanistic link between detoxification processes in general and different epigenetic mechanisms, i.e., DNA methylation and histone modifications. Similar to the regulation and allocation of methyl groups, histone modifications by acetyl groups have been found to have an impact on detoxification pathways in the honeybee *Apis mellifera* (Hu et al., 2017). Inhibition of histone deacetylation triggers histone tail acetylation and leads to gene activation and changes in gene expression (Jenuwein and Allis, 2001). In honeybees, low concentrations of sodium butyrate act as histone deacetylase inhibitor.

By regulating histone acetylation with sodium butyrate, it was possible to increase the honeybee's tolerance towards imidacloprid (Hu et al., 2017). Expression of genes of the cytochrome P450 pathway (CYP genes), a superoxide dismutase gene, and the glutathione-*S*-transferase gene was found to be enhanced after sodium butyrate treatments (Hu et al., 2017). All these genes code for enzymes that are known to be involved in detoxification of neonicotinoid insecticides, as is imidacloprid, or reactive oxygen species as general response to cellular stress.

More empirical studies are necessary to provide a fundamental statement to the role of epigenetics in insecticide resistance. Nevertheless, at the interface of genotype and phenotype, epigenetics displays a major regulatory unit and modifications to this end can even be inherited to future generations.

4. EPIGENETICS CAN ENHANCE EARLY ON ADAPTATION

In principal, insecticides pose a strong selection pressure on insect populations; thus, preadaptations (i.e. a genetic key adaptation already being present in the gene pool) or rapid responses due to phenotypic plasticity are necessary to survive the toxic exposure. Epigenetic responses, such as changes in DNA methylation, may contribute to environmentally induced phenotype variability by modifying gene expression (Angers et al., 2010). This is exactly what was shown for *M. persicae* (see above and Field et al., 1989). Irrespective of a specific selection pressure, Kronholm and Collins (2016) used a simulation approach to investigate the effects of epigenetic mutations (as we could call changes of epigenetic marks at a specific site) on the dynamics of adaptive evolution. Given that an epigenetic mark affects a fitness-related trait and is inherited to subsequent generations (Jablonka and Raz, 2009; Johannes et al., 2009), they found that initial stages of adaptation can be faster when epigenetic variation is involved. Furthermore, there is evidence for genetic assimilation, where epigenetic modifications can eventually be replaced by genetic changes (Kronholm and Collins, 2016). The translation into genetic changes is driven by the revertability and thus instability of epigenetic changes (Johannes et al., 2009) that will not allow to reach the highest final fitness at the long term (Kronholm and Collins, 2016). In other words, epigenetics may provide phenotypic flexibility early on in adaptation, and if the selection regime stays constant, these plastic responses are eventually replaced by genetic evolution to maximise fitness. Empirical data support this idea, indicating that epigenetic effects have contributed to the induction and transmission of tolerance to the synthetic pesticide emamectin benzoate in the diamondback moth during the first generations, until genetic changes were established (Rahman et al., 2010). Similar evidence of an epigenetic basis of the adaptive response to climate selection was shown in a fish species (*Leucoraja ocellata*, Lighten et al., 2016).

Until now it can only be speculated that this mechanism also shapes the rapid development of insecticide resistance. Nevertheless, with the advancing insight in molecular, epigenetic, transcriptomic, and genomic systems, challenges to answer these questions can easily be overcome.

5. EPIGENETIC SYSTEM AS PROMISING TARGET FOR NEW INSECTICIDES

An in-depth understanding of mechanisms involved in insecticide resistance is essential to design new chemical tools for insecticide resistance management in pest control and vector control. On the one hand, this involves the understanding of how insecticide resistance might be based on epigenetic mechanisms. On the other hand, vector genetics and biochemistry can most probably make use of the epigenetic system as targets for the development of novel insecticidal chemicals (Jenkins and Muskavitch, 2015). Many epigenetic modifiers have been shown to be involved in the regulation of developmental processes, mediation of homeostasis, induction of responses to the external environment, and even in transgenerational inheritance of gene expression patterns (Cantone and Fisher, 2013; Greer et al., 2011; Kiefer, 2007; Lunyak and Rosenfeld, 2008; Meissner, 2010). The investigation of a gene ensemble that underlies epigenetic control in *D. melanogaster* revealed a high level of conservation also among anopheline mosquitoes (Jenkins and Muskavitch, 2015). It therefore seems promising to use epigenetic genes underlying those processes to develop insect-selective or even further species-selective pesticides by inhibiting functions of metamorphic development (Jenkins and Muskavitch, 2015). Functional conservation of epigenetic genes among insects in general thus also means that mass-administered small molecule antagonists may also affect nontarget insects as *Apis* spp. (Jenkins and Muskavitch, 2015). Therefore, a better understanding of epigenetic systems, its machinery, functioning, and dynamics among insects sets the fundamental basis for the advancement of new and more sustainable insecticidal interventions that only target respective vector and pest species.

REFERENCES

Angers, B., Castonguay, E., Massicotte, R., 2010. Environmentally induced phenotypes and DNA methylation: how to deal with unpredictable conditions until the next generation and after. Mol. Ecol. 19, 1283–1295.

Assogba, B.S., Djogbénou, L.S., Saizonou, J., Milesi, P., Djossou, L., Djegbe, I., Oumbouke, W.A., Chandre, F., Baba-Moussa, L., Weill, M., Makoutodé, M., 2014. Phenotypic effects of concomitant insensitive acetylcholinesterase (ace-1 (R)) and knockdown resistance (kdr (R)) in *Anopheles gambiae*: a hindrance for insecticide resistance management for malaria vector control. Parasit. Vectors 7, 548.

Bass, C., Puinean, A.M., Zimmer, C.T., Denholm, I., Field, L.M., Foster, S.P., Gutbrod, O., Nauen, R., Slater, R., Williamson, M.S., 2014. The evolution of insecticide resistance in the peach potato aphid, *Myzus persicae*. Insect Biochem. Mol. Biol. 51, 41–51.

Blackman, R.L., Spence, J.M., Field, L.M., Devonshire, A.L., 1995. Chromosomal location of the amplified esterase genes conferring resistance to insecticides in *Myzus persicae* (Homoptera: Aphididae). Heredity 75, 297–302.

Brown, A., 1986. Insecticide resistance in mosquitoes: a pragmatic review. J. Am. Mosq. Control Assoc. 2, 123–131.

Cantone, I., Fisher, A.G., 2013. Epigenetic programming and reprogramming during development. Nat. Struct. Mol. Biol. 20, 282–289.

Carson, R., 2002. Silent Spring. Houghton Mifflin Harcourt, Boston, MA (first published in 1962).

Davis, F.R., 2017. Insecticides, agriculture, and the Anthropocene. Glob. Environ. 10, 114–136.

Després, L., Stalinski, R., Faucon, F., Navratil, V., Viari, A., Paris, M., Tetreau, G., Poupardin, R., Riaz, M.A., Bonin, A., Reynaud, S., David, J.P., 2014. Chemical and biological insecticides select distinct gene expression patterns in *Aedes aegypti* mosquito. Biol. Lett. 10, 1–7.

Devonshire, A.L., Field, L.M., Foster, S.P., Moores, G.D., Williamson, M.S., Blackman, R.L., 1998. The evolution of insecticide resistance in the peach–potato aphid, *Myzus persicae*. Philos. Trans. R. Soc. Lond. B Biol. Sci. 353, 1677–1684.

Devonshire, A.L., Moores, G.D., 1982. A carboxylesterase with broad substrate specificity causes organophosphorus, carbamate and pyrethroid resistance in peach-potato aphids (*Myzus persicae*). Pestic. Biochem. Physiol. 18, 235–246.

EFSA, 2012. Scientific opinion on the science behind the development of a risk assessment of plant protection products on bees (*Apis mellifera, Bombus spp.*, and solitary bees). EFSA J. 10, 2668.

EU—European Union, 2017. Pesticides and Bees. https://ec.europa.eu/food/animals/live_animals/bees/pesticides_en.

Ffrench-Constant, R.H., 2013. The molecular genetics of insecticide resistance. Genetics 194, 807–815.

Field, L.M., Devonshire, A.L., 1998. Evidence that the E4 and FE4 esterase genes responsible for insecticide resistance in the aphid *Myzus persicae* (Sulzer) are part of a gene family. Biochem. J. 330 (Pt. 1), 169–173.

Field, L.M., Devonshire, A.L., Ffrench-Constant, R.H., Forde, B.G., 1989. Changes in DNA methylation are associated with loss of insecticide resistance in the peach-potato aphid *Myzus persicae* (Sulz.). FEBS Lett. 243, 323–327.

Field, L.M., Blackman, R.L., Tyler-Smith, C., Devonshire, A.L., 1999. Relationship between amount of esterase and gene copy number in insecticide-resistant *Myzus persicae* (Sulzer). Biochem. J. 339, 737–742.

Glare, T.R., Gwynn, R.L., Moran-diez, M.E., 2016. Development of biopesticides and future opportunities. In: Glare, T.R., Moran-Diez, M.E. (Eds.), Microbial-Based

Biopesticides: Methods and Protocols, Humana Press, New York, Copyright: Springer Science+Business Media, 211–221.

Goulson, D., 2013. An overview of the environmental risks posed by neonicotinoid insecticides. J. Appl. Ecol. 50, 977–987.

Greer, E.L., Maures, T.J., Ucar, D., Hauswirth, A.G., Mancini, E., Lim, J.P., Benayoun, B.A., Shi, Y., Brunet, A., 2011. Transgenerational epigenetic inheritance of longevity in *Caenorhabditis elegans*. Nature 479, 365–371.

Haynes, K.F., 1988. Sublethal effects of neurotoxic insecticides on insect behavior. Annu. Rev. Entomol. 33, 149–168.

Hemingway, J., Ranson, H., 2000. Insecticide resistance in insect vectors of human disease. Annu. Rev. Entomol. 45, 371–391.

Hick, C.A., Field, L.M., Devonshire, A.L., 1996. Changes in the methylation of amplified esterase DNA during loss and reselection of insecticide resistance in peach-potato aphids, *Myzus persicae*. Insect Biochem. Mol. Biol. 26, 41–47.

Hirata, K., Komagata, O., Itokawa, K., Yamamoto, A., Tomita, T., Kasai, S., 2014. A single crossing-over event in voltage sensitive Na^+ channel genes may cause critical failure of dengue mosquito control by insecticides. PLoS Negl. Trop. Dis. 8. e3085.

Hu, Y.T., Wu, T.C., Yang, E.C., Wu, P.C., Lin, P.T., Wu, Y.L., 2017. Regulation of genes related to immune signaling and detoxification in *Apis mellifera* by an inhibitor of histone deacetylation. Sci. Rep. 7, 41255.

Ihara, M., Buckingham, S., Matsuda, K., Sattelle, D., 2017. Modes of action, resistance and toxicity of insecticides targeting nicotinic acetylcholine receptors. Curr. Med. Chem. 24, in press.

Jablonka, E., Raz, G., 2009. Transgenerational epigenetic inheritance: prevalence, mechanisms, and implications for the study of heredity and evolution. Q. Rev. Biol. 84, 131–176.

Jenkins, A.M., Muskavitch, M.A.T., 2015. Evolution of an epigenetic gene ensemble within the genus *Anopheles*. Genome Biol. Evol. 7, 901–915.

Jenuwein, T., Allis, C.D., 2001. Translating the histone code. Science 293, 1074–1080.

Johannes, F., Porcher, E., Teixeira, F.K., Saliba-Colombani, V., Simon, M., Agier, N., Bulski, A., Albuisson, J., Heredia, F., Audigier, P., Bouchez, D., Dillmann, C., Guerche, P., Hospital, F., Colot, V., 2009. Assessing the impact of transgenerational epigenetic variation on complex traits. PLoS Genet. 5, e1000530.

Kasai, S., Ng, L.C., Lam-Phua, S.G., Tang, C.S., 2011. First detection of a putative knockdown resistance gene in major mosquito vector, *Aedes albopictus*. Jpn. J. Infect. Dis. 64, 217–221.

Kiefer, J.C., 2007. Epigenetics in development. Dev. Dyn. 236 (4), 1144–1156.

Kronholm, I., Collins, S., 2016. Epigenetic mutations can both help and hinder adaptive evolution. Mol. Ecol. 25, 1856–1868.

Lee, D.H., Jacobs, D.R., Porta, M., 2009. Hypothesis: a unifying mechanism for nutrition and chemicals as lifelong modulators of DNA hypomethylation. Environ. Health Perspect. 117, 1799–1802.

Lighten, J., Incarnato, D., Ward, B.J., van Oosterhout, C., Bradbury, I., Hanson, M., Bentzen, P., 2016. Adaptive phenotypic response to climate enabled by epigenetics in a K-strategy species, the fish *Leucoraja ocellata* (Rajidae). R. Soc. Open Sci. 3, 160299.

Lunyak, V.V., Rosenfeld, M.G., 2008. Epigenetic regulation of stem cell fate. Hum. Mol. Genet. 17, R28–R36.

Lyko, F., Ramsahoye, B.H., Jaenisch, R., 2000. DNA methylation in *Drosophila melanogaster*. Nature 408, 538–539.

Mazzi, D., Bravin, E., Meraner, M., Finger, R., Kuske, S., 2017. Economic impact of the introduction and establishment of *Drosophila suzukii* on sweet cherry production in Switzerland. Insects 8, 18.

Meissner, A., 2010. Epigenetic modifications in pluripotent and differentiated cells. Nat. Biotechnol. 28, 1079–1088.

Namountougou, M., Simard, F., Baldet, T., Diabaté, A., Ouédraogo, J.B., Martin, T., Dabiré, R.K., 2012. Multiple insecticide resistance in *Anopheles gambiae* s.l. Populations from Burkina Faso, West Africa. PLoS One 7, e48412.

Nauen, R., Denholm, I., 2005. Resistance of insect pests to neonicotinoid insecticides: current status and future prospects. Arch. Insect Biochem. Physiol. 58, 200–215.

Nkya, T.E., Akhouayri, I., Kisinza, W., David, J.-P., 2013. Impact of environment on mosquito response to pyrethroid insecticides: facts, evidences and prospects. Insect Biochem. Mol. Biol. 43, 407–416.

Oppold, A.-M., Kreß, A., Vanden Bussche, J., Diogo, J.B., Kuch, U., Oehlmann, J., Vandengehuchte, M.B., Müller, R., 2015. Epigenetic alterations and decreasing insecticide sensitivity of the Asian tiger mosquito *Aedes albopictus*. Ecotoxicol. Environ. Saf. 122, 45–53.

Palmquist, K., Salatas, J., Fairbrother, A., 2012. Pyrethroid insecticides: use, environmental fate, and ecotoxicology. In: Perveen, D.F. (Ed.), Insecticides—Advances in Integrated Pest Management. INTECH Open Access Publisher, Rijeka, pp. 251–278.

Rahman, M.M., Baker, G., Powis, K.J., Roush, R.T., Schmidt, O., 2010. Induction and transmission of tolerance to the synthetic pesticide emamectin benzoate in field and laboratory populations of diamondback moth. J. Econ. Entomol. 103, 1347–1354.

Rider, S.D., Wilde, G.E., Kambhampati, S., 1998. Genetics of esterase-mediated insecticide resistance in the aphid *Schizaphis graminum*. Heredity 81, 14–19.

Roush, R.T., Tabashnik, B.E., 1990. Pesticide Resistance in Arthropods. Chapman and Hall, New York and London. 241 pp.

Sheikh, A.A., Wani, M.A., Bano, P., Un Nabi, S., Bhat, T.A., Bhat, M.A., Dar, M.S., 2017. An overview on resistance of insect pests against Bt Crops. J. Entomol. Zool. Stud. 5, 941–948.

Srinivasan, R., Shelton, A.M., Collins, H.L., 2011. The Sixth International Workshop on Management of the Diamondback Moth and Other Crucifer Insect Pests.

Tabashnik, B.E., 1994. Evolution of resistance to *Bacillus thuringensis*. Ann. Dent. 39, 47–79.

Vandegehuchte, M.B., Janssen, C.R., 2011. Epigenetics and its implications for ecotoxicology. Ecotoxicology 20, 607–624.

WHO, 2012. Global Plan for Insecticide Resistance Management. http://www.who.int/malaria/vector_control/gpirm_executive_summary_en.pdf?ua=1.

WHO, 2016. Factsheet Vector-Borne Diseases. http://www.who.int/mediacentre/factsheets/fs387/en/.

Wood, O., Hanrahan, S., Coetzee, M., Koekemoer, L., Brooke, B., 2010. Cuticle thickening associated with pyrethroid resistance in the major malaria vector *Anopheles funestus*. Parasit. Vectors 3, 67.

Woodwell, G., Craig, P., Johnson, H., 1971. DDT in the biosphere: where does it go? Science 174, 1101–1107.